STATISTICS

Difficult concepts, understandable explanations

STATISTICS
Difficult concepts, understandable explanations

DALE E. MATTSON, Ph.D.

Professor of Biometry, School of Public Health,
University of Illinois, Chicago, Illinois

The C. V. Mosby Company

ST. LOUIS • TORONTO • LONDON   1981

MOSBY

1906 **75** 1981
YEARS

A TRADITION OF PUBLISHING EXCELLENCE

Copyright © 1981 by The C. V. Mosby Company

Printed in the United States of America

The C. V. Mosby Company
11830 Westline Industrial Drive, St. Louis, Missouri 63141

Library of Congress Cataloging in Publication Data

Mattson, Dale E   1934-
  Statistics: difficult concepts, understandable
explanations.

  Includes index.
  1. Statistics. 2. Medical statistics. I. Title.
QA276.12.M37   519.5   80-24947
ISBN 0-8016-3173-4

AC/D/D 9 8 7 6 5 4 3 2 1   01/D/028

To
Mary Ellen, Greg, Bruce, and Minnie

Preface

It is usually not very long before a new acquaintance asks the question: "What do you do for a living?" Persons who answer, "I teach statistics courses," soon learn that an almost universal response can be expected. The new acquaintance, in almost an attitude of awe, confesses that he or she once struggled to comprehend statistics, either as a student in an introductory statistics course or as part of a course that required the application of statistics. The thought that someone actually understands statistics well enough to attempt to teach the subject to others is mind boggling. Often the new acquaintance who is so intimidated by statistics is well educated and may in fact be successful in an area requiring advanced training in a very complex area.

What is it about statistics that engenders such awe? I am convinced that, more than anything else, it is the way the subject is taught. Both teachers of statistics courses and authors of statistics textbooks are individuals who have enjoyed and done well in mathematics classes over a period of 20 or more years of formal schooling. It is difficult for such individuals to recognize that they are members of a minority group. The majority of students in statistics classes and the readers of statistics books are persons who have struggled as average students in mathematics classes. They are not likely to be greatly moved by an eloquent mathematical proof for a statistical principle. They are satisfied to know what works to solve a problem and to have an intuitive belief that the solution is logical or reasonable.

This textbook has been written with the majority of students in mind. The mathematics major who wants to learn statistics will not be very satisfied with the presentation of the subject in this text. Such a person will find the text wordy and repetitive. Concepts that could be explained with a single page of formulas will instead be explained in several pages of text, with worked examples to illustrate all calculations. My hope is that the very characteristics that make the book unsatisfactory to the mathematics major will make it useful to the person who could profit from an understanding of statistics but who is intimidated by pages of formulas. The reaction of most students who have studied from early drafts of the book has encouraged this hope.

Throughout the writing of the text, I have tried to keep in mind the goal of making the concepts and principles understandable to the *average* reader. The use of symbols and formulas has been kept to a minimum, especially in early sections of the book. By the time the reader has progressed to later sections, certain symbols and concepts should be so familiar that their use in formulas should not cause a problem. Appendix D contains a listing of all formulas of general importance in the order in which they appear in the text.

The material is organized in a sequential fashion. When first introduced, concepts are presented at a very basic level; more sophisticated interpretations are made with successive use. Starting with a chapter from the middle of the book will be similar to attempting to do the same with a complex mystery novel.

The fact that the *presentation* in the text is intended for the average reader does not mean that the *content* of the book represents a superficial introduction to statistics. The book is divided into three parts. Part One contains material generally included in the most basic of texts, such as self-study, programmed texts. Part Two describes the principal concepts and rationale of inferential statistics. To keep things as uncomplicated as possible, the only statistics considered are proportions and means. Attention is restricted to the one-sample case. Part Three of the text contains descriptions of methods of inferential statistics usually included in second-level courses and intermediate textbooks. The concepts and calculations are complex. Worked examples including all calculations are provided, making it possible for all readers to learn how to make the tests.

Parts One and Two and selected lessons from Part Three represent a good introductory course for the person who needs to be an intelligent reader of the scientific literature. I have used these parts of the text for this purpose in teaching introductory courses. The decision to include all of the material in Part Three in the text was made in the belief that a statistics book ought to be a reference book that is retained and used after a course has been completed. As such, it should be as complete as possible. It is frustrating to discover that the text with which you are familiar does not cover a topic, but instead refers you to another text that uses a completely different style and a different set of symbols. Of course, no textbook in statistics can ever be "complete." The content of this text is based on my experience with the kinds of questions most frequently asked by graduate students and others who come for help with the analysis of data.

The three parts of the book are divided into chapters, which are further divided into lessons. I have found that the content of a lesson is approximately the amount of new material that can be read in advance and then thoroughly discussed in a 50-minute class period. Lessons near the end of the book contain more information than lessons near the beginning. This is intentional. There is value in starting at an easy pace, which assures that students will be successful from the very beginning. Each lesson contains exercises designed to provide an opportunity for the student to use the concepts from that lesson. Answers to all exercises are provided in Appendix B.

I presently teach courses in a school of public health and have spent most of my career as a teacher, consultant, and administrator in health-related organizations. My graduate training was in an educational psychology program. My leisure-time activities are centered on athletic activities. All of these experiences have contributed to the choices of examples and problems that are included in the text. The data used in these examples and problems, however, are artificially constructed to serve as good teaching examples. Though I believe that each data set is reasonable for the circumstances described, the reader needs to be aware that the data are artificial.

A word about the applicability of this book is in order. The concepts included in the text are applicable not only in the health sciences, but also in education, psychology, sociology, business, and other fields. Appendix A, which contains additional lessons of particular relevance to public health, can be omitted by persons who will not need to know about special statistical techniques used in this field.

It is traditional to conclude a preface with acknowledgment of all of the individuals who have contributed to the long, tedious project of producing a textbook. I am indebted to a host of individuals, including instructors, students, and long-suffering secretarial personnel. Particular mention should be made of colleagues at the School of Public Health of the University of Illinois: Drs. Paul Levy, Edwin Chen, Saul Rosenberg, and Robert Anderson. These individuals provided help with the initial organization of the material and have served as sounding boards for ideas for the book. They deserve credit for its positive features; I alone am responsible for any weaknesses. Robert Anderson deserves a special word of thanks for his effort in preparing detailed answers to many of the exercises. I also wish to express my appreciation to Howard Hait, who as a graduate student critically reviewed significant portions of the text. Secretaries who have patiently typed sections of the drafts of the text include Liz Shelby, Bessie Nunn, and Vanessa Sturkey.

Dale E. Mattson

Contents

Appendixes

PART ONE

Descriptive statistics

CHAPTER 1

Scope and content of statistics

Lesson 1

The majority of students in a course dealing with statistics or biostatistics are enrolled in the course to meet a requirement of a curriculum. Most have heard that an understanding of the material in the course is important to their careers or at least to other courses they will be taking. Few have any clear idea about the content of the course except that "you need to know math."

Before we look at the content of biostatistics, we may find it helpful to identify the kinds of questions that a study of biostatistics will help to solve. The wide variety of these questions will demonstrate the scope of applications of the knowledge you will gain. It is hoped that the listing will also suggest a question in which you might be interested, making the study of biostatistics valuable beyond the completion of a course requirement. The following listing then is merely an illustration of the wide range of questions for which biostatistics skills are appropriate. The first grouping includes general questions involving the common experiences of most readers. The second contains questions relating to health topics.

General questions

1. If your child seems to exhibit a preference for using the left hand and you attempt to encourage the use of the right hand, are you increasing the child's chances of developing speech and reading problems?
2. Your uncle has been asked to consider seeking the Democratic nomination for state senator. What are his chances of gaining the nomination?
3. Suppose your brother's child is having some difficulty in school and it is discovered that he is slightly below average in learning ability. The option of putting him in a class of slow learners is available. Is this likely to result in greater or less educational progress than if he is left in the regular classroom?
4. You have a bet with your brother-in-law that more people watch the news nightly on Channel 2 than on Channel 7. Who wins the bet?
5. You are about to purchase a new car and have narrowed the choice down to two models that are comparable so far as you are concerned. The deciding factor in your mind is potential reliability. Which of the two cars is likely to have the lower cost of repairs and service during the first 5 years of operation?

3

6. Someone offers to bet you that in the majority of games that your favorite baseball team wins this year, it will score as many or more runs in its best inning as the team it beats will score in the whole game. Should you take the bet?

Health-related questions*

1. An outbreak of measles occurs in a school district enrolling over 100,000 children. It is apparent that the forms in the central office that are supposed to include information on inoculation are inaccurate and incomplete. What is the real proportion of children who do not have protection?
2. Is the presence of an abnormal electroencephalogram reading taken 48 hours after a head injury indicative of a future occurrence of epileptic seizures?
3. Rates for cancer of the small intestine are higher in industrial societies than in rural societies. Are dietary patterns a factor in these differences?
4. Recently many hospitals have tried alternatives to traditional methods of organizing nursing care. Are the attitudes of nurses about their jobs more favorable under these alternative approaches?
5. Your organization of 1000 employees decides to set up its own individual retirement plan. For every dollar set aside yearly by each employee during the length of employment, how much can be paid out yearly to an employee who retires at age 60, 65, or 70?
6. Among the several devices available for use in hip replacement surgery, which has the best results in terms of patient satisfaction with the outcome?
7. In what ways do the physical growth patterns of children with sickle cell anemia differ from those of their normal siblings?

The promise that you will be able to solve all of these problems after a single course in biostatistics would be unrealistic. It is not unrealistic, however, to promise that if you understand the material in this textbook, you will be able to understand the approach used to analyze the data from such studies. In other words, you are not going to be expected to design the data-gathering process needed to answer the questions. You will understand the language and appreciate the rationale used to examine the data.

REQUIRED SKILLS

The promise made in the foregoing paragraph is a big one, and it contains a big "if." What are your chances of understanding the material contained in this book? The answer is: "Better than you think, either now or later, as each new topic is presented." Understanding will take place *gradually* as concepts are first introduced and then used repeatedly throughout the book. Also, set aside any feelings that your math skills are inadequate for the study of biostatistics. Many educated, intelligent persons who have proved their abilities in a wide range of fields approach the study of biostatistics with fear and trembling because of a belief that they will need the skills of a math major to succeed. Actually a very basic level of algebra skills is all that is required for understanding the concepts in this book. You are expected to know the basic rules of algebra. There is a test (Exercise

*Lesson 1 of Appendix A describes the kind and sources of data that are particularly relevant to persons interested in public health.

A) at the end of this lesson that will help you to assess your algebra skills. If you have difficulty with the items on this test, you will need to do some reviewing. Even then, do not plan to take a course in college algebra before proceeding. There are several books that present, in capsule form, what you will need to know. An example is Chapter 1 of the 1954 edition of *Statistical Methods for the Behavioral Sciences,* by Allen Louis Edwards.

Although you will be expected to know basic algebra, actually you will not even do much arithmetic. Instead, you should rely on electronic calculators. Modern electronic calculators make possible a wide range of calculations on large sets of numbers with very minimal effort. Exercise B, at the end of this chapter, involves becoming familiar with what one of these calculators can do. Be sure that you solve the problems on a relatively sophisticated calculator that allows you to solve problems 15 and 16 by entering each number only once. The instruction manuals accompanying these calculators should be self-explanatory. If you are considering purchasing a hand calculator, you might want to ask whether it can do correlations in a single operation. This will be an indication of its potential for saving you time when dealing with the kind of problems you will encounter in this book.

THE SUBJECT MATTER OF BIOSTATISTICS

The mention of statistics evokes an image of reams of numbers and long, complicated formulas. Actually, however, the goal of the statistician is to replace reams of numbers by a few concepts that provide meaning for the reams of numbers. In the first part of the book, this process of reduction will result in averages and pictures to represent whole distributions of numbers. This part of the material will be referred to as *descriptive statistics.* It is the subject matter of Chapters 1 to 3 and Appendix. A.

The second and third parts of the book involve what is called *inferential statistics.* Here the goal is also to make a descriptive statement about a population of numbers. The problem will be that we will have to infer what the population is like without ever having access to all the numbers in the population. We will do so based on samples taken from the population. This sampling process is not without its risks, but we will learn how to minimize these risks.

CONCEPTS OF MEASUREMENT

As has been stated, most people recognize that statisticians are people who work with numbers. Before we proceed to topics involving how a statistician deals with these numbers, it is in order to consider their origin and meaning. In the remainder of this chapter the topic of numbers will be discussed from a theoretical point of view and will contain information usually included in textbooks on educational measurement.

Definition of measurement

When asked where numbers come from, most people would probably conclude that numbers result when things are counted; for example, 12 apples, 68 heartbeats per minute, or 20 ml of serum. Often a number represents a quantity even though the objects have not actually been counted; for example, a check for 1 million dollars. Confusion sometimes arises when an arithmetic operation is performed on a set of numbers, each of which itself represents a count; for example, the number of children per family. The result of the arithmetic operation may be a new number (such as the average number of children per family) that is difficult to conceptualize (1.87 children per family). Yet this new number may be very useful for purposes of comparisons and still can be thought of as representing quantity.

In addition to numbers used to represent quantity, there are numbers that represent other concepts, such as size, quality, or color (for example, size 11AAA shoe, grade number 1 pistachio nuts, or shade number 4 on a particular dental shade guide). It is important to recognize that numbers do not always represent quantity, and in fact are often used simply as a substitute name for a group of things that have some characteristic in common. In this last instance, the numbers really represent a code name for a class of things. The operations that can be logically performed on such numbers are quite different from those for numbers that represent quantity. We will return to this topic later when describing scales of measurement.

Before describing the various scales of measurement, we should arrive at a definition of what the term *measurement* means. We will use the term in a very broad sense, as virtually synonymous with the term *classification*. Formally, we will think of measurement as any process that involves classifying objects, concepts, or phenomena into two or more classes that are both exhaustive and mutually exclusive. By saying that the classes must be mutually exclusive, we mean that each "thing" to be measured must fall into only one class. By saying that classes must be exhaustive, we are saying that there must be a class for all of the "things" we wish to measure. The basis for classification may be a single attribute (such as size), or it may involve two or more attributes (such as size and color). Although the mutually exclusive classes are usually referred to by numerical symbols, this is not an essential part of the basic definition of measurement; in fact, letter symbols are often used.

As an illustration of this definition of measurement, consider a class of 24 dental students. This class of 24 students may be divided into smaller groups in nearly an infinite number of ways, depending on the attributes that are considered. Suppose it is decided to divide the total class into smaller groups on the basis of age. There may be three 20-year-olds, fourteen 21-year-olds, four 22-year-olds, two 23-year-olds, and one 27-year-old. The students could also be divided into groups on the basis of other attributes, such as the number of nerves they are able to identify correctly on a diagram of some part of the human anatomy. The grouping of the class might then be as follows: two persons, none correct; seventeen persons, 1 correct; one person, 2 correct; and four persons, 3 correct. A third grouping might be made on the basis of gender. The class would then fall into only two groups: male and female.

In each of the three instances given, the students in the class have been "measured" according to a different attribute, and as a result each time the class has been divided differently. It follows that the first step in measurement must be to decide which attribute or attributes are to be measured. As an illustration of measurement on the basis of two attributes simultaneously, the class might be divided in such a way that each group is made up of persons who are both the same age and the same sex (24-year-old men, 24-year-old women, 23-year-old men, and so forth).

It might be noted that in each instance in which the class of 24 students is partitioned on the basis of one or more attributes, the resulting smaller groups are mutually exclusive (no person in the class could belong to more than one group). For example, a student could not be both a 23-year-old and a 24-year-old; nor could he be a 23-year-old able to identify no anatomic structures and a 23-year-old able to identify three anatomic structures.

Notice also that, in each of the three instances cited, the process of measurement would differ. In the first case, students might be asked to give their dates of birth; by so doing, they would in essence be measuring themselves on the attribute age. In the second instance, a written test might be given. In the third case, students might be divided into male and female on the basis of their appearance.

Value of measurement

Perhaps the chief value of measurement is that it facilitates communication. Can you imagine the problems of communicating with another person about a patient's condition without measurement? What are some of the common measurements used in describing a patient's condition?

Another value of measurement is that it makes possible what psychologists refer to as logical operations. At a very simple level an example might be comparisons. Without an established process of measurement, even simple physical measurements would not be comparable from time to time, place to place, and physician to physician. Established procedures of measurement allow a physician in Seattle to compare the metabolism rate of his patient with that of 20 other patients observed by another physician in New York over the past 20 years. At the next level of complexity, we can perform arithmetic operations on the measurements and compare averages to reach conclusions about similarities and differences among *groups* of things.

Limitations of measurement

Measurement reduces data. Once a complex entity, such as a human being, is measured for a single attribute (for example, age) other attributes of the individual are often lost or ignored. The result may be to think of all the persons placed in the same class through measurement to be similar for other variables. As an example, all 67-year-olds, who have one thing in common (having lived 67 years), may be thought to be similar in other attributes: energy, mental acuity, attitudes, and so forth. Such stereotyping is unfair and leads to faulty conclusions.

Types of variables

It is useful to divide the variables to which measurement is to be applied into two types: discrete variables and continuous variables. In some instances, distinct classes are readily available as soon as an attribute is decided upon. For example, within a classroom, if sex of the student is used as the variable in measuring (classifying) the student, two classes (male and female) become obvious. Sex is, therefore, an example of a discrete variable. On the other hand, suppose we decide to group the class according to the attribute height. We must arbitrarily define the classes to be used. We can use inches, half inches, quarter inches, or some other unit of measure. The precision of the measuring device or the precision needed in the results will determine how wide each class is to be. Given more precise measuring instruments, the width of classes can be continuously reduced until virtually an infinite number of classes are used. In the case of height, measurement may be accurate to the nearest inch, the nearest tenth of an inch, or the nearest hundredth of an inch. Variables for which the boundaries of classes are arbitrary are said to be continuous variables.

Scales of measurement

Now let us return to a discussion of the type of measurements that are made so that we can consider the meaning of the numbers or symbols that result from the measurement process. Four types of measurement scales are widely used: nominal, ordinal, interval, and ratio. As one moves from the lowest scale to the highest, the result of measurement

Table 1-1. Scales of measurement

Scale	Characteristics	Examples
Nominal	Objects are assigned to classes that may be denoted by numbers. That the number for one class is greater or less than another number reflects nothing about the properties of the objects other than that they are different.	Racial origin, eye color, numbers on football jerseys, sex, clinical diagnosis, automobile license numbers, Social Security numbers
Ordinal	The relative sizes of the numbers assigned to the objects reflect the relative amounts of the attribute the objects possess. Equal differences between the numbers do not imply equal differences in the amounts of the attributes.	Hardness of minerals, grades for achievement, ranking of personality traits, military ranks
Interval	Numbers are assigned so that equal differences on the scale reflect equal differences in the amounts of the attribute measured. The zero point of the interval scale is arbitrary and does not reflect absence of the attribute.	Calendar time, Fahrenheit and Centigrade temperature scales
Ratio	The numbers assigned to objects have all the properties of the intrval scale. In addition, an absolute zero point exists on the scales. A measurement of zero indicates absence of the property measured. Ratios of the numbers assigned in measurement reflect ratios in amounts of the property measured.	Height, weight, temperature on the Kelvin (absolute zero) scale

gets closer to a pure measure of a count of quantity or amount. This can be seen in Table 1-1, which describes each of the scales.

As mentioned earlier, the four scales of measurement differ in terms of the types of analyses that are appropriate. For nominal data, about all that can be done is to report frequency counts for each class. For ordinal data, comparisons are meaningful (for example, the horse that finished first ran faster than the horse that finished third); however, arithmetic operations such as addition are not meaningful. As an example, consider two race horses, one with a second-place and a sixth-place finish and another with two fourth-place finishes. Adding the finishes would give a total of eight for both, but there would be a significant difference in the results of betting on both horses to win, place, or show.

Arithmetic operations *are* meaningful when the data are on an interval scale or a ratio scale. The total length of two boards, one of which is 1 foot long and the other 3 feet, is the same as the total length of two boards each 2 feet long. Ratios, however, are meaningful only for data on a ratio scale. It is not twice as hot when it is 80° F as when it was 40° F.

Attributes of satisfactory measurement

Three attributes are required for a satisfactory system of measurement. First, the measurements must be *reliable*. Suppose we are concerned about infant mortality. If our system for measuring infant mortality in various populations is to be of value, different investigators looking at the infant mortality in the same population during the same period of time must come up with the same result. In other words, the measurements must be reproducible, or reliable. The second attribute of an acceptable measurement system is that it must be *valid*. This means the measurements must represent what they claim. If the system for measuring infant mortality systematically excludes deaths of infants that do not occur in hospitals, the results are reproducible but invalid. In a later chapter you will learn a procedure that results in a correlation coefficient that can be used as an index of reliability or validity.

Finally, a measurement system must be *usable*. In other words, the system must be feasible and workable. Although a system of measuring infant mortality in a population that is based on individual questioning of each adult member of the population might result in valid and reliable measures, it would be so costly as to be unusable.

Suggestions for improving measurement

Most problems of measurement could be avoided if an attempt were always made to specify exactly the process used in making measurements. There would then be no doubt about what attributes were considered or how classification was determined. In instances in which the reliability of the measurement procedure is in doubt, a technique that is often used is to make several measurements and then to take an average of these repeated measurements.

Distinction between measurement and evaluation

The term *evaluate* is often confused with the term *measure* as defined here. A process of measurement that is largely subjective in nature is often referred to as evaluation. Actually, evaluation is the process of placing a value judgment on a measurement.

An example may help to distinguish between the two concepts. Suppose a dental clinic instructor is inspecting five cavity preparations. The attribute of interest might be the extent of the preparations. Another person, such as an engineer, could do just as well as the clinic instructor at measuring the physical extent of the preparation. This might be done either subjectively or by means of precise instruments. Suppose, however, that the object is to determine whether the extent of each cavity preparation is satisfactory. In this case a value judgment is to be placed on the measurements. Although both dentist and engineer could *measure* the extent of a cavity preparation, only the dentist would be qualified to *evaluate* the extent of the preparation.

SUMMARY

This introductory lesson began with a listing of some of the kinds of questions that an understanding of the concepts in this text can help to answer. It was then said that a knowledge of basic algebra would be a sufficient mathematics background for success in learning the material. The content of the book involves two general subjects: descriptive statistics, (Chapters 1 through 3 and Appendix A) and inferential statistics (Chapters 4 through 14). The procedures of descriptive statistics provide concise descriptions of complicated or extensive data. The procedures of inferential statistics result in statements about populations on the basis of an analysis of the observations made in a sample of the population.

The second part of the lesson contained a description of some basic concepts of measurement theory. Measurement was defined as a process of placing things into two or more classes that are both exhaustive and mutually exclusive. The difference between discrete and continuous variables was explained. Four scales of measurement were described: nominal, ordinal, interval, and ratio. The lesson concluded with suggestions for improving measurements and an explanation of the distinction between measurement and evaluation.

LESSON 1 EXERCISES
Exercise A: Test of numerical skills

$$
\begin{array}{llllll}
\textbf{1.}\quad \begin{array}{r}(-9)\\ +(-4)\end{array} &
\textbf{2.}\quad \begin{array}{r}0\\ +(-12)\end{array} &
\textbf{3.}\quad \begin{array}{r}22\\ +(-14)\end{array} &
\textbf{4.}\quad \begin{array}{r}-9\\ -(-4)\end{array} &
\textbf{5.}\quad \begin{array}{r}0\\ -(-12)\end{array} &
\textbf{6.}\quad \begin{array}{r}22\\ -(-14)\end{array}
\end{array}
$$

7. $(-9) \times (-4) =$ _____   **9.** $(0.1) \times (0.1) =$ _____

8. $(22) \times (-4) =$ _____   **10.** $(-18) \div (2) =$ _____

11. $(0.006) \div (0.0002) = $ _____

12. $(-0.862) \div (-0.02) = $ _____

13. $(7 + 2)^2 = $ _____

14. $-6 + 4 - 3 = $ _____

15. $\frac{3}{4} \div (-2)/8 = $ _____

True or false

16. $\dfrac{36 + 9}{6} = 6 + \frac{9}{6}$

17. $\dfrac{36 \times 9}{6} = 6 \times \frac{9}{6}$

18. $(X - Y)^2 = X^2 + Y^2 + 2XY$

Solve for X

19. $\dfrac{X}{Y} = Z$

$X = $ _____

20. $X^2 = (Y^2)(12 + 4)$

$X = $ _____

21. $\dfrac{2X}{3} = X - 2$

$X = $ _____

22. If 70% of 300 cases have trait A, what number of cases have trait A? _____

Find the square root

23. $\sqrt{0.09} = $ _____

24. $\sqrt{3600} = $ _____

Exercise B: Use of electronic calculators

Solve each of the following problems by using an electronic calculator in the most efficient method possible.

Addition

1.	**2.**	**3.**
527	16.432	23,462,179
624	19.715	1,212,107
+371	+ 3.641	+ 16,869

Subtraction

4. $8141 - 756 = $ _____
5. $872 - 946 = $ _____
6. $759 - 114 = $ _____

Multiplication

7. $67 \times 112 = $ _____
8. $1.5 \times 3.7 \times 14 = $ _____
9. $1.472 \times 0.31 = $ _____

Division

10. $147 \div 7 = $ _____
11. $0.00820 \div 4 = $ _____
12. $18 \div 900 = $ _____

Squares and square roots

13. $(143)^2 = $ _____

14. $\sqrt{0.000225} = $ _____

15. For the following series of numbers find both the sum and the sum after squaring each number. You should be able to get both results in a single operation.

$$17, \ 21, \ 32, \ 11, \ 14, \ 53, \ 12, \ 37, \ 9, \ 28$$

The first result will be designated as ΣX and the second as ΣX^2 where Σ represents "the sum of" and X is simply a letter name given to signify any of the numbers.

16. Often a group of numbers contains repetitions of the same numbers. Some calculators allow you to enter these numbers only once, along with the frequency of occurrence and still get ΣX and ΣX^2. Find the ΣX and ΣX^2 for the following numbers with the associated frequencies (listed under f). For example, there are 7 numbers equal to 24.

X	f
28	1
26	3
24	7
22	4
20	2

CHAPTER 2

Presentation of data

Lesson 1 □ Frequency distributions

Data sets that contain many measurements can result in a confusing jumble of numbers unless one has some method of organizing and condensing the information. In this chapter you will learn of several ways to show visually the information present in large data sets. The purpose in each case will be to condense the information while still retaining the meaning of the original data. One of the most commonly used methods, called a frequency distribution, will be described in the first lesson. The second lesson will describe graphic methods of showing a frequency distribution. It will also present ways of showing how one variable (such as population counts) changes relative to another (such as time).

The concept of a frequency distribution is straightforward. Instead of reporting every measurement separately, the analyst groups the measurements into classes and then reports the number (frequency) within each class. This technique can be used with either continuous or discrete variables and for measurements on any of the scales of measurement. Counting how many of the measurements fall in each class is easy once the classes have been unambiguously defined. The only thing to learn, then, is how to define the classes.

When working with nominal data, each class or combination of several classes becomes a class in the frequency distribution. The table showing the frequency in each class is easy to construct. An example appears in Table 2-1. Notice that since the scale is a nominal one and the order of the classes has no meaning, the usual practice of listing classes in descending order of frequency has been followed.

When the scale of measurement is either an interval or ratio scale, the question the analyst needs to answer is "How wide should my classes be in relation to the total spread of the set of observations?" If very narrow classes are used, a large number of classes will be needed to extend throughout the range of data. Carried to an extreme, a large number of very small classes will defeat the purpose of the frequency distribution. At the other extreme, making classes large in relation to the range of the observations will result in a frequency distribution that has only a few classes. Such a frequency distribution achieves

Table 2-1. Numbers of fish (by species) resulting from a sample survey of Holland Lake

Species	Frequencies
Brook trout	472
Yellow perch	126
Chubs	27
Suckers	14
TOTAL	639

the goal of condensing information, but fails because too much of the meaning of the original set of data is lost.

What is needed is a compromise that condenses the data enough to make a table of the frequency distribution easy to comprehend, while still retaining the essential meaning of the original measurements. Most statisticians suggest that the width of the class interval should be chosen in such a way as to result in somewhere between ten and twenty classes. Since the classes must extend the entire range of the data, this means that the width of the class should be somewhere between one tenth and one twentieth of the total range of the observations. Generally, the table will have an uncluttered appearance and be easy to read if the width of the class is a convenient number, such as 0.1, 0.5, 1, 2, 5, 10, or 20. Once a class width is chosen, the next problem is to decide where to start the first class. Again, the table will be easier to read if the lower limit of each class is a multiple of the class size. Of course, you will want to choose the lower limit of the first class in such a way that the lowest number in your data set falls within the first class. The lower limit of each subsequent class is then obtained by adding the class size to the lower limit of the current class.

It is a little harder to describe the process for choosing the upper limit for each class. A common error is to make the lower limit of one class correspond to the upper limit of the preceding class. This leads to ambiguity when dealing with a measurement that corresponds to the upper limit of one class and the lower limit of another. To avoid this problem, the upper limits are chosen to correspond to the largest measurement that could have been made without falling into the next class. If the data measurements are reported to the nearest one unit, this means that the upper limit of one class and the lower limit of the next class will be one unit apart. If data measurements are reported to the nearest five units, the upper limit of one class and the lower limit of the next class will be five units apart.

For data recorded to the nearest ten units with a range that extends from 230 to 750, a frequency distribution might look like the one in Table 2-2. Notice that a class size of 50 has resulted in 12 classes, that the lower limit of the first class is a multiple of 50, and that the lowest class includes the lowest score. Although there is not absolute consistency in the matter, the table usually starts with the lowest class at the top, as in Table 2-2.

A special case arises when data accurate to the nearest one unit are reported and the

Table 2-2. Verbal scores of applicants to the fall 1979 Master of Public Health program

Class limits	Frequencies
200-240	1
250-290	0
300-340	2
350-390	0
400-440	3
450-490	12
500-540	39
550-590	47
600-640	23
650-690	7
700-740	0
750-790	1
TOTAL	135

size of the class interval is also chosen to be one unit wide. In this case the upper and lower limits of each class are the same, and only the one number is listed. The same is true when data are reported to the nearest tenth of a unit and the width of the class interval is chosen to be 0.1, when data are reported to the nearest hundredth of a unit and the width of the class interval is chosen to be 0.01, or in any other case in which the class interval and the accuracy of the reported data are represented by the same unit of measurement.

For situations in which the data represent a continuous variable, it is often useful to consider *real boundaries* for each class as well as class limits. The class limits correspond to the recorded data, whereas class boundaries represent theoretical dividing points between classes. Suppose for example, that the weights of a group of individuals range from 81 to 235 pounds, reported to the nearest 1 pound. The observation reported as a weight of 81 might actually have represented a weight anywhere between 80.5 and 81.5. Similarly, the observation reported as 235 might have been an actual weight somewhere between 234.5 and 235.5. The actual range of the weights might, therefore, extend from 80.5 to 235.5, or 155 units. This is one unit greater than the distance between the highest and lowest observations.

In a similar manner, if each class interval is ten units wide and the limits of the lowest interval are chosen as 80 to 89, what actual weights of individuals would fall in this class? Possible reported weights would be 80, 81, 82 . . . 89. However, an observation reported as an 80 would represent a weight somewhere between 79.5 and 80.5, and the observation reported as 89, somewhere between 88.5 and 89.5. The real boundaries for the class are, therefore, 79.5 and 89.5. Notice that the distance between real boundaries is exactly the size of the class width. Class boundaries, unlike class limits, *do* overlap, with the upper boundary of one class identical to the lower boundary of the next. This causes no problems, since none of the reported scores will fall on a boundary. Boundaries contain one more decimal place of accuracy than the data that are reported.

A frequency table with both class boundaries and class limits for data reported to the

Table 2-3. Months of survival of 100 patients following diagnosis
of cervical cancer

Class boundaries	Class limits	Frequencies
0-5.5	0-5	1
5.5-11.5	6-11	1
11.5-17.5	12-17	3
17.5-23.5	18-23	7
23.5-29.5	24-29	8
29.5-35.5	30-35	5
35.5-41.5	36-41	12
41.5-47.5	42-47	9
47.5-53.5	48-53	7
53.5-59.5	54-59	5
59.5+	60+	42
TOTAL		100

nearest 1 month appears in Table 2-3. The class size in this instance is 6 months. Notice that the distance between the upper limit of one class and the lower limit of the next is one unit, since the data are reported to the nearest 1 month. Also notice that for this particular frequency distribution the real boundary of the first class is listed as zero (a number below zero would be meaningless) and the last class is open ended. The last class is left open ended because 5 years is generally considered a "cure."

In summary, a frequency distribution can be constructed by completing the following eight steps.

1. Find the highest and lowest observation in the data set.
2. Find the differences between the highest and the lowest observation.
3. Divide this difference by 10 and also by 20. Choose as a class size a convenient number somewhere between the two answers you get from these divisions.
4. Start the lowest class with a lower limit that is smaller than the lowest observation in your data set and that is a multiple of class size.
5. To get subsequent lower limits, add the class size to each current lower limit.
6. Determine the accuracy of the data being reported; for example, nearest hundredth, nearest tenth, nearest one unit, or nearest ten units. Now subtract this number from the class size. Adding the result to each lower limit provides upper limits.
7. If the data represent a continuous variable, you can get the real boundaries by finding the midpoint of the distance between the upper limit of one class and the lower limit of the next contiguous class.
8. Once the class limits are listed, consider each measurement. Mark the class each measurement falls into by placing a check mark beside the class. When check marks have been entered for all measurements, count the check marks beside each class to obtain the appropriate frequencies.

By following these steps, you should now be able to construct a frequency distribution appropriate for the data in the exercise for this lesson. The next lesson will use this frequency distribution as an example in showing some ways to illustrate a frequency distribution.

SUMMARY

This lesson described the construction of a table called a frequency distribution. The purpose of the table is to condense the information contained in a large data set, while retaining the meaning of the original measurements. Instead of reporting each individual measurement, a frequency distribution reports the frequency of measurements in each of a set of classes that extend throughout the range of the measurements. It was suggested that somewhere between 10 and 20 classes would be adequate. A set of steps that results in the construction of a satisfactory frequency distribution was provided.

LESSON 1 EXERCISE

Construct a frequency distribution to represent the following data, which are measures of the weights of laboratory animals of various ages following 6 months of a special diet. Weights are rounded to the nearest gram.

454	436	531	189	281
396	405	478	391	352
280	476	648	512	189
221	428	366	397	254
549	376	349	365	362
146	447	361	468	319
428	308	241	310	369
345	311	530	331	414
230	272	199	285	489
370	319	289	406	484

Lesson 2 □ Graphic methods

In this second lesson on presentation of data we will start with two further concepts to be used with frequency distributions: *cumulative* columns and *percentage* columns. First, let us consider a frequency distribution from the exercise in the previous lesson, as contained in Table 2-4. Notice that using 50 as the class size has resulted in 11 classes. Using the steps suggested, you may have chosen some other convenient number between 50.2 and 25.1 (648 − 146 divided by 10 and by 20) and constructed a different frequency distribution that also satisfactorily represents the data. Notice that the lower limit of the first class is a multiple of 50, that the first class contains the lowest number, and that the upper limit of one class and the lower limit of the next class do not overlap. The real class boundaries for each class are also shown. They do overlap and are reported to one more decimal place of accuracy than are class limits.

In addition to the columns already discussed, the table also contains three additional columns. The first is headed "*cf*," for cumulative frequency. It represents the cumulative number of observations in the class under consideration and all preceding classes. It is obtained by adding. The second column, headed "%," lists the percentage of the total numbers of observations falling in each class and is obtained by dividing each entry in the "*f*" column by the total number of observations (50 in this case) and then multiplying by 100. The third column, headed "*c%*" for cumulative percentage, tells the total percentage of observations falling below the upper boundary of each class. It is obtained by dividing each entry in the *cf* column by the total number of observations and then multiplying by 100. Notice that the last column is especially useful for describing how the weights are distributed; for example, 26% of the weights are below 299.5 grams, and 90% are below 499.5 grams.

Table 2-4. Weights (in grams) of laboratory animals following 6 months of special diet

Class boundaries	Class limits	Midpoint	f	cf	%	c%
99.5-149.5	100-149	124.5	1	1	2	2
149.5-199.5	150-199	174.5	3	4	6	8
199.5-249.5	200-249	224.5	3	7	6	14
249.5-299.5	250-299	274.5	6	13	12	26
299.5-349.5	300-349	324.5	8	21	16	42
349.5-399.5	350-399	374.5	11	32	22	64
399.5-449.5	400-449	424.5	7	39	14	78
449.5-499.5	450-499	474.5	6	45	12	90
499.5-549.5	500-549	524.5	4	49	8	98
549.5-599.5	550-599	574.5	0	49	0	98
599.5-649.5	600-649	624.5	1	50	2	100
TOTAL			50		100	

HISTOGRAMS AND POLYGONS

Two types of pictures are often used to portray a frequency distribution: histograms and frequency polygons. A histogram for the data in Table 2-4 appears as Fig. 2-1, and a frequency polygon appears as Fig. 2-2.

Fig. 2-1 contains a conventional histogram with a vertical scale for frequency and a horizontal scale made up of the boundaries of the classes. Since the visual concept of area is used to represent frequency, all columns are the same width. (This is easily accomplished using graph paper.) A column twice as tall as another would have twice the area, representing twice the frequency. The staggered line at the start of the horizontal scale is used to indicate that some classes have been left out. In this case, the two unneeded classes below 99.5, for which there were no weights, have been omitted. The same technique is used at the extreme ends of the scale for classes when there are several classes without any frequencies, followed by a class that does have an entry. Another way of handling this latter case is to spread the frequencies in the last class across the preceding empty classes. This is done by averaging; in other words, dividing the frequency in the last class by the number of empty classes plus the class with the entry. When this is done, the concept of area remains valid. In our example in Fig. 2-1, following this technique would result in 0.5 for the frequency for each of the last two columns. The last two columns of the histogram would then have shown a frequency as illustrated by the dotted line.

The frequency polygon appearing in Fig. 2-2 has the same purpose as the histogram and is similar in construction. It is formed by joining the midpoints of the tops of the

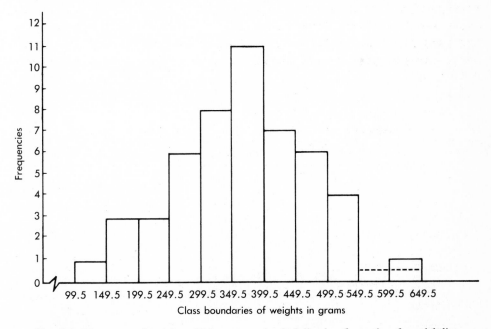

Fig. 2-1. Histogram of weights of laboratory animals following 6 months of special diet.

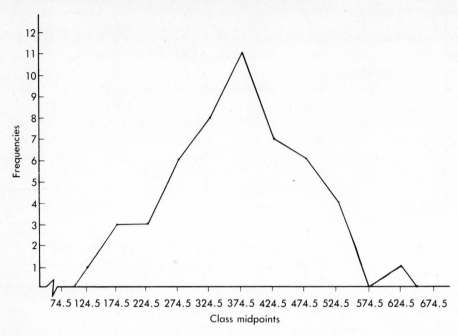

Fig. 2-2. Frequency polygon of weights of laboratory animals following 6 months of special diet.

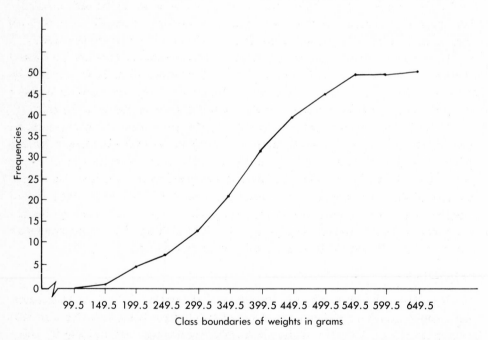

Fig. 2-3. Cumulative frequency polygon of weights of laboratory animals following 6 months of special diet.

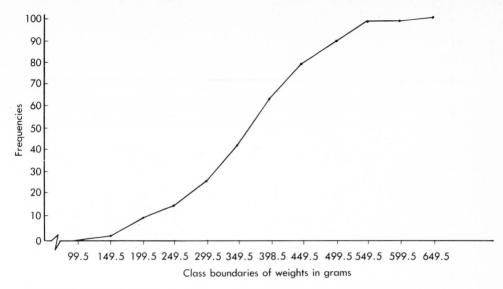

Fig. 2-4. Cumulative percentage polygon of weights of laboratory animals following 6 months of special diet.

histogram. Accordingly, the horizontal scale contains the midpoints of the classes rather than the boundaries. It is conventional to bring the line down to the base line at the beginning and the end, so that it will not appear to be floating in space. The usual points of contact with the base line are the appropriate boundaries of the first and last classes.

It is also rather common to see cumulative frequency polygons and cumulative percentage polygons (Figs. 2-3 and 2-4). These are constructed like the frequency polygon except that: (1) the entries in the cf and $c\%$ columns are used as the basis for height on the vertical scale, and (2) the points on the horizontal scale are class boundaries.

It can be seen that the cumulative percentage polygon is a useful way to present the data from a distribution of scores. It makes estimating the percentage of scores below any given number easy. You do this by going vertically from any desired point on the horizontal scale to the cumulative percentage line and then horizontally from that point to the vertical scale of percentages. For example, looking at Fig. 2-4, we see that a vertical line from 449.5 on the horizontal scale appears to meet the cumulative percentage line at a height corresponding to a percentage of approximately 80. Therefore, one might estimate that approximately 80% of the scores would fall below 449.5.

QUANTILES

As an aside before turning our attention to line diagrams, let us digress to clarify percentages and related concepts. Almost everyone uses percentages, having learned in elementary school that "you divide the number by the total and multiply by 100." Percentiles are not as often used and are sometimes misunderstood. Percentiles are used to designate the number in the frequency distribution below which a certain percentage of scores will fall. The ninety-fifth percentile is the number that divides the distribution into the lower 95% of the scores and the upper 5%. A closely related concept is the percentile rank. In this instance you start with the number and ask what its percentile rank is in the distribution. You might learn, for example, that a score of 43 on a biostatistics examination has a percentile rank of 72. This means that 72% of the students received scores below 43.

It is not necessary to divide the distribution into 100 units (centiles) as is done in computing percentages. In some instances the distribution is divided into ten units (deciles) or four units (quartiles). The general name applied to these numbers (centiles, deciles, quartiles) is *quantiles*. Both deciles and quartiles are in rather common use for describing the position of a score within a score distribution, and they are used in the same way as percentiles. For example, the second percentile would exceed two one-hundredths of the distribution, the second decile two tenths of the distribution, and the second quartile two fourths of the distribution.

SCATTER DIAGRAMS AND LINE DIAGRAMS

Graphs are often used to show how one variable changes in relation to another. One variable is plotted on the horizontal scale and the other on the vertical scale. There must, of course, be pairs of measurements; for example, weight and height, year and population, or age and blood pressure. When using regular graph paper, you should plan the vertical and horizontal scales so that the scale will fit nicely on the paper. For example,

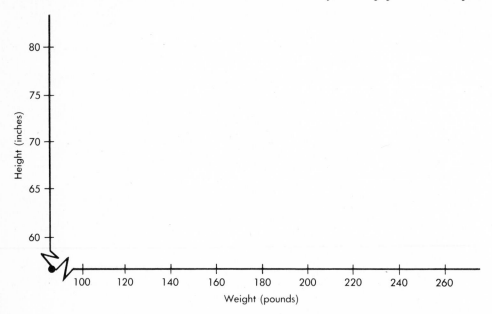

Fig. 2-5. Sample graph scales for height and weight.

suppose weights ranging from 100 to 250 pounds are to be plotted on a horizontal scale on graph paper with 18 squares across. Leaving some room for margins, you may decide to use 15 of the squares for the scale. Each square on the horizontal scale would represent 10 pounds. Similar computations for heights to be represented on the vertical scale might result in a scale in which each square represents 2.5 inches. You then would construct the scales starting at the lower left-hand corner of the paper. The scales might appear as in Fig. 2-5.

Once you have constructed the scales, you then place a dot on the graph for each pair of measurements. For some types of data each point on the horizontal scale may have more than one point on the vertical scale. For example, persons weighing 180 pounds may vary in height from 65 to 75 inches. When you are plotting data in which these multiple measurements are possible, the result will be a scatter diagram, as shown in Fig. 2-6. Remember that each dot represents a pair of scores for an individual. The height on the vertical scale indicates the person's height, and the distance to the right on the horizontal scale represents the person's weight. When in general the tall persons are also the heavy persons, the dots will form an elongated cluster from the lower left to the upper right. When there is no relationship between the variables, the cluster of dots in the scatter diagram will not show any clear tendency to angle either upward or downward, but may instead simply have a circular shape. In a later lesson you will learn to compute an index that measures the strength of the linear relationship between two variables.

There are also many situations in which you wish to show the relationship between two variables where only a single measurement exists for each point on the horizontal

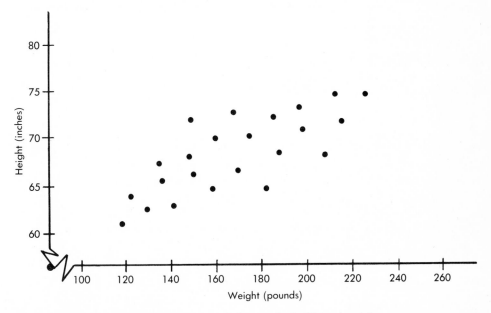

Fig. 2-6. Sample scatter diagram for height and weight.

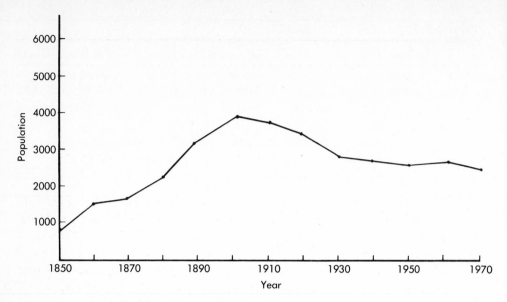

Fig. 2-7. Line diagram: population of Newberry, Pa. (1850-1970).

scale. The most usual example is when time is plotted on the horizontal scale and a measurement for some other variable (such as population) exists at regular intervals over the period of interest. In this instance the dots plotted for the pairs of measurements (such as year and size of population) are usually connected by straight lines, and the result is called a line diagram (Fig. 2-7).

In many instances it is desirable to compare the changes over time in a variable (for example, population) for two different populations (for example, two cities, or a city and its suburbs). In this case, a single graph contains two lines; one may be solid, the other broken.

LOGARITHMIC GRAPH PAPER

Ordinary graph paper contains squares of equal size throughout and is useful for a wide range of situations. There is also a special kind of graph paper called logarithmic paper, in which the squares are not of equal size, but instead are arranged so as to represent a logarithmic scale. If both the horizontal and vertical scales are logarithmic, the paper is full-log paper. If only one scale is logarithmic and the other is ordinary, the paper is semilog paper. The number of times the logarithmic scale is repeated on either the horizontal or vertical scale determines whether the paper is two- or three-cycle paper. For the exercise you will be asked to do, you will need three-cycle semilog paper.

Two major reasons for plotting graphs on logarithmic paper can best be illustrated by a greatly simplified example. Suppose you wish to compare population changes in city *A* with those of city *B* and the data are as reported in Table 2-5. If you try to plot the data of

Table 2-5. Populations of two cities from 1900 to 1960

Year	City A	City B
1900	1000	100,000
1910	2000	110,000
1920	4000	120,000
1930	8000	130,000
1940	16,000	140,000
1950	32,000	150,000
1960	64,000	160,000

Table 2-5 on ordinary graph paper, you will immediately be confronted with a problem. The scale for population size appropriate for B is not appropriate for A. Using a scale that will accommodate the great disparity in city sizes makes the changes in the population of city B indistinguishable. In addition, even if you were able to plot the numbers, the result might be misleading, in that both cities would appear to be growing at about the same rate. Although it is true that each increased by about 60,000 over 60 years, the proportional change is much greater in city A. It would be possible to do the calculations necessary and then plot *percentage* of increase every 10 years. A simple solution, which also takes care of the disparity in population sizes, is to use three-cycle semilog paper.

When using semilog paper you scale time just as before on the ordinary scale along the horizontal (Fig. 2-8). The logarithmic scale used for the vertical may be a bit confusing at first. At the very bottom you will see a 1, then a 2, a 3, and so forth up to 9, and then the cycle starts over again. Each of the cycles can be used to represent a decimal place. Three-cycle paper can be used, therefore, for data with a size disparity of three decimal places; for example, 1000's to 100,000's, or 100's to 10,000's. In our example the data for city A, which start in the 1000's and end in the 10,000's will be plotted in the first cycle and extend into the second. All that must be remembered is that when you get into the second cycle each number on the scale then represents 10,000. The number 16,000 will, therefore, be between the 1 and the 2 of the second cycle. If every number equals 10,000, it is clear that the ten divisions between the 1 and the 2 of the second cycle must each represent 1000. The number 16,000 will, therefore, be six divisions above the 1 of the second cycle. The data for city B are all in the 100,000's, and so will be plotted entirely in cycle three. Again, note that in the third cycle each number represents 100,000, so each of the ten divisions between 1 and 2 must equal 10,000. The number 110,000, therefore, is marked at the first division above the 1 in the third cycle.

When we look at Fig. 2-8, it is apparent that the rate of growth is much greater in city A than in city B. Note also that the rate of change in city A is a constant doubling of the population every 10 years. This constant proportional change is shown on the graph by a straight line for city A. The population of city B increases by a constant 10,000 each year, but the proportional increase gets smaller as the population base increases. Although it is difficult to see, this is reflected by a line that is slightly curved rather than straight.

The use of the logarithmic scales can probably be understood only when you have had the experience of using them. You will get to do this as part of the exercises that follow. Remember that when you wish to compare or measure *proportional* change between consecutive time points, logarithmic paper is appropriate. When *absolute* change is being compared, ordinary graph paper is the choice.

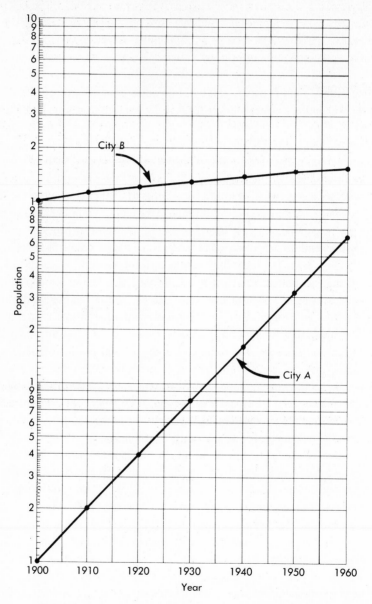

Fig. 2-8. Log scale for population of two cities (1900-1960).

SUMMARY

This lesson began with a description of three additional columns often included as part of a frequency distribution: (1) cumulative frequency, (2) percentage, and (3) cumulative percentage. Next, graphic methods for portraying the information contained in a frequency distribution were illustrated. Histograms, frequency polygons, cumulative frequency polygons, and cumulative percentage polygons were included. The last part of the lesson described scatter diagrams and line diagrams as a means of showing the relationship between two variables. The use of semilog paper in conjunction with line diagrams was shown to be an easy way to examine proportional rate of change.

LESSON 2 EXERCISES

1. Construct a histogram, a frequency polygon, and a cumulative percentage polygon for the data appearing in the following frequency distribution. The data represent grade point averages (GPA) for Master of Public Health (MPH) graduates in a school of public health.

Frequency Table of the GPA for the 82 MPH graduates in 1975

Class boundaries	Class limits	f	%	cf	c%
3.595-3.695	3.60-3.69	2			
3.695-3.795	3.70-3.79	0			
3.795-3.895	3.80-3.89	0			
3.895-3.995	3.90-3.99	2			
3.995-4.095	4.00-4.09	7			
4.095-4.195	4.10-4.19	7			
4.195-4.295	4.20-4.29	11			
4.295-4.395	4.30-4.39	9			
4.395-4.495	4.40-4.49	13			
4.495-4.595	4.50-4.59	7			
4.595-4.695	4.60-4.69	10			
4.695-4.795	4.70-4.79	6			
4.795-4.895	4.80-4.89	6			
4.895-4.995	4.90-4.99	2			

2. Construct a line diagram to show the number of mopeds sold annually in Chicago and Paris. Do it once with ordinary graph paper and once with semilogarithmic paper. What conclusions do you reach regarding growth of sales in the two cities?

Year	Chicago	Paris
1964	142	30,212
1966	360	32,982
1968	940	34,490
1970	2104	37,721
1972	4271	41,289
1974	8012	44,003
1976	11,271	46,572

CHAPTER 3

Summary statistics

Lesson 1 □ Symbols and measures of central tendency

In the preceding chapter you learned how to organize data into frequency distributions and how to represent these frequency distributions in histograms and frequency polygons. Often the information from a frequency distribution can be conveyed even more concisely by describing three characteristics of the frequency distribution or its corresponding frequency polygon: (1) central tendency, (2) variability, and (3) shape. As an illustration of how these characteristics differ, consider frequency polygons for the age distributions of the following groups: (1) students in an elementary school, (2) major league baseball players, and (3) residents of a nursing home. The polygons would appear as shown in Fig. 3-1. The three frequency polygons have obvious differences: they are located in different places on the age scale, the amount of spread in the polygons varies from about 7 years for the elementary school students to over 40 years for the nursing home residents, and the shapes of the three frequency polygons differ considerably. It is possible to arrive at measures of each of these characteristics so that we can describe any frequency distribution by stating its central tendency, its variability, and its shape. You will learn to make extensive use of these three characteristics of a frequency distribution. Later, we will distinguish between distributions that represent entire populations and distributions representing samples from a population. When the measures of central tendency, variability, and shape describe a distribution of a *sample,* the measures are called *statistics.* When they describe a *population* distribution, they are called *parameters.*

In this first lesson, we will start with measures of central tendency, since they are the most familiar. Then in subsequent lessons we will deal with the other two characteristics of a distribution and learn of some uses for the concepts. Before proceeding, it will be necessary to introduce some symbols to make explanations more concise.

SUMMATION SYMBOLS

So much of the work that statisticians do involves adding sets of numbers that it is convenient to have a shorthand way of saying "add the weights of all the persons in group *A*" or "add the hospitalization costs of all the persons in group *B*." We begin by using a

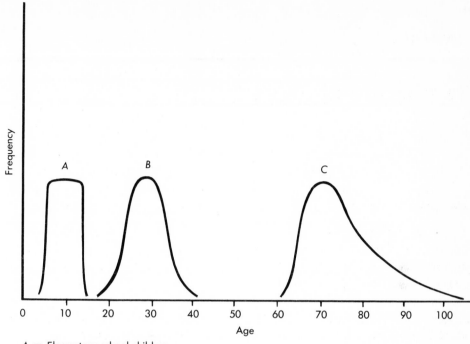

A = Elementary school children
B = Major league baseball players
C = Nursing home residents

Fig. 3-1. Sample frequency polygons for three age distributions.

letter to symbolize the measurement for weight, or hospitalization cost, or whatever it is we are interested in. Any letter can be chosen, so long as we specify what it means. Often X, Y, or Z are used, but there is nothing sacred about these choices. The characteristic represented by the letter is called a *variable,* as opposed to a *constant.* As you might easily guess, a variable is any characteristic that when measured gives varying results for the different members of the group being studied. Age, weight, and height might all be variables of interest in a given study, and we might decide to refer to them as variables A, B, and C; or X, Y, and Z; or any other letters we might wish to designate.

Suppose we have ten individuals and for each we have measures of the variables age, weight, and height. The data might be shown as in Table 3-1. A subscript is used to designate a particular individual. It is then possible to designate a measure for any of the variables for any of the individuals. For example, referring to Table 3-1: $X_2 = 18$, $Y_9 = 173$, and $Z_8 = 68$.

Once we have decided to use a letter to designate a variable, with a subscript for a particular individual, all we need is a symbol that means to add the designated numbers. The symbol is Σ, which is referred to as a summation sign. When all of the numbers in a group for variable X are to be added together, we can simply write ΣX. It is also possible

Table 3-1. Ages, weights, and heights of ten individuals *(i)*

i	Age = X	Weight = Y	Height = Z
1	20	112	65
2	18	147	67
3	13	95	61
4	27	180	72
5	22	195	74
6	15	108	64
7	33	130	62
8	29	148	68
9	24	173	69
10	25	132	70

to designate that only the numbers for certain members of the group are to be added. This is done by stating limits for the addition above and below the summation sign. For example, $\sum_{i=2}^{4} X$ means to add the measures on variable X for individuals two through four. When $\sum X$ is used alone, it is understood that what is meant is $\sum_{i=1}^{n} X$ where *n* stands for the total number.

There are three rules for algebraic manipulation of summation signs. Later, knowing these rules will make it possible to transform several forms of the same formula when dealing with measures of variability. However, we may as well introduce them now. All three will appear self-evident, when you think about them.

1. $\Sigma K = nK$ where K stands for any constant

To say that a measure is a constant for any group means that all individuals in the group have the same score on that measurement; for example, each of the ten people in a group might be the same age, 20 years. If this were the case, you could add the age, 20, ten times to get a total of 200 years; or you could multiple ten by twenty to get the same result.

2. $\Sigma(X + Y) = \Sigma X + \Sigma Y$ where X and Y are both variables

This formula says that if you have two variables measured for each individual and you want to know the sum of all the scores, you can either add the two measures for each individual, and then add all the results; or you can add all the scores for each variable to get two totals, and then add the two totals. Obviously, the result is the same either way.

3. $\Sigma KX = K\Sigma X$ where K is a constant and X is a variable

This formula says that if you have a single variable measured for each individual and you want to find what the total would be if all the measurements were multiplied by a constant, you can either multiply each of the measurements by a constant and then add, or you can add all the measurements first and them multiply by the constant.

To be sure you understand these rules, think through the following questions: (1) If each of ten people earns $5000 per year, what are two ways to compute their total annual income? (2) What are two ways to compute the total number of correct answers on the midterm and final examination by a class of biostatistics students? (3) What are two ways to figure out what the total weight of a group of laboratory animals will be if each animal doubles its weight?

MEASURES OF CENTRAL TENDENCY

The mean. Now that you understand summation notation, it will be easy to write the formula for one measure of central tendency, the mean. The mean is the arithmetic average that you get by adding all of the measurements in a group and then dividing by the total number. It is what most people think of when they refer to an "average." One way to describe the different frequency polygons in Fig. 3-1 would be to state the mean for each distribution. The mean would give an indication of the location of each group on the scale for age. It is traditional in statistics books to use two different symbols for the mean, depending on whether it is for an entire population of scores or for a sample. For the population mean the Greek letter for *m* is used. It looks like this, μ, and is pronounced "mew." For samples, it is traditional to designate the mean by placing a bar over the symbol for the variable, for example, $\overline{X}$ or $\overline{Y}$. For now we will consider our examples to be populations; therefore we will use μ as our symbol for the mean. If two variables, X and Y, are being considered, we will use a subscript with μ as an identifier, for example, μ_X or μ_Y.

The formula for the computation should be easy to write, using summation notation. It is:

$$\mu = \frac{\Sigma X}{N} \qquad \text{3.1-1}$$

When data have been placed in a frequency table, the class midpoint is then the representative value for every observation in that class. When computing a mean from a frequency table, you must multiply the *midpoint* of each class by the frequency *(f)* for that class and then add the results. If we let M equal the midpoint for each class, the formula is:

$$\mu = \frac{\Sigma f M}{N} \qquad \text{3.1-2}$$

where $N = \Sigma f$.

Notice that in this case both f and M are variables, since the frequencies vary from class to class and the midpoints are all different. Some exactness in finding the mean is lost when you multiply the midpoints by the frequencies rather than add all the scores in each class. However, with ten or more classes the errors will average out from one class to another and the result will be quite close. An exercise at the end of the lesson will demonstrate this.

The mean is the most used measure of central location or tendency. It is easy to understand and, with electronic calculators, rather easy to compute. As we will see later, however, there are some situations in which use of the mean as a measure of central tendency may be misleading.

The median. Another measure of the central tendency of a distribution of observations is called the median. It is generally thought of as the number that divides the distribution into two equal parts. The median will, therefore, be the same as the fiftieth percentile. Computing the median would appear to be a rather straightforward task: arrange the scores in order, then count up until you have reached the halfway point. Questions can arise, however. For example, what would be the median of the following numbers?

$$1, 2, 2, 3, 3, 3, 3, 5$$

If you answer 3, there are three numbers below 3 and only one above. If the numbers represent a discrete variable, there is really no good solution to the question. When the numbers represent a continuous variable, however, there is a logical procedure for defining the median. The numbers recorded as 3's are really numbers somewhere between 2.5 and 3.5. Since we have no way to know their real locations, we will assume they are evenly distributed throughout this interval. Under this assumption, going halfway into the interval would include half the numbers; three fourths of the distance would include three fourths of the numbers, and so forth. Since in our example there are eight observations in all, four should be below the median and four above. There are three below the interval 2.5 to 3.5. We need one more. We want to go far enough into the interval to get one more observation. Since there are four observations in the interval (assumed to be evenly distributed) and we want one of them, we will need to go one fourth of the distance into the interval. Since the interval is one unit wide, we will add $1/4$ of 1, or $1/4$, to the lower boundary, 2.5, resulting in a median of 2.75.

The preceding logic is also applied to find the median of data that have been grouped into a frequency distribution. Consider for example Table 2-4 in Lesson 2 of Chapter 2. Altogether there were 50 scores, so 25 should be above the median and 25 below. There were 21 scores below the class of 349.5 to 399.5 (represented in class boundaries), and 11 scores in that class interval. The median is somewhere within the class. Since we already have 21 of the 25 scores needed, we want to choose a point that includes 4 of the 11 scores. Again, we assume the scores are evenly spread throughout the interval. We go $4/11$ of the distance into the interval. Since the class interval is 50 units wide, we multiply $4/11$ times 50, add the result to the lower boundary, and call this the median.

$$\text{Median} = (4/11)(50) + 349.5 = 18.18 + 349.5 = 367.68$$

You will have an exercise for this lesson that requires you to find the median under similar circumstances.

The median is fairly easy to compute and is often used as a measure of central tendency in situations where the mean is difficult to compute or, perhaps, inappropriate.

The mode. The mode is another measure of central tendency that is sometimes used, largely because of ease of computation. It is the measurement that appears most frequently. In the case of observations in a frequency distribution, the mode becomes the modal class, the one with the greatest frequency. In some instances, two different values that are widely separated may be tied for highest frequency, or the same may be true of two classes in a frequency distribution. In this instance, the distribution is said to be bimodal.

COMPARISON OF MEASURES OF CENTRAL TENDENCY

As stated earlier, the mean is by far the most common measure of central tendency. In the days before electronic calculators, the median was often used as a substitute because it was easier to calculate for large data sets. In some instances the median is the more appropriate measure because it better represents the measurements. An example occurs when a few extreme observations would greatly influence the mean. When computing the median, you would merely count these extreme observations as additional values above the median. The mode is the least-used measure of central location, largely because of its instability. A change of a single number can greatly affect the mode.

SUMMARY

The lessons of this chapter describe indexes that can be used to condense even further the information contained in a frequency distribution. The indexes describe three attributes of a distribution of measurements: (1) central tendency, (2) variability, and (3) shape. This first lesson was concerned with the first attribute, central tendency. To make it easy to describe the formulas for descriptive statistics, a symbol called the summation sign (Σ) was introduced. Three rules for the algebraic manipulation of terms that include summation signs were given. The use of a letter with subscripts to designate measurement for an individual for a specified variable was also described. The remainder of the lesson explained the calculation and use of three measures of central tendency: (1) the mean, (2) the median, and (3) the mode.

LESSON 1 EXERCISE

The weights, in ounces, of 48 babies born at a community hospital are shown below. The letter X is used to represent the variable weight.

i	X	i	X	i	X	i	X	i	X
1	93	11	114	21	115	31	123	41	131
2	117	12	114	22	131	32	136	42	98
3	120	13	106	23	114	33	112	43	134
4	122	14	94	24	107	34	130	44	126
5	98	15	109	25	107	35	138	45	109
6	122	16	117	26	86	36	123	46	114
7	115	17	142	27	128	37	152	47	47
8	110	18	122	28	123	38	107	48	101
9	117	19	110	29	133	39	144		
10	146	20	133	30	101	40	105		

1. Find X_{23}.
2. Find $\sum\limits_{i=10}^{20} X$.
3. Find ΣX.
4. Find μ.
5. Find the median of X.
6. Now make a frequency distribution with a class size of ten.
7. Find μ from the frequency distribution.
8. Find the median from the frequency distribution.

Lesson 2 □ Measures of variability

In the previous lesson you learned about several measures of central tendency. These measures tell where, "on the average," the observations are located. However, two groups of observations may have the same central tendency and be quite different on the other two attributes of a distribution: variability and shape. As an example, the average daily highs for temperature over a year of observation, might be very similar for Seattle and Kansas City. However, the distribution of daily highs would be very dissimilar. This lesson deals with variablity of a distribution. The next will deal with shape.

RANGE AND INTERQUARTILE RANGE

You already used one measure of the variability of observations when you figured the range in the construction of frequency distributions. Remember that the range was obtained by subtracting the smallest observation from the largest observation. In a number of statistics books you will find that a modification is suggested when you are finding the range for measurements of a continuous variable that have been rounded to some given degree of accuracy. If the data are reported to the nearest whole unit, it is suggested that you add one to the difference you get when subtracting the lowest reported score from the highest; if data are rounded to the nearest one tenth, you add one tenth, and so forth. This suggestion is based on logic similar to that used in finding the class boundaries of a frequency distribution.

The range is easy to compute, and it does give an idea of the spread of the observations. Unfortunately, since the range depends on only the two most extreme numbers, it does not represent the rest of the numbers very well and is not very stable. Dropping either the highest or the lowest observation might have a great effect on the reported range. A measure that represents more of the data and is still relatively easy to compute is called the interquartile range. It is the distance between the first and third quartiles, or the twenty-fifth and seventy-fifth percentiles. Accordingly, the interquartile range indicates how far apart the middle 50% of the observations are spread.

AVERAGE DEVIATION

Both the range and interquartile range represent only part of the data. A measure of variablity that makes use of all of the data is the average deviation. The word *average* implies that you will "add something and divide by *n*." In the case of average deviation, what is averaged is the distance of each observation from the center of the distribution, as represented by the mean. Closely related to the idea of "distance from the mean" is the concept of "deviation from the mean." When statisticians speak of a deviation from the mean, they add a sign to the distance. If the score is above the mean, the deviation is positive; if the score is below the mean, the deviation is negative. Deviations from the mean are used often enough in statistics to warrant a special symbol. It is the small letter corresponding to the capital letter used to designate a variable. If X is used to represent the variable "weight" for a population of 25 individuals, μ_X represents the mean weight of the 25 individuals, X_{12} represents the weight of individual number 12, and x_{12} represents the deviation from the mean for individual number 12. The formula for a deviation score appears as Formula 3.2-1.

$$x_i = X_i - \mu \qquad \qquad \textbf{3.2-1}$$

where i stands for any individual.

Note that if the individual's score is lower than the mean, his deviation score will be negative.

We could use Formula 3.2-1 to compute each person's deviation score and then average these scores to find the average deviation. If we try this, an interesting thing will happen. Adding all the deviation scores, some of which are positive and some negative, will result in a total equal to zero. In other words, the plus scores will exactly balance the minus scores. This is one of the properties of the mean: the sum of the deviations from the mean is zero.

To compute the average deviation, we will have to consider just the distance from the mean, without worrying about the direction of the distance. Mathematicians use two vertical parallel lines for this purpose and call the value included within the parallel lines an *absolute value*. The formula for average deviation can be stated as in Formula 3.2-2.

$$\text{Average deviation} = \frac{\Sigma|X_i - \mu|}{N} \quad \text{or} \quad \frac{\Sigma|x|}{N} \qquad \textbf{3.2-2}$$

where N = the number of scores.

Since we are actually averaging absolute deviations, perhaps the name should have been average absolute deviation. As it happens, it is of little consequence, since the use of the average deviation has become very limited.

VARIANCE AND STANDARD DEVIATION

The variance and its square root, the standard deviation, have largely replaced the average deviation as a measure of variability. The variance also uses deviation scores, as can be seen in Formula 3.2-3.

$$\text{Variance} = \frac{\Sigma(X - \mu)^2}{N} \quad \text{or} \quad \frac{\Sigma x^2}{N} \qquad \textbf{3.2-3}$$

Notice that instead of taking the absolute value of each deviation score, the deviation scores have been squared, making all the negative deviation scores positive values. In other words, the variance is the average squared deviation from the mean. The variance will usually be a larger number than the average deviation, since squaring deviation scores results in larger numbers (unless the absolute deviations are between 0 and 1). Taking the square root of the variance results in a number called the standard deviation, which is closer to the average deviation. The formula is as follows:

$$\text{Standard deviation} = \sqrt{\frac{\Sigma x^2}{N}} \qquad\qquad \textbf{3.2-4}$$

Later you will learn to use different symbols for variance and standard deviation, depending on whether you are considering the variability of a whole population of observations or the variability of a sample of observations. You will also learn to use a capital N when referring to the number in a population and a small n when referring to the number in a sample. The formulas for variance and standard deviation also differ slightly, depending on whether they are being computed for a sample or a population. For now, let us confine our discussion to populations and use the symbol σ^2 for variance and the symbol σ for standard deviation. Since we are talking about populations, we will use N in the denominator of formulas. Although σ and σ^2 may not be the easiest symbols to use, their use is rather standard, so you may as well get used to them. The name for the symbol is the Greek letter sigma. You will sometimes see the standard deviation referred to as sigma. In words, the standard deviation is the square root of the average squared deviation from the mean. At first you will have great difficulty conceptualizing the meaning of the standard deviation. For now, just think of it as a measure of variability that is related to the average deviation, which you should be able to conceptualize.

SUM OF DEVIATIONS SQUARED AND COMPUTATIONAL FORMULAS

What you have seen so far may be referred to as definitional formulas for the variance and standard deviation. To compute these measures of variability using these formulas would be a tedious job. For example, to find the variance you would have to do the following:

1. Find the mean.
2. Subtract the mean from each score to get deviation scores.
3. Square all the resulting deviation scores.
4. Add all the squared deviation scores.
5. Divide by N.

If the mean turned out to be an uneven number, the process would become very tedious indeed.

Fortunately there are computational formulas that are exactly equivalent to the definitional formulas and that make computations easier. These formulas are based on an alternate method of finding the sum of deviations squared. In the process just described, steps 1 through 4 would be the usual way to compute the sum of deviations squared. A process that will yield exactly the same result is described in Formula 3.2-5.

$$\Sigma x^2 = \Sigma X^2 - \frac{(\Sigma X)^2}{N} \qquad\qquad \textbf{3.2-5}$$

The x to the left of the equal sign is a small x, signifying a deviation score, and those to the right are large X's, signifying raw scores. By using the rules you learned earlier relating to summation signs, you may be able to prove Formula 3.2-5. If not, see the technical note at the end of the lesson.

The concept of the sum of deviations squared and the computational formula for the sum of squares are extremely important in statistics. The sum of deviations squared might be called the building block or the ''brick'' of statistics, because it appears in so many formulas. Be certain that you know what it means. It might even be a good idea to make a red circle around Formula 3.2-5 and name Σx^2 the ''brick.''

Notice that to compute Σx^2 using Formula 3.2-5 it is not necessary to compute the mean. The right-hand part of the formula says you only need to find the sum of all the observations and the sum of the squares of each observation. These values, along with the number of observations, are substituted in the proper places in the formula, providing the value of Σx^2 without ever calculating a single deviation score. To clarify things, let us look at an example. Take the scores 1, 2, 4 and find the Σx^2 by two methods. Using the definitional formula, you follow steps 1 through 4 listed earlier. Since the mean turns out to be 2.33, the calculations become rather involved, but you should get a total sum of deviations squared equal to 4.6667. Using the computational formula, you find $\Sigma X = 7$, $\Sigma X^2 = 21$, $N = 3$ and

$$\Sigma x^2 = \Sigma X^2 - \frac{(\Sigma X)^2}{N} = 21 - \frac{(7)^2}{3} = 4.6667$$

Now that you are familiar with the brick and its computational form, you should be able to write the computational formulas for variance and standard deviation by substituting the computational form of the brick in the proper places in the definitional formulas. The resulting formulas are given as Formulas 3.2-6 and 3.2-7.

$$\text{Variance or } \sigma^2 = \frac{\Sigma X^2 - \dfrac{(\Sigma X)^2}{N}}{N} \qquad \textbf{3.2-6}$$

$$\text{Standard deviation or } \sigma = \sqrt{\frac{\Sigma X^2 - \dfrac{(\Sigma X)^2}{N}}{N}} \qquad \textbf{3.2-7}$$

Formulas 3.2-6 and 3.2-7 look a bit imposing, but you will use them so often that they will become familiar. Many calculators provide a method of calculating ΣX, ΣX^2, and N by entering each raw score only once. The exercise you are asked to do at the end of this lesson will require very little time if you use such a calculator. Some advanced model calculators will not only find the three values required for the brick (Σx^2) but will even substitute this value in the appropriate place in Formula 3.2-7 and display the result. You should be aware that most such calculators use $n - 1$ in place of the bottom N in Formula 3.2-7. The reason for this will be explained later.

When finding the variance or standard deviation for data from a frequency distribution, we must do as we did for the mean. Actually we find the standard deviation of the midpoints of the classes. Remember to take the frequencies within each class into account. The formula for variance is stated as Formula 3.2-8.

$$\sigma^2 = \frac{\Sigma f(M^2) - \dfrac{(\Sigma fM)^2}{N}}{N}$$

3.2-8

where M = midpoints,
f = frequencies within each class, and
$N = \Sigma f$.

Many calculators have special functions that allow you to find $\Sigma f(M^2)$, ΣfM, and N in a single operation.

Now that you know how to compute the standard deviation and you know that it is an indication of the amount of variability in a set of scores, you are probably ready to say, ''so what?'' or ''who cares?'' Be patient and be sure you know how to compute these measures now. In the fourth lesson of this chapter you will begin to see some of the usefulness of the concepts. Later you will use the concepts extensively when studying inferential statistics.

To provide a bit more feel for the meaning of the standard deviation, something should be said about the meaning of the size of the standard deviation. When is the standard deviation large, indicating great variability, and when is it small? The answer is that it depends on the size of the measurements you are dealing with. A standard deviation of ten may be very large in one setting or very small in another. A concept that relates the variability of a set of scores to the average size of a set of scores is the *coefficient of variation*. In the form of a percentage, it states the ratio of the standard deviation to the mean. The formula is given as Formula 3.2-9.

$$\text{Coefficient of variation} = \frac{\text{standard deviation}}{\text{mean}} \times 100$$

3.2-9

When two distributions have different means, comparison of the respective coefficients of variation is more meaningful than comparison of the respective standard deviations.

SUMMARY

This second lesson on descriptive statistics discussed measures of variability within a distribution of scores. The range and interquartile range were mentioned first. You had previously used the range when deciding what size of class interval to use when constructing a frequency distribution. You had also learned previously about the meaning and calculations for a quartile. The interquartile range was described as the distance between the first and third quartile.

Next, a concept called a deviation score was described. The formula was $X_i - \mu$. A measure of variability called the average deviation was then explained. This measure is seldom used any more. It was described because it provides a reference for understanding two indexes of variability that are widely used: variance and standard deviation. Both definitional and computational formulas were given for these indexes of variability. A term that appears in these computational formulas and many other statistical formulas was given a name, the "brick." The brick is the sum of squared deviations. The lesson concluded with a description of a concept called the coefficient of variation. The coefficient of variation relates the standard deviation of a set of scores to the mean of the scores.

LESSON 2 EXERCISE

All questions refer to the 48 weights listed for the exercise of the preceding lesson. Consider these to be a population of observations.
1. Find the range of the 48 scores.
2. Find the average deviation of the first ten scores.
3. Find the variance of the 48 scores.
4. Find the standard deviation of the 48 scores.
5. Find the coefficient of variation.
6. Now use the frequency distribution you constructed as part of the previous exercise and find the standard deviation and the variance.
7. Find the interquartile range from the frequency distribution.

TECHNICAL NOTE

If you remember the three rules given earlier for algebraic manipulation of summation signs, you should be able to follow the seven steps below. The seven steps demonstrate that the definitional form of sum of deviations squared (the brick) is the same as the computational form.

$$\text{Definitional form: } \Sigma x^2 = \Sigma (X - \mu)^2$$

$$\text{Computational form: } \Sigma x^2 = \Sigma X^2 - \frac{(\Sigma X)^2}{N}$$

1. $\Sigma x^2 = \Sigma (X - \mu)^2$ but $(X - \mu)^2 = X^2 - 2X\mu + \mu^2$
2. $\therefore \Sigma x^2 = \Sigma (X^2 - 2X\mu + \mu^2)$
3. or $\Sigma x^2 = \Sigma X^2 - \Sigma 2X\mu + \Sigma \mu^2$
4. or $\Sigma x^2 = \Sigma X^2 - 2\mu\Sigma X + N\mu^2$ but $\mu = \dfrac{\Sigma X}{N}$
5. $\therefore \Sigma x^2 = \Sigma X^2 - \dfrac{2\Sigma X}{N}(\Sigma X) + N\dfrac{\Sigma X}{N}\dfrac{\Sigma X}{N}$
6. or $\Sigma x^2 = \Sigma X^2 - \dfrac{2(\Sigma X)^2}{N} + \dfrac{(\Sigma X)^2}{N}$
7. or $\Sigma x^2 = \Sigma X^2 - \dfrac{(\Sigma X)^2}{N}$

Lesson 3 □ **Shape of a distribution and the normal curve**

Two distributions can have the same mean and the same standard deviation and still be quite different because of the shapes of the distributions. For example, some distributions are symmetrical about the mean. By saying they are symmetrical, we are referring to the condition in which a vertical line drawn through the mean of the distribution would divide the distribution into two equal-appearing halves. In contrast to symmetrical distributions, others are skewed (not symmetrical). The degree of skewness may vary from slightly skewed to markedly skewed. There are ways of measuring the amount of skewness, but we will not be concerned with this. It is traditional to say that a distribution is positively skewed if the tail is to the right and negatively skewed if the tail is to the left (Fig. 3-2).

When we speak about the shape of a distribution, it is not sufficient simply to describe its symmetry or any skewness that might be present. It is also possible to differentiate between distributions in which observations occur in approximately equal numbers throughout the range and distributions in which there is a tendency for observations to cluster about the mean and occur with less frequency the further you go from the mean in either direction. The first is an even distribution; the second has a degree of peakedness. The peak results from the larger proportion of scores near the mean. Again, there are measures of peakedness in a distribution, but we will not be concerned with this either.

Instead of measuring degree of skewness and degree of peakedness when considering the shape of a distribution of scores, we will generally compare the distribution with a theoretical distribution with known characteristics. One of the most important of these is the normal distribution.

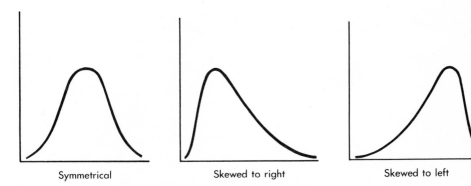

Symmetrical   Skewed to right   Skewed to left

Fig. 3-2. Shapes of distributions.

NORMAL DISTRIBUTIONS

If you were to make up a frequency polygon for heights of 1000 randomly chosen men, birth weights of 1000 newborn guinea pigs, or lengths of the dorsal fins of 1000 mature salmon, you would discover that the shapes of the distributions would have much the same appearance. Strangely enough, if each of 1000 people flips a coin 20 times and records the number of heads, a frequency polygon of the results will have an appearance similar to those of the examples above. Further, suppose you instruct 1000 elementary school principles to choose 30 students at random, record their ages, and compute the mean age for those 30 students. Within each elementary school, the distribution of *ages* will be rather evenly distributed from 6 to 13 years. However, when you now plot the 1000 *mean ages,* the same familiar shape emerges.

This familiar bell-shape curve that occurs in all of these circumstances is called a normal distribution. It is symmetrical and has a clustering of observations near the mean. Real data sets do not exactly correspond to a normal curve but are close enough to make the curve useful for a wide variety of applications. Tables are widely available that give information as to the percentage of observations that fall in various parts of the normal curve. We will look at one of these and show how to read it later. Let us begin by getting an overall understanding of what normal curves look like. The appearance is something like the distribution in Fig. 3-3.

As with a histogram, the area under the curve of Fig. 3-3 represents frequency. If the whole area under the curve represents 100% of the observations, then we can talk about the area under part of the curve as representing some percentage of the observations. The

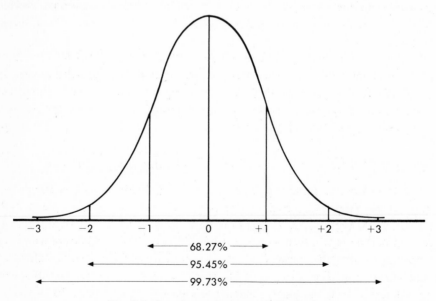

Fig. 3-3. Sample standard normal distribution.

scale along the bottom requires some explanation. The numbers indicate distances in terms of standard deviations either above or below the mean. You will recall that the standard deviation is a measure of variability and changes from one distribution to another. The +1 indicates a point one standard deviation above the mean. For example, if we were referring to Graduate Record Examination (GRE) scores, the mean happens to be 500 and the standard deviation 100. The score that corresponds to +1 would be a score of 600.

Notice that a large part of the total area under the curve falls between −1 and +1 (roughly 68% of the scores) and that a very large part falls between −2 and +2 (actually slightly over 95%). Although theoretically the curve never comes down to touch the baseline, it is very close by the time you get to three standard deviations from the mean. To summarize, when the distribution of a variable is approximately normal, about two thirds of the observations are within one standard deviation of the mean, about 95% are within two standard deviations, and almost all the observations are within three standard deviations.

The normal curve is not a recent discovery. In fact, it goes back to the sixteenth century. Its uses and applications have continued to expand since that time. There is an exact mathematical formula for the curve that gives the height of the curve for any point on the horizontal axis. The formula contains μ and σ as variables, resulting in different curves for different values of μ and σ. All normal curves have several things in common. There is a peaking of the distribution at the mean and a steep decline on either side, followed by a slower decline toward each of the tails. The point at which the curve changes from concave to convex is called the point of inflection. This point of inflection occurs at a distance one standard deviation from the mean in either direction.

z SCORES

The measurements making up the scale at the bottom of Fig. 3-3 report the number of standard deviations away from the mean and are referred to as standard scores. The symbol usually used is a small z. Often the scores are simply referred to as z scores. Since z scores indicate the number of standard deviations away from the mean, the formula for changing any observed score X into a z score is:

$$z_i = \frac{X_i - \mu}{\sigma} \quad \text{or} \quad \text{standard score} = \frac{\text{observed score} - \text{mean}}{\text{standard deviation}} \qquad \textbf{3.3-1}$$

If we know that a variable is approximately normally distributed and we transform all the scores representing that variable into z scores, the result will be a new set of scores called *standard normal scores*. The mean of these standard normal scores will be zero, about $^2/_3$ of the scores will be between −1 and +1, about 95% will be between −2 and +2, and virtually all will be between −3 and +3. We can find the exact percentile for any standard normal score by referring to a table developed for this purpose. We will describe the procedure for doing so shortly. First, it is necessary to be certain that a common misconception is avoided.

Fig. 3-3 contains an approximate drawing of a standard normal curve. Such an illustration is commonly referred to as a bell-shaped curve. It should be pointed out that there is not a *single* normal curve. The appearance of the normal curve will depend on the mean and standard deviation of the distribution and how they are scaled on the paper. A *standard* normal curve is a particular normal curve that is based on standard scores (z scores). Standard scores are scores with a mean of zero and a standard deviation of one. Some standard scores are normally distributed; others are not. Those that are, are called standard normal scores. Similarly, some scores that are normally distributed have a mean of zero and a standard deviation of one; others do not. Those that do are called standard normal scores. As you have just learned, any set of scores can be changed to z scores. If the original set of scores is normally distributed, the result will be a set of standard normal scores.

READING A TABLE OF THE STANDARD NORMAL CURVE

Tables of the standard normal curve can be presented in various ways, and it can be confusing to switch from one to another. All of the tables contain entries for z scores and corresponding proportions or percentages. However, the tabled proportions may represent the proportions of scores relative to different reference points. You may have the proportions of scores between the mean and the tabled z score, you may have the proportions of scores beyond the z score in the positive direction, or you may have the proportions of scores beyond the z score in the negative direction. In addition, there are tables that report values in both tails of a distribution and tables that report percentages in terms of a single tail.

The table of the standard normal curve, appearing as Table E-6 in Appendix E, should make it possible to avoid much of the confusion. It lists a set of z scores from 0 to beyond 4 and has separate columns for one-tail and two-tail proportions. The single set of z scores can be used for either positive or negative z scores. Under each column (one tail and two tail) there are two columns, headed "π beyond" and "π remainder," where π stands for proportion. To change these proportions to percentages, you would move the decimal point two places to the right. To illustrate the meaning of these two columns, look at the entries corresponding to a z score of 2.00. Do you remember that it was stated earlier that approximately 95% of all scores in a normal distribution are within two standard deviations of the mean, with the remaining 5% divided between the two tails? If you look at the two columns for the two-tail proportions, it indicates that the exact percentage within two standard deviations is 95.45% and that 4.55% is divided between the two tails. Now, looking at the one-tail proportion, you find that 2.28% of the scores are in each tail, more than two standard deviations from the mean. This means that 2.28% of all z scores in a standard normal population are greater than +2.00 and that 97.72% are less than +2.00. Similarly, 2.28% are less than −2.00 and 97.72% are greater than −2.00.

To find the z score that will leave exactly 5% in the two tails of the normal curve (2.5% in each), look for 0.05 under the π beyond column of the two-tail section of the table. The z score for this row of the table is 1.96. The meaning of this number is as follows. If a set of measurements is normally distributed, 95% of all measurements will be within 1.96 standard deviations of the mean. How many standard deviations would you have to go above and below the mean to include 99.9% of all the measurements? Again, looking at the two-tail section of the table, you will look for 0.001 in the π beyond column or 0.999 in the π remainder column. The corresponding z score is 3.291. To find the z score that will leave a certain percentage in *one* tail of the curve, the procedure is the same, except you look in the one-tail section of the table.

Now here are some examples to be certain that you understand how to read the table. Suppose a person gets a GRE score of 650. Where would this score fall in the distribution of all GRE scores, which are nomally distributed with a mean of 500 and a standard deviation of 100? First, we change the 650 to a standard normal score of $+1.50$. We then look this z score up on the table and find that 6.68% of the scores are above $+1.50$ and 93.32% are below $+1.50$. If for some reason we wanted to know how many persons had received scores this far from the mean in either direction, we would look at the two-tail probabilities and find that 13.36% of the scores are either below -1.5 or above $+1.5$.

Suppose a man weighs 105 pounds. Where does his weight fall among the population of persons whose weights are normally distributed with a mean of 180 and a standard deviation of 30? His weight converted to a z score would be -2.5. Looking at the table for a z score of 2.5, we find that less than 1% (actually 0.62%) of the scores would be less than -2.50.

THE CORRELATION COEFFICIENT

Just as we considered the two-variable (bivariate) case in the lesson on graphic display of data, we will pause in our discussion of the characteristics of frequency distributions to consider a descriptive statistic that is an index of the strength of the linear relationship between two variables. The index most frequently used is the correlation coefficient, which has a very formidable looking computational formula, but which becomes more comprehensible when broken down into its components. The symbol usually used to denote the correlation coefficient is a small r. Let us start with the formula in its *definitional* form.

$$r_{XY} = \frac{\Sigma z_X z_Y}{n} \qquad \qquad \text{3.3-2}$$

where z_X and z_Y are standard scores for X and Y and
n is the number of pairs of scores.

The formula says that you could compute the correlation coefficient by the following steps:
1. Change each X and Y score to a standard score.
2. Multiply each pair of standard scores together.
3. Add these cross-products.
4. Divide the result by the number of pairs of scores.

Although it is by no means obvious, the maximum value you could get by following this process would be $r = +1.00$; the minimum would be $r = -1.00$. If there is no relationship between the variables, you would get a value of zero. An r of either $+1.00$ or -1.00 means there is a perfect linear relationship between the variables. In this case, it is possible to perfectly predict the value of one variable for any individual by knowing his measurement on the other. For example, suppose blood pressure were perfectly correlated with the percentage of fat in the body. This would mean that if you knew the percentage of fat in the body of a person, you could perfectly predict his blood pressure. Similarly, if you knew his blood pressure, you could perfectly predict the percentage of fat in his body. Of course, many other factors are interrelated with blood pressure, so the relationship will not be near perfect. Such is usually the case with real data in the biological sciences. The result is that r values seldom approach either $+1.00$ or -1.00. Before we go on, it should be explained that negative correlation coefficients occur when the relationship is such that high values on one variable tend to be paired with low values on the other and low values on one variable tend to be paired with high values on the other. An example of a negative relationship would be skin elasticity and age in human beings.

As with the standard deviation, in addition to the definitional formula there is a computational formula for the correlation coefficient. Let us get to it in steps. By using the formula for a z score, you may be able to demonstrate that Formula 3.3-2 can also be written:

$$r_{XY} = \frac{\Sigma xy}{\sqrt{(\Sigma x^2)(\Sigma y^2)}} \qquad \text{3.3-3}$$

where   the small x's and y's are deviation scores.

Now if you think back, you will recognize the entries inside the parentheses. These are simply the sums of deviations squared for X and Y. Remember the brick, first given as Formula 3.2-5.

$$\Sigma x^2 = \Sigma X^2 - \frac{(\Sigma X)^2}{n}$$

The numerator in Formula 3.3-3 is called the sum of cross-products of deviations. Its computational form is:

$$\Sigma xy = \Sigma XY - \frac{(\Sigma X)(\Sigma Y)}{n} \qquad \text{3.3-4}$$

Now, plugging all these components into one intimidating formula, we get the computational formula for a correlation coefficient.

$$r_{XY} = \frac{\Sigma XY - \frac{(\Sigma X)(\Sigma Y)}{n}}{\sqrt{\left[\Sigma X^2 - \frac{(\Sigma X)^2}{n}\right]\left[\Sigma Y^2 - \frac{(\Sigma Y)^2}{n}\right]}} \qquad \text{3.3-5}$$

Notice that to use Formula 3.3-5 you do not need any deviation scores. You work entirely with raw scores and need six values to substitute in the appropriate places in the formula: n, ΣX, ΣY, ΣX^2, ΣY^2, ΣXY. Many modern electronic calculators will give all six values just by entering each pair of numbers once. Some will even substitute the values in the formula for you.

SUMMARY

The third characteristic of a distribution, in addition to central tendency and variability, is shape. When discussing shape of a distribution, we consider both degree of *skewness* and degree of *peakedness*. Observed distributions are often compared with theoretical distributions that have known characteristics for central tendency, variability, and shape. One of the most important of these is the normal curve. When a distribution of scores is normally distributed, it is possible to estimate the percentile value of any observed score by transforming the observed score to a standard score (z score). A table of the normal curve, such as Table E-6 in Appendix E, can then be used to obtain the desired percentiles.

In addition to the normal distribution, there are many other theoretical distributions that statisticians use. This book will use primarily the normal curve in the introductory chapters dealing with inferential statistics. Later, you will also become familiar with tables of the t distribution, the chi-square distribution, and the F distribution.

At the end of the lesson you learned about an index that measures the relationship between two variables X and Y. It is called the correlation coefficient and is referred to by the symbol r. Both a definitional and a computational formula were provided for r.

LESSON 3 EXERCISES

All of the problems refer to the data in the following table. The measurements are birth weights in pounds (X) and birth lengths in inches (Y) for 50 baby girls.

X	Y	X	Y	X	Y	X	Y	X	Y
5.8	18.7	7.1	19.5	7.2	19.4	7.0	18.9	8.2	21.1
7.3	19.5	7.1	20.5	8.2	19.9	8.5	20.3	6.1	19.4
7.5	19.6	6.6	19.5	7.1	20.3	7.0	19.8	8.4	20.4
7.6	20.3	5.9	19.3	6.7	19.3	8.1	21.3	6.2	19.3
6.7	19.4	6.8	19.1	6.1	19.5	8.1	20.5	6.8	19.7
7.6	19.6	7.3	19.9	5.4	18.6	7.7	20.0	7.1	20.6
7.2	19.5	8.9	20.4	6.9	19.5	9.4	21.7	4.2	17.9
6.9	20.0	7.6	20.0	7.7	19.5	6.7	19.9	6.3	18.9
7.3	19.8	6.9	19.5	8.3	20.2	9.0	21.7	7.4	19.9
9.1	21.0	8.3	21.4	6.3	18.9	7.4	19.7	5.5	19.0

1. What are n, ΣX, ΣY, ΣX^2, ΣY^2, ΣXY?
2. What are Σx^2, Σy^2, Σxy?
3. Find the means and standard deviations for the X and Y scores.
4. What is the z score for $Y = 21$? For $Y = 18.9$? If the Y scores were normally distributed what would be the percentile rank for $Y = 21$? For $Y = 18.9$? What proportion of scores would be between $Y = 21$ and $Y = 18.9$?
5. What is r_{XY}? What is the meaning of this index?

Lesson 4 □ Score conversions

You have learned the meaning of and the computations for the mean, the median, the standard deviation, and the correlation coefficient. You have also learned about frequency polygons, which describe the shape of a distribution, and about one particular distribution called the normal distribution. The uses of these concepts to describe large data sets should be obvious. If the only use to be made of these concepts were to be to describe large populations of scores, your efforts would be worthwhile. Applications of the concepts go much beyond this limited use, however. Later, in lessons dealing with inferential statistics, you will use these concepts extensively when learning how to make inferences about populations by examining the results of samples. There is also an important application of the concepts that relates to the comparability of various measurements. This application will be the subject of this lesson.

Often, the same attribute is measured by two or more different measuring devices. The results of the various measuring devices may be reliable and valid, but they may still not be directly comparable. Outdoor temperatures may be measured with thermometers giving either Celsius or Fahrenheit readings. The weights of newborn babies may be measured by scales that give readings in either ounces or grams. The distance to the nearest hospital may be measured in miles or kilometers. Before these measurements can be compared, some conversion is necessary. For example, either ounces must be converted to grams, grams to ounces, or both to some other common unit.

When the measurements being compared are physical measurements such as size, weight, or temperature, each scale is based on a standard unit that is precisely defined, and the conversion involves application of a standard formula. For example, to change a Fahrenheit reading to Celsius, you subtract 32 and then multiply by $^5/_9$; to change ounces to grams you divide by 0.035.

When the measurements being made are not physical measurements, the problem is not quite as straightforward, but the idea is the same. Before comparing two different measurements of IQ, made with different measuring devices, we must convert both to the same unit of measurement. Unfortunately, the National Bureau of Standards does not have a standard, unchanging unit for measuring IQ. Nevertheless, some common reference is necessary. The solution is to use a defined group of persons as a common reference. If a large number of adults from many different backgrounds take both tests, we can use the results to determine how scores from the two test compare. We could do this, by finding for each test the score that corresponds to each percentile. For example, if the ninetieth percentile for test *A* is 120 and the ninetieth percentile for test *B* is 580, then a score of 120 on test *A* is comparable to a score of 580 on test *B*. In effect we are using position of persons within the distribution of scores on the two tests as a common reference for making comparisons.

The process just described can be used to make up conversion tables for comparing the results of two different tests. A more efficient procedure is possible if the shapes of the score distributions of the tests are similar. Instead of using percentiles as an indication of position within each score distribution, we can use z scores. You remember that a z score indicates the number of standard deviations above or below the mean. If the score distributions have similar shapes, a score that is one standard deviation above the mean on test A would be comparable to a score one standard deviation above the mean on test B. Changing the scores from both tests to z scores would therefore result in comparable scores. Remember that to convert a raw score to a z score, you subtract the mean and divide by the standard deviation. These means and standard deviations for each test are computed from the score distributions resulting from giving the tests to the large reference group.

To see how this process works, let us take a hypothetical example. Suppose you were given the responsibility of developing a national testing program to be used in recertification of practical nurses every 5 years. Suppose in addition that a person who fails the examination is allowed to retake it after a period of 1 month. You will probably want to develop at least two forms of the examination and be sure that the scores are comparable. Let us call the two tests form A and form B. Again, it is possible for both forms to be reliable and valid but not comparable. Scores may tend to be higher or to spread out more on one than on the other. To solve this problem, you arrange to have a large group of practical nurses take both forms A and B. You then make a frequency distribution and a frequency polygon for each score distribution. If the score distributions have different shapes, you will need to make up a conversion table showing, for each percentile, the corresponding raw scores on forms A and B. These conversion tables will then have to be used when scores from the two different forms are to be compared.

Suppose, however, that the shapes of the two score distributions are similar and that the means and standard deviations are as follows.

	Test A	Test B
μ	100	80
σ	10	20

Instead of deriving a conversion table based on percentiles, we can accomplish the same purpose by reporting each score as a z score. To change each score from form A to a z score, we will subtract 100 and then divide by 10. To change each score from form B to a z score, we will subtract 80 and divide by 20. If these conversions are made before scores are reported, the scores for forms A and B will be directly comparable.

So far, what we have described is what is actually done when tests are standardized. There is usually one additional step. When scores are changed to z scores, all those scores below the mean will be negative. Scores both above and below the mean will be decimal numbers. Decimal numbers, especially negative decimal numbers, are not easily understood by the general public. To avoid this difficulty, the z scores are modified. To avoid the reporting of decimals, the z scores are multiplied by 100 and then rounded to a whole number. A z score of $+1.55$ would become a score of 155. This still will not solve the problem of negative numbers, however, since a score 1.55 standard deviations below the mean would become a -155. The negative numbers can be eliminated, however, if a constant is added to each score. The size of the constant chosen will, to a certain extent, be arbitrary; the only requirement is that it must be larger than the lowest z score multiplied by 100. Test programs like the Graduate Record Examination (GRE) and the Medical College Admissions Test (MCAT) add the constant 500.* Since the shapes of the score distributions from these tests are approximately normal, most of the scores are within plus or minus three standard deviations of the mean. If you have understood what has been said in this section, you should be able to see why the range of scores reported for the MCAT and the GRE is approximately from 200 to 800.

CODING AND SCORE CONVERSIONS

Underlying the process just described for converting scores is a concept that we shall call coding. Coding means performing the same arithmetic operation (addition, subtraction, multiplication, or division) on all of the scores in a distribution. You code by adding a constant to or subtracting it from each score, or by multiplying or dividing each score by a constant.

The effects of each coding operation on the mean, standard deviation, variance, and correlation coefficient are shown in Table 3-2.

The entries in the first two columns should make sense. Adding, subtracting, multiplying, or dividing would all be expected to affect the average of a set of scores. Multiplying or dividing by a constant would also affect the spread of a set of scores, making them either closer together (coding by division) or further apart (coding by multiplication). On the other hand, adding or subtracting a constant would not affect the spread of scores but would simply move the total distribution in one direction or another.

The fact that coding has no effect on the correlation coefficient is a little harder to understand. If you look at the definitional formula for the correlation coefficient, you will see that the scores used are standard scores. Although the fact is by no means obvious, if you try some test numbers you may be able to see that the use of standard scores in the formula ensures that coding will not affect the correlation coefficient. If you cannot see why this is true, do not worry about it, just accept it. Later in the lesson you will see a demonstration of the principle.

The information in Table 3-2 is of more than academic interest, since it allows us to convert any score distribution to arrive at a new distribution with any desired mean and standard deviation. You should be able now to see why z scores have a mean of 0.0 and a standard deviation of 1.0. Think it through. What is the coding that takes place when you calculate a z score?

Previously we said that GRE and earlier MCAT scores are obtained by first converting raw scores to z scores, then multiplying these z scores by 100, and finally adding 500.

---

*This was true for MCAT scores before a recent revision. Scores for the revised MCAT are reported on a scale with a mean of approximately 8 and a standard deviation of approximately 2.

Table 3-2. Effects of coding

Coding operation	Statistics			
	Mean	Standard deviation	Variance	Correlation coefficient
Addition and subtraction	Comparable*	None	None	None
Multiplication and division	Comparable	Comparable	Comparable²†	None

*Comparable means that increasing or decreasing all the scores by adding or subtracting a constant increases or decreases the mean by the same constant.

†Comparable squared is a shorthand method of saying that multiplying or dividing all the scores by a constant affects the variance by multiplying or dividing it by the square of the constant.

Using the information in Table 3-2, you should now be able to figure out what the means and standard deviations of these distributions are—500 and 100 respectively. Right?

Suppose you have wanted both forms of your practical nurse accredition tests to have means of 100 and standard deviations of 15. Can you describe the coding that would be necessary for the scores from each form? Can you do this without ever changing the scores to z scores? Think about the problem in light of the information in Table 3-2. Try solving the problem before reading on.

When any score distribution is to be converted to one with a desired mean and a desired standard deviation, it is advantageous to first code the scores to the desired standard deviation by multiplication or division. Of course, the original mean will also be affected by this coding. A second coding by addition or subtraction will result in the desired mean without affecting the standard deviation. For form A, the mean and standard deviation of the raw scores are respectively 100 and 10. If all raw scores are coded by multiplying by 1.5, the resulting score distribution will have a mean of 150 and a standard deviation of 15. If the resulting scores are then coded by subtracting 50, the resulting score distribution will have the desired mean of 100 and standard deviation of 15. The total coding process can be described by the following formula:

$$A_i' = (A_i)(1.5) - 50$$

where A_i is any raw score for form A.

Can you derive a similar formula for coding the scores of form B? A similar process of reasoning leads to the following formula for converting scores from form B:

$$B_i' = (B_i)(0.75) + 40$$

Earlier, the problem of comparing scores from two tests measuring the same variable was posed. You should now be able to see how the problem can be solved if the shapes of the score distributions are comparable. By coding, either one distribution can be made comparable to the other (same mean and standard deviation) or both can be coded to have some arbitrarily chosen mean and standard deviation. The latter approach is being used when both are converted to z scores (mean equals 0.0, standard deviation equals 1.0).

It should be noted that coding does not affect the shape of the distribution in terms of how closely it approximates the normal curve. If the raw score distribution is normal, the coded scores will also be normal; if it is skewed, the coded scores will be skewed. There is a process for normalizing a distribution of scores. It is referred to as score transformation, as opposed to score conversion. You will see an example of score transformation in a later chapter.

CODING AND COMPUTATION OF STATISTICS

In the days before electronic calculators, all statistics books contained a section showing how coding could be used to make computations easier when you are finding the mean and standard deviation from frequency distributions or when you are finding the correlation coefficient. Now that almost everyone has access to a pocket calculator, the explanation is given only for its instructional value.

Coding to find the mean and standard deviation. You will remember that when scores have been placed in a frequency distribution, the midpoints are used to find the mean and standard deviation. To make computations easier, it is possible to code the midpoints by subtraction and division, making the numbers smaller. The coding is a two-step process. First, choose one of the midpoints near the center of the distribution and use this as the constant to subtract from each midpoint. Let us call these "once-coded midpoints." Since all classes are the same size, the once-coded midpoints are multiples of the class size. The second step is to code by dividing all once-coded midpoints by the class size. You can then find the mean and standard deviation of the resulting "twice-coded midpoints." These twice-coded midpoints will be smaller numbers, starting with zero for the class whose midpoint was used for the first coding and increasing or decreasing by one as you go in either direction from that class. Once you have found the mean and standard deviation of these twice-coded midpoints, you will have to reverse the process to find the mean and standard deviation of the original score distribution. Remember the information in Table 3-2 when doing this. Why does the first coding not have to be considered when computing the standard deviation?

Coding to find the correlation coefficient. In some instances coding can also be used to make finding the correlation coefficient without the use of a calculator easier. This works when the scores are large and they are multiples of some number. Take, for example, the following scores for five persons who took both the GRE and the earlier version of the MCAT. What is the correlation between these scores?

i	GRE	MCAT
1	400	500
2	400	500
3	500	600
4	700	700
5	600	800

Using the computational formula for correlation will be a lot of work without a calculator and without coding. However, suppose we decide to code by dividing all scores by 100 (it is not necessary to use the same coding for both sets of scores). We then have the following:

i	GRE	MCAT
1	4	5
2	4	5
3	5	6
4	7	7
5	6	8

We could further code by subtracting five from each GRE score and six from each MCAT score. We would then have the following scores:

i	GRE	MCAT
1	−1	−1
2	−1	−1
3	0	0
4	2	1
5	1	2

Now, finding ΣX, ΣY, ΣX^2, ΣY^2, ΣXY and n would not be difficult without a calculator. Use a calculator and find r_{XY} for the three different sets of scores. You will find it to be 0.85 in each case. As indicated in Table 3-2, coding has not affected the correlation coefficient.

SUMMARY

This lesson described a process for comparing scores from two different score distributions. When the score distributions have different shapes, a score conversion table is required. For each score distribution, the conversion table provides the score corresponding to each percentile rank. When the two score distributions have the same shape, a more efficient procedure is available. Through a process of coding, it is possible to convert scores to make them directly comparable. In the last lesson you learned of one such conversion: coding by subtracting the mean and dividing by the standard deviation. The resulting scores are called standard scores, or z scores. A table was provided that describes the effect of types of coding on the statistics you have learned to calculate. Once the information in the table is understood, it is possible to convert any score distribution to one with any desired mean and standard deviation. Such a conversion will not affect the basic shape of the original distribution.

This lesson concludes the information on descriptive statistics that is usually covered in statistics books. Be sure you are confident of your ability to handle the concepts covered. They will be used extensively in the following chapters dealing with inferential statistics.

LESSON 4 EXERCISES

1. Consider the following population of numbers:

   10
   10
   18
   18

   a. Convert it into a standard score distribution.
   b. Convert it into a distribution having mean of 10 and standard deviation of 9.

2. Consider the data set:

   100,001
   100,101
   100,201

   Without a desk calculator, use coding to aid in finding the mean and standard deviation of the above set of data.

3. Without a calculator, use coding to aid in finding r_{XY}.

X	Y
1050	110
1100	115
1200	125

PART TWO

Rationale of inferential statistics

CHAPTER 4

Probability

Lesson 1 □ Definitions and combined events

With this lesson you will begin the study of inferential statistics. A drum roll would not be inappropriate accompanying this announcement. What may sound like a dry, theoretical topic is actually a powerful tool with applications in virtually every human activity. The idea is a simple one. The world we live in is one of uncertainty. We do not know for certain whether it will rain tomorrow, whether the president will be reelected, whether the outstanding spades are split seven and zero, whether our next child will be male or female, whether IBM's earnings will be up or down, whether a birth control method will prevent conception, or whether a drug will leave an undesirable side effect. At the other extreme, neither is the world a place of total disorder. It probably will not rain tomorrow with a high pressure area moving into the region, a popular president doing a creditable job will probably not be defeated, the spades will probably not split seven and zero.

In this world of uncertainty, people are constantly asking, "What will happen if . . .?" The seer and others possessing mystical powers provide definitive answers. The inferential statistician says, "Give me enough information about past experiences with similar circumstances, and we may be able to identify all the possible outcomes and tell how likely each is." In words the seer says, "No, it will not rain tomorrow" and "No, the spades are not split seven and zero." The statistician says, "On other days when the situation was like today, it rained only 10% of the time" or "The likelihood of the seven spades all being in one hand is less than 1 in 100."

The statistician looks at events in the world as sample events from a whole population of similar events. The statistician's work involves trying to learn what populations are like by examining samples; or alternatively, by knowing what a population is like, he or she tries to determine the likelihood of a particular sample's occurring. In either case the answer will involve a statement of probability. For example, when comparing the effects of two drugs, the statistician will look at the results of samples and conclude that the probability of drug A's being more effective than drug B on the average over all possible comparisons is less than 1 in 20. Or, knowing the distribution of weight losses among obese patients who participate in a behavior modification therapy group, he or she will determine that the probability that a patient will succeed in losing at least 10% of body weight during the next year is less than 1 in 5.

Since the concept of probability is basic to all of inferential statistics, this will be the subject of the first two lessons. This first lesson will explain what a statistician means when he or she uses the word *probability* and will provide information as to how probabilities are computed for some relatively uncomplicated situations. The second lesson includes a discussion of how probabilities are computed in three complicated situations. You will probably not have difficulty with the material in the first lesson. The second will require some real study to master. Fortunately, a complete understanding of the material in Lesson 2 is not essential for the study of the remainder of the lessons in the book. The section entitled the binomial expansion does provide continuity for what will follow later. The section under Bayes' theorem has particular applications to disease diagnosis and is the basis for an approach to statistical inference called Bayesian statistics. We will use the concept only tangentially.

DEFINITION OF PROBABILITY

Before we define probability it is necessary to clarify some terms. Suppose we are interested in the probability of a person with a certain disease exhibiting symptom A. Each person with the disease will be considered to be an occurrence of an event. The absence or presence of the symptom for any event will be called an outcome. The presence of the symptom will be called a positive outcome and the absence, a negative outcome. To make it easier to present this material we will use *P(A)* as a shorthand for saying: "The probability of a positive outcome when looking for trait A in any event." The probability of a negative outcome will be written *P(NA)*.

With these clarifications in mind, we can now state that statisticians define the probability of any event's exhibiting a particular trait as a decimal value arrived at by dividing the number of all events exhibiting that trait by the total number of all events.

In symbols:

$$P(A) = \frac{N_A}{N}$$

4.1-1

where $P(A)$ is the probability that any event will exhibit trait A,
N_A is the number of events with trait A, and
N is the total number of events.

Using this definition, you can calculate the probability of a randomly selected card's (event's) being an ace (positive outcome) in either of two ways: empirically or theoretically. On an empirical basis, you would randomly draw a card, note its identity, replace it in the deck, thoroughly shuffle the deck, and then repeat the process. After thousands of repetitions you could then compute the probability of drawing an ace, using Formula 4.1-1, where N_A would be the number of aces observed and N would be the number of observations made.

On a theoretical basis, you would go through the deck of cards counting the number of aces and the total number of cards. The probability of an ace's being drawn would again be calculated from Formula 4.1-1, with N_A being the number of aces in the deck and N being the number of cards. The theoretical probability is based on the assumption that the random card is picked in such a way that all of the cards have an equal chance of selection. We will use this definition of random in future lessons.

Earlier, when learning to use the table of the normal curve, you found that the table lists *proportions* of areas of the curve for different values of z. These proportions are often referred to as *probabilities*. You can see now why statisticians often use the two terms interchangeably. In Formula 4.1-1, the probability of outcome A is defined as the proportion of events that have trait A among all events under consideration.

To be sure you understand the definition of probability and how it can be calculated both empirically and theoretically, think through the calculations for the following questions. "What is the probability of an even number's appearing when one die from a pair of dice is thrown?" Would you expect to get exactly the same result from the empirical method as from the theoretical method? Which is more valid?

There are many circumstances where we want to know what will happen over a series of events. For example, what is the probability of *all* the outcomes' being positive or perhaps their all being negative? What is the probability of *at least* one positive outcome? What is the probability of *exactly one* positive outcome? What is the probability of 80% positive outcomes?" We will examine each of these questions in the remainder of this lesson and the one to follow.

The multiplicative rule

There are times when it is necessary to find the probability that a single event will exhibit all positive outcomes for several traits; for example, the probability of a single patient's having all of the symptoms A, B, and C. At other times one wants to know the probability that all of a number of events will exhibit positive outcomes for a single trait; for example, the probability that all of the patients will have symptom A. In either case, the solution is to use what is generally called the multiplicative rule. The general form of the multiplicative rule is usually stated for the case of two traits, A and B, for a single event. It is as stated in Formula 4.1-2.

$$P(A \text{ and } B) = P(A) \times P(B|A) \qquad \textbf{4.1-2}$$

where $P(A \text{ and } B)$ equals the probability of positive outcomes for both traits A and B and
$P(B|A)$ equals the probability of a positive outcome for B when the outcome for A is positive.

Note the meaning of the vertical line. It can be read "given that there is." When the probability of B is totally unrelated to the outcome for A, then A and B are said to be independent traits. In symbols, traits A and B are independent when $P(B) = P(B|A)$. In this case, the last term in Formula 4.1-2 would be just $P(B)$.

The multiplicative rule can be used for three or more traits for a single event, but it gets a bit complicated when the traits are not independent. For the case of three or more traits *where all traits are independent,* you find the product of all of the trait probabilities. For example, for a certain disease symptom A has a probability of 0.50; symptom B, 0.20; and symptom C, 0.10. The probability of any symptom is independent of any other. In this case, by use of the multiplicative rule, the probability of a patient with the disease having all three symptoms is:

$$(0.50)(0.20)(0.10) = 0.01$$

As stated earlier, the multipicative rule can also be used to find the probability of all positive outcomes for several events when considering a single trait. The usual application is for the case where the probability of the trait remains the same from event to event. In this case the multiplicative rule can be stated as follows.

$$P \text{ (all } n \text{ positive outcomes for trait } A) = [P(A)]^n \qquad \textbf{4.1-3}$$

where n is the number of events.

Use this rule to find the probability of tossing a die three times and getting a six all three times. You find the probability to be $1/216$, or 0.0046. Right?

Can you write the formula for *no* positive outcomes for A for any chosen number of events? Note that $P(A) + P(NA) = 1.00$. What is the probability of getting no sixes in three tosses of a die? You should get $125/216$, or 0.5787.

Understanding how to use the multiplicative rule is not difficult. The difficulty lies in deciding when it is appropriate. It will help you in this regard to get used to looking for the key words *all* or *none,* remembering that the rule can be used either for multiple traits for a single event or for multiple events for a single trait. Here are some sample questions.

1. What is the probability of throwing a coin five times and getting all heads? Answer: $(1/2)^5$, or $1/32$.
2. What is the probability of drawing two cards from a deck of cards and getting all spades? No spades? Careful now. Are the outcomes independent? Answer: $(13/52)(12/51)$ and $(39/52)(38/51)$.
3. A certain disease exhibits symptoms A and B. The $P(A) = 0.8$, $P(B) = 0.4$, $P(B|A) = 0.3$. What is the probability that a patient will have both symptoms A and B? Answer: $(0.8)(0.3) = 0.24$.

The additive rule

In many circumstances we want to know the probability of *at least one* positive outcome. As with the multiplicative rule, our interest may be in considering several traits for a single event or several events for a single trait. Let us take the former as an example, since it is the most obvious. Let us make the throwing of a die an event. Trait A will be the number 2 and trait B the number 3. What is the probability of at least one positive outcome for a single throw? In other words, what is the probability of a 2 or 3? It is apparent that if we add the $P(A)$ and $P(B)$ we will get $2/6$. This checks with our definition of probability, since there are six sides to the die and two of them are the 2 and 3. The complete additive rule is shown in Formula 4.1-4.

$$P(A \text{ or } B \text{ or both}) = P(A) + P(B) - P(A \text{ and } B) \qquad \textbf{4.1-4}$$

Notice that in our example of a single die, traits A and B were mutually exclusive and the last term in the formula was equal to zero. Suppose you were asked to give the probability of drawing a single card that is either an ace or a spade. An event is the drawing of a card; a positive outcome for A is drawing an ace, and a positive outcome for B is the drawing of a spade. What is the probability of at least one favorable outcome?

$$P(A) = 4/52 \qquad P(B) = 13/52 \qquad P(A \text{ and } B) = 1/52$$
$$\therefore P(A \text{ or } B) = 4/52 + 13/52 - 1/52 = 16/52$$

This checks with our knowledge that there are 13 spades and 3 aces that are not spades, for a total of 16 cards that meet *at least one* of our conditions.

As stated at the start of this section, the additive rule can also be used to figure the probability of at least one positive outcome of a single trait among several events. Usually, positive outcomes are not mutually exclusive from event to event, so that there must again be some subtraction for this factor. The problem is that when there are more than two events the computation of the probabilities for all the overlapping situations becomes complex. Fortunately, there is a much easier solution. You learned earlier to use the multiplicative rule to find the probability of no positive outcomes of a certain trait over several instances. How would you use this information to find the probability of at least one positive outcome? The alternate to finding no positive outcomes is to find at least one positive outcome. Therefore, to find the probability of at least one positive outcome, first use the multiplicative rule to find the probability of no positive outcomes. Then subtract the result from 1.00.

Here is a question to test your understanding of the additive rule. What is the probability of throwing a pair of dice and getting at least one six? We can find this in two different ways. First, using the additive rule we calculate as follows:

$$P(\text{at least one } 6) = P(6 \text{ on first die}) + P(6 \text{ on second die}) - P(6 \text{ on both dice})$$
$$= (1/6) + (1/6) - (1/6)(1/6)$$
$$= 11/36$$

Or, using the multiplicative rule to find the probability of no sixes, and then subtracting from 1.00, we get:

$$P(\text{at least one } 6) = 1 - (P \text{ of no } 6\text{'s})$$
$$= 1 - (5/6)(5/6)$$
$$= 11/36$$

As with the multiplicative rule, the difficult part of using the additive rule is in deciding when it is appropriate. The key words to look for are *or* and *at least one*. Usually you will make use of the additive rule when considering several traits for a single event.

Table 4-1. Example data for smoking history and running speed ($n = 300$)

Smoking history*	Time required to run 1 mile			
	< 6 min	6 to 8 min	> 8 min	Total
None	30	10	2	42
<1 pack per week	20	15	8	43
<1 pack per day	7	25	20	52
<2 packs per day	3	32	50	85
≥2 packs per day	0	8	70	78
TOTAL	60	90	150	300

*Information obtained through self-reports for the last year.

As has been pointed out, there is an easier solution when considering several events and a single trait.

Conditional probabilities

We have already introduced the idea of conditional probabilities when talking about independence and the multiplicative rule. Remember we used the vertical line to indicate "given that" some condition exists. Before concluding this first lesson on probability, let us expand the example to include more than just absence or presence of a trait and instead include levels of a trait. Let us consider a population in which there are two traits of interest: smoking experience and time required to run a mile. The data might be as shown in Table 4-1. The total number of persons represented in Table 4-1 is 300. Going back to our definition of probability, what is the probability that a person chosen at random from the 300 is able to run the mile in less than 6 minutes? What is the probability that a person is a nonsmoker and can run the mile in less than 6 minutes?

Neither of these questions puts any restrictions on the population from which an event is drawn. However, suppose we ask, "What is the probability that a person can run the mile in less than 6 minutes given that he is a nonsmoker?" This is called a conditional probability. It places a condition on the events that can occur. Instead of including all 300 persons for the denominator of our fraction, we narrow the group down to some portion of the total through the condition that is stated. In this case, 42 persons are nonsmokers. Of these, 30 are able to run the mile in under 6 minutes. The answer, therefore, is $^{30}/_{42}$, or 0.714. You will be given a chance to test your understanding of conditional probabilities as well as your understanding of the multiplicative and additive rules in the exercises that follow.

SUMMARY

This first lesson dealing with probability provided a basic definition of probability. In symbols the definition is:

$$P(A) = \frac{N_A}{N}$$

where $P(A)$ is the probability that any event will exhibit trait A,
N_A is the number of events with trait A, and
N is the total number of events.

Two rules were given for calculating combined probabilities. The multiplicative rule was said to be applicable when calculating the probability of *all* positive outcomes. The additive rule was said to be applicable when calculating the probability of *at least one* positive outcome. Examples were given for each application.

The last section of the lesson explained the meaning of conditional probability and provided an example of the use of the concept.

LESSON 1 EXERCISES

1. Let A and B denote two genetic characteristics. Suppose that the probability is $1/2$ that an individual chosen at random will exhibit A and $3/4$ that he will exhibit B. Assume that these characteristics occur independently. What is the probability that an individual chosen at random will exhibit:
   a. Both?
   b. Neither?
   c. Exactly one?
   d. At least one?
2. If the probability that any child born is a boy is equal to $1/2$, and if sex is independent from child to child, what is the probability that of five children born on a given day in a given hospital:
   a. All will be boys?
   b. None will be boys?
   c. At least one will be a boy?
3. Persons of appropriate age can qualify for Social Security benefits either because they have worked under Social Security or because their deceased spouses have. In a population of women, 40% have worked under Social Security and 70% are widows of men who worked under the system. The two events are independent in the population. What portion is qualified for benefits?
4. Four hundred men with angina pectoris are classified by age and weight as follows:

Age (in years)	Weight (in pounds)			
	130-149	**150-169**	**170-189**	**≥190**
30-39	10	20	20	40
40-49	10	15	50	70
50-59	5	15	50	40
60-69	5	10	15	25

Using the above table, find for an individual the probability that he:
a. Is in the age interval 40-49.
b. Is in the age interval 40-49 and weighs 170-189 lb.
c. Is in the age interval 40-49 or 60-69.
d. Is in the age interval 40-49 or 60-69 and weighs 150-169 lb.
e. Is in the age interval 40-49, given that he weighs 150-169 lb.
f. Weighs less than 170 lb.
g. Weighs less than 170 lb and is less than 50 years old.
h. Weighs less than 170 lb, given that he is less than 50 years old.

Lesson 2 □ The binomial distribution, the Poisson distribution, and Bayes' theorem

In the preceding lesson you learned to use the multiplicative rule to find the probability of *all* positive outcomes or *no* positive outcomes. You learned to use the additive rule to find the probability of *at least one* positive outcome. During the first section of this lesson you will learn how to find the probability of *exactly one, exactly two, exactly three,* or any other number of positive outcomes for any designated number of events. In the second section you will learn a method for finding combined probabilities for rare events. Then in the third section you will make use of what you learned about conditional probabilities in studying Bayes' theorem.

THE BINOMIAL DISTRIBUTION

Suppose a friend of yours claims that he can predict the outcome of professional football games by consulting biorhythm charts for the opposing quarterbacks. You are rather skeptical, but he insists that you put him to the test. Deciding to humor him, you choose two games for the following Sunday that are rated as toss-ups. You challenge him to predict the winners. To your surprise he correctly predicts both games. Being an accomplished statistician after your mastery of the material in the last lesson, you tell him you are not very impressed, because his feat could easily be the result of guessing. You know there are only three possibilities for the number of positive outcomes: none correct, one correct, or two correct. If we consider a correct prediction to be A, and your friend is merely guessing, $P(A)$ for each game would be $1/2$ and $P(NA)$ would be $1/2$. Using the multiplicative rule, we find that the probability of two positive outcomes would be $P(A)^2$, or $1/4$, and the probability of no positive outcomes would be $P(NA)^2$, or $1/4$. Since the person must get either none, one, or two correct, the total probabilities must add to 1.00. Therefore, the probability of one correct must be $1/2$.

Your friend, of course, is indignant, saying "My record was perfect, what more do you want?" You reply, "Your result could easily be a chance happening. Just guessing, you could be expected to get both games right 1 out of 4 times. Let us see what happens next Sunday when you try to pick the winners of five games rated as toss-ups." You have already figured that his probability of getting all five correct by guessing is $(1/2)^5$, or only $1/32$. If he does this, you are prepared to believe he may be on to something.

Your friend studies his charts all week and confidently hands you his list of predictions on Sunday morning. You listen to the news Sunday night and learn that he was right on four out of five predictions. Now what will you say to your friend over coffee Monday morning? You know his probability of guessing all five right and his probability of guessing all five wrong, but what about four right, or for that matter three, or two, or one? Somehow, overnight you are going to have to extend your knowledge of probabilities to include the whole distribution of possible results. You are going to have to learn about something called the *binomial distribution*. Be warned ahead of time that it is a bit complicated.

Combinations

The formula for the binomial distribution makes use of an expression you are probably not familiar with. It is the expression $_nC_k$ and stands for "the combination of n things taken k at a time." An example will clarify the meaning of the expression. How many different ways can you select two courses from among four that are offered? You could call these four courses a, b, c, and d. Then by a process of enumeration you could list all possible pairs: ab, ac, ad, bc, bd, and cd. There are six ways to select two things from among four.

Suppose, however, you wanted to know how many ways there are to select two courses from among six being offered or three courses from among eight being offered. Although it can still be done, the process of listing and then counting all the combinations becomes difficult as the number selected and the group from which selections are to be made increase.

There is a formula for finding how many ways there are to select any designated number of things at a time (k) from among any designated size of group (n). It is Formula 4.2-1.

$$_nC_k = \frac{n!}{(n - k)!\, k!}$$ **4.2-1**

where n is the number to be chosen from and
         k is the number chosen each time.

Formula 4.2-1 contains exclamation marks. You may not be familiar with the use of this symbol in a mathematical equation. The symbol signifies "factorial." For example, a number 3 followed by an exclamation mark is read 3 factorial, a number 6 followed by an exclamation mark is read 6 factorial, and so forth. Factorial means that you are to find the product of that number and all whole numbers less than it all the way down to 1; for example:

$$3! = (3)(2)(1) = 6$$
$$6! = (6)(5)(4)(3)(2)(1) = 720$$

Although it will not make much sense at this time, you should also be aware that by definition $0! = 1$.

To find the number of ways to select two courses from among four offered, our computation is as follows:

$$_4C_2 = \frac{4!}{(4 - 2)!\, 2!} = \frac{(4)(3)(2)(1)}{(2)(1)(2)(1)} = 6$$

This is the same answer we got through a process of enumeration. Now try the formula for finding the number of ways to select two courses from six, and four courses from eight. You should get 15 and 70.

The formula for combinations can be of use apart from the use we are going to make of it in computing probabilities for the binomial distribution. Knowing that there are 70 different combinations possible when selecting four things from among eight means that any particular combination has one chance out of 70 of occurring when four selections are made at random. Exercises at the end of this chapter will demonstrate the use that can be made of this information.

Just as an aside, it should be noted that if each different ordering of any combination is to be counted separately, these different orderings of the combinations are called *permutations*. The expression for permutation is $_nP_k$ and the formula is Formula 4.2-2.

$$_nP_k = \frac{n!}{(n-k)!} \qquad \text{4.2-2}$$

Now that you know about combinations and factorials, we can return to our original question, "What is the chance of your friend's simply guessing and getting four positive outcomes out of five events?" Again, his chance of guessing any game correctly is $1/2$, so $P(A) = 1/2$ and $P(NA) = 1/2$. The probability of getting exactly k positive outcomes out of n instances is given by Formula 4.2-3.

$$P \text{ (exactly } k \text{ out of } n) = [_nC_k][P(A)]^k[P(NA)]^{n-k} \qquad \text{4.2-3}$$

It has become rather common in other texts to use a lower case p to represent the probability of a positive outcome and a lower case q to represent the probability of a negative outcome. As a result, in other statistics books the right-hand side of Formula 4.2-3 looks like this: $_nC_k\, p^k q^{n-k}$. We will continue to use $P(A)$ and $P(NA)$ to avoid confusion with other uses for the symbol p.

Using our knowledge of combinations and Formula 4.2-3, we can now compute the probability of guessing four out of five right when the probability of guessing any one right is $1/2$.

$$
\begin{aligned}
P(4 \text{ out of } 5) &= \left[\frac{5!}{(5-4)!\ 4!} \right][1/2^4][1/2^1] \\
&= (5)(1/2)^5 \\
&= 5/32 \\
&= 0.15625
\end{aligned}
$$

Someone could be expected to get exactly four out of five right about 15% of the time when simply guessing. However, we are not interested just in the probability of exactly four out of five correct. Getting all five right would be even more impressive. We want to know what the probability is of getting *at least* four out of five correct. This means either four correct or five correct. Since these results are mutually exclusive, we can add the probability of four correct ($5/32$) and the probability of five correct ($1/32$) and get $6/32$, or 0.1875.

Although you may still not be convinced that your friend is doing anything more than guessing, you will probably take him a bit more seriously in future discussions of his theory.

It should be apparent that Formula 4.2-3 can be used to figure the probabilities associated with any of the possible numbers of positive outcomes. The only requirement for the use of Formula 4.2-3 is that $P(A)$ remain constant for all events. When the probabilities are computed, a picture similar to a histogram can be drawn, with the number of positive outcomes as the horizontal scale and probability as the vertical scale. These pictures are referred to as probability distributions for the binomial distribution. Fig. 4-1 contains some probability distributions for the binomial distribution for $P(A) =$

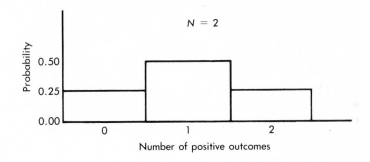

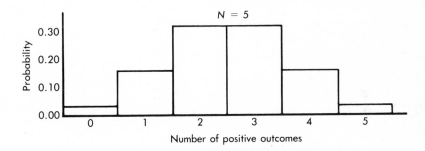

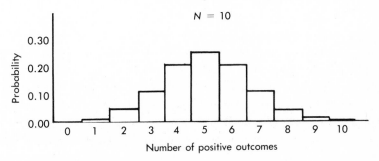

Fig. 4-1. Binomial distributions; $P(A) = \frac{1}{2}$.

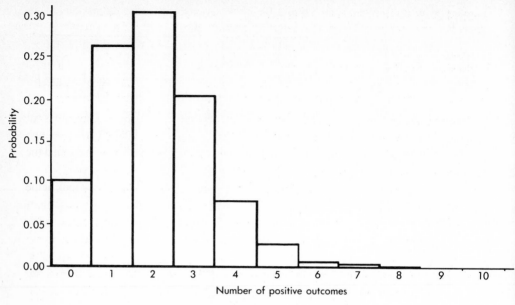

Fig. 4-2. Binomial distribution; $P(A) = 0.2, N = 10$.

$^1/_2$, showing what happens as n increases. Notice that the shape of the distribution begins to take the shape of the normal curve by the time $n = 10$.

Fig. 4-2 is similar to the bottom example in Fig. 4-1 in that it is also for ten events. The differences are that $P(A) = 0.2$ rather than 0.5 and that it is drawn to a larger scale. Notice that when $P(A)$ is not 0.5, the distribution is skewed. This will make sense if you consider the increased difficulty of getting a large proportion of positive outcomes when the probability of a positive outcome is small for any single event.

You will have an exercise at the end of this lesson that requires you to use a binomial distribution. In a later lesson you will learn to use the normal curve as an approximation for the binomial distribution when n is sufficiently large relative to the $P(A)$.

TABLES FOR COMBINATIONS AND THE BINOMIAL THEOREM

You have learned to use Formula 4.2-1 to determine the possible number of different combinations of k things selected from among n objects. In the example given, we found that $_4C_2 = 6$. Now, look at the computer-generated Table E-2 in Appendix E. This table provides information for any combination of n and $k \leq 15$. Looking at the left-hand column and locating $n = 4$, we then move horizontally to the column for $k = 2$. The number reported is 6, verifying the results obtained earlier. The table eliminates the need for doing calculations for small values of n and k. It is a good idea to use both the formula and Table E-2 when doing exercises at the end of the lesson to be sure you understand both. Incidentally, Appendix E also contains a table for the number of possible permutations of k things selected from among n objects (Table E-3).

Table 4-2. Possible outcomes for five events

	0	1	2	3	4	5
Number of survivors ($\pi = 0.9$)	0	1	2	3	4	5
Number of deaths ($\pi = 0.1$)	5	4	3	2	1	0

You have also learned to calculate the probability of exactly k positive outcomes for n events using Formula 4.2-3. The formula is applicable when the probability of a positive outcome remains the same from one event to another. In the example given, we found the probability of four out of five correct guesses for the case where the probability of a correct guess for each event was 0.5. The probability found was 0.15625. Now look at the computer-generated Table E-4 in Appendix E. The elements in the table give probabilities for the binomial formula for cases in which the probability for each single event is: 0.10, 0.20, 0.30, 0.40, and 0.50. (The probability for each single event is represented in the table by the Greek letter π. We shall have more to say about this use of the symbol π later.) The fifth element in Table E-4, where $\pi = 0.50$, contains the value 0.1562 for $n = 5, k = 4$, confirming the result of our calculations using the formula for the binomial distribution.

Actually, for the problem given we were interested in the probability of *at least* four out of five positive outcomes. We, therefore, added the probabilities for four out of five *and* five out of five positive outcomes. The result was 0.1875. Now look at Table E-5. The elements in this table give probabilities for *at least* k positive outcomes for n instances. The fifth table contains information for $\pi = 0.5$. For $n = 5, k = 4$, the value listed is 0.1875, confirming the calculations made earlier.

Tables E-4 and E-5 contain only elements for $\pi = 0.10, 0.20, 0.30, 0.40$, and 0.50. Suppose you are interested in the probability of at least four out of five patients surviving when $\pi = 0.90$ rather than 0.50. We do not have a table for $\pi = 0.90$. However, we can change the question to one involving the number of deaths rather than the number of survivors. If the probability of survival is 0.90, then the probability of deaths is 0.10. Table 4-2 lists the possible outcomes in terms of deaths versus survivors. Instead of asking the question "What is the probability of at least four surviving," we ask the equivalent question: "When $\pi = 0.10$, what is the probability of one or fewer deaths?"

We could use Table E-4 to find that the probability of no deaths is 0.5904 and that the probability of one death is 0.3280. Adding these two probabilities, we find that the probability of one or fewer deaths is 0.9184. Another alternative, which makes use of Table E-5, is to ask: "What is the probability of at least two deaths?" The tabled value for $\pi = 0.1$ and $n = 5$ is 0.0814. Since the alternative to at least two deaths would be one or fewer deaths, we can subtract 0.0814 from 1.00 and get 0.9186, which, except for rounding off, is the same value just obtained. The value 0.9186 is the probability of one or fewer deaths, or four or more survivors.

For even values of π and small values of k and n, the tables of binomial probabilities included as Tables E-4 and E-5 eliminate the need to do the many calculations required by the binomial formula. To be sure you understand the use of the formula, you should use both the formula and the tables to solve the problems in the exercises.

THE POISSON DISTRIBUTION

You learned earlier that when π is near 0.5, the binomial probability distribution is nearly symmetrical. As the sample size gets larger, the distribution begins to take on the form of the normal curve. What happens to the distribution as π approaches zero or 1.00? In this case the distribution becomes more and more skewed, and for very large samples the probability distribution approximates another theoretical distribution, called the Poisson distribution.

To illustrate the Poisson distribution, let us consider the incidence of a rare birth defect. The data available include the counts of the birth defect for a given year for 100 cities with populations over 10,000. The counts range from 0 to 12 with a mean of 3. If occurrences of the birth defect are independent events, the distribution of counts will follow the Poisson distribution. The probability of k occurrences of the event in a Poisson distribution can be obtained from Formula 4.2-4.

$$P(k \text{ occurrences}) = \frac{(\mu)^k}{k!e^\mu} \qquad \textbf{4.2-4}$$

where μ is the average number of occurrences in samples of very large size and
         e is the base for natural logarithms, or 2.7183.

For our example the probability distribution would be found by substituting 3 for μ in the formula. If the birth defects are independent events, we would expect the distribution to be as shown in Table 4-3.

Notice that if the occurrence of the birth defect is indeed independent from one case to another, approximately 5% of the cities will have no occurrences, 15% will have one occurrence, approximately 22% will have two occurrences, another 22% will have three occurrences, and more than eight occurrences will happen in only 35 of every 10,000 cities.

Can you guess what the distribution will look like if occurrences of the birth defect are not independent? Suppose, for example, that there is a clustering based on environmental factors. In this case the relative frequencies for the two ends of the distribution will exceed those of the Poisson distribution. Since it was stated that the observed frequencies for our example ranged from 0 to 12, there is some evidence that occurrences of the birth defect are not independent events. In a later chapter you will learn about a technique called the *goodness of fit test,* which can be used to test whether data from an experiment approximate a set of expected frequencies such as those derived from the Poisson distribution.

It should be noted that if the binomial formula is used to calculate the probabilities for $\pi = 0.0003, n = 10,000$, and $k = 0$ to 10, you will get almost the exact probabilities

Table 4-3. The Poisson distribution for $\mu = 3$

Number of occurrences (k)	Relative frequency
0	$(3)^0/[0 ! \ (e)^3] = 0.0498$
1	$(3)^1/[1 ! \ (e)^3] = 0.1494$
2	$(3)^2/[2 ! \ (e)^3] = 0.2240$
3	$(3)^3/[3 ! \ (e)^3] = 0.2240$
4	$(3)^4/[4 ! \ (e)^3] = 0.1680$
5	$(3)^5/[5 ! \ (e)^3] = 0.1008$
6	$(3)^6/[6 ! \ (e)^3] = 0.0504$
7	$(3)^7/[7 ! \ (e)^3] = 0.0216$
8	$(3)^8/[8 ! \ (e)^3] = 0.0081$
9	$(3)^9/[9 ! \ (e)^3] = 0.0027$
10	$(3)^{10}/[10 ! \ (e)^3] = \underline{0.0008}$
	0.9996

listed in Table 4-3. As stated at the beginning of this discussion, when π approaches zero and n is very large, the binomial distribution approaches the Poisson distribution.

Notice that to use the formula for the binomial distribution it is necessary to have values for both π and n. In contrast, the use of Formula 4.2-4 requires only μ, the average number of occurrences for samples of very large size (in theory infinite). In our example we did not worry about the actual populations of each city. We merely stated that each was in excess of 10,000. The Poisson distribution is often used to test for independence in situations in which the event being studied is rare and sample sizes are all large but not necessarily equal.

BAYES' THEOREM

Earlier, when explaining the multiplicative rule, we made use of the concept of independence. We said if A and B are independent, then:

$$P(A) = P(A|B) \quad \text{and} \quad P(B) = P(B|A)$$

When A and B are not independent, $P(A) \neq P(A|B)$ and $P(B) \neq P(B|A)$. It cannot be assumed in such a case that $P(A|B) = P(B|A)$. In fact, there are many situations, especially in medical practice, where we know $P(A|B)$ and want to determine $P(B|A)$ or vice versa. An example is the skin test for tuberculosis. We may know the prevalence of tuberculosis in the general population, *P(A);* the probability of positive reactions in the general population, *P(B);* the probability of positive reactions among persons with tuberculosis, $P(B|A);$ and the probability of a positive reaction among persons without tuberculosis, $P(B|NA)$. What we really want to know is the probability that a person has tuberculosis, given that the skin test is positive, *P(A|B)*. The solution can be found by applying Formula 4.2-5, which is called Bayes' theorem.

$$P(A|B) = \frac{P(A) \cdot P(B|A)}{[P(A) \cdot P(B|A)] + [P(NA) \cdot P(B|NA)]} \qquad \textbf{4.2-5}$$

Let us suppose that we know that approximately 5% of a population has tuberculosis, that the skin test is positive for 99% of those with tuberculosis, and that it is positive for 10% of those without tuberculosis. What is the probability that a person has tuberculosis, given that his skin test is positive? The data and computations are as follows:

$$P(A) = 0.05$$
$$P(B|A) = 0.99$$
$$P(B|NA) = 0.10$$

$$P(A|B) = \frac{(0.05)(0.99)}{(0.05)(0.99) + (0.95)(0.10)} = \frac{0.0495}{0.1445} = 0.342$$

Bayes' theorem has broad applications, not only in medicine, but in many statistical situations. The logic behind the formula is rather difficult to follow; I will not try to go into an explanation, but will leave it to you to puzzle over. One of the exercises will test your ability to apply the rule and should make the rationale behind the theorem understandable.

One word of comfort is in order. If you have managed to struggle through these two lessons on probability and understand the material, you have made it past a major hurdle. Students who have difficulty with the material in this book generally find the content of these two lessons to be among the most difficult of all the material.

After seeing that you can do the following exercises, you will proceed to a more detailed discussion of the idea of sampling and sampling distributions.

SUMMARY

This second lesson on probability began with a discussion of the problem of finding the probability of exactly k positive outcomes for a total of n events. The solution involves the use of the formula for the binomial expansion. The formula includes an expression for calculating the number of ways in which k things can be selected from among n things ($_nC_k$). The binomial formula is used to answer questions like the following: "If the probability of survival is 0.8, when is the probability that out of 20 patients with a disease, 19 will survive?"

The second section of the lesson described the Poisson distribution. This distribution occurs when π is near zero and samples are very large.

The third section of the lesson described a formula referred to as Bayes' theorem. It is used to find conditional probabilities and has applications in the diagnosis of diseases as well as in many other situations.

LESSON 2 EXERCISES

1. A medicine cabinet contains five similar bottles (A, B, C, D, and E) containing tablets. Single tablets from any of the bottles would have no adverse effect. However, if a person ingests both a tablet from bottle A and a tablet from bottle B within a 24-hour period, there will be a severe reaction. Suppose a child randomly ingests a tablet from each of two bottles on a paticular day. What is the probability of a severe reaction?
2. Suppose for the problem above the reaction occurs only if the tablet from A is ingested first,

followed by the one from B. The child again ingests a tablet from one bottle, then, an hour later, a tablet from a different bottle. Now what is the probability of a severe reaction?

3. A person whose credentials are somewhat suspect claims to have developed a saliva test that will identify whether a patient has a malignant tumor. You are dubious and set up a test as follows. You present him with eight unidentified samples, four from patients known to have malignant tumors and four from healthy patients. He is asked to identify whether each comes from a healthy patient or from one with cancer. He gets six correct and makes an error on two.
   a. What is the trait A?
   b. What is $P(A)$?
   c. What is $P(NA)$?
   d. How many events are there?
   e. Will the probability distribution be symmetrical?
   f. List all possible results in terms of the number of favorable outcomes.
   g. What is the probability of exactly six right?
   h. What is the probability for each of the rest of the results?
   i. What is the probability of six or more correct?
   j. What do you conclude?

4. Suppose you are told that among residents of a retirement community, 62.5% are female and 60% of the females are over 70 years old. You are also told that only one third of the male residents are over 70. What is the probability that a person chosen at random will be female, given that the person selected is over 70? Use Bayes' theorem where A = being female and B = being over 70.

$$P(A) = \underline{\hspace{4cm}}$$
$$P(NA) = \underline{\hspace{4cm}}$$
$$P(B|A) = \underline{\hspace{4cm}}$$
$$P(B|NA) = \underline{\hspace{4cm}}$$

$$P(A|B) = \frac{P(A) \cdot P(B|A)}{[P(A) \cdot P(B|A)] + [P(NA) \cdot P(B|NA)]}$$

$$P(A|B) = \underline{\hspace{4cm}}$$

5. Now suppose you were told that the community of Problem 4 had 10,000 residents. Use the same percentages given in Problem 4 to complete squares (1), (2), (3), and (4) of the following box.

	F	M
>70	(1)	(2)
≤70	(3)	(4)

Now, to answer the question of Problem 4, you will divide the number for square (1) by the total number of residents over 70, or squares (1) and (2). In other words:

$$P(A|B) = \frac{\text{square (1)}}{\text{square (1)} + \text{square (2)}}$$

If you will now examine how you computed the numbers for squares (1) and (2), you will see the logic for Bayes' theorem.

CHAPTER 5

Sampling and sampling distributions

Lesson 1 □ Random samples

It was stated earlier that the purpose of inferential statistics is to learn something about a population by looking at a sample. Before we proceed any further, it makes sense to clarify exactly what is meant by a sample versus a population and to examine several methods for choosing a sample.

POPULATIONS VERSUS SAMPLES

Statisticians do not use the word *population* in the same way it is used by the general populace. Instead of referring to a group of people residing in a given area, the word *population* as used by statisticians refers to any "group of things of interest." The group of things may be some finite number of physical objects, such as automobiles registered within a country at a certain time, hospital patient charts for a given hospital for a given day, or persons who received Medicaid payments during a certain year. On the other hand, a population may be a defined group of physical or nonphysical objects of undetermined number, such as all patients with carcinoma, all treatments for the common cold, or all biometry examination questions. In the first grouping it would be possible, albeit time-consuming, to list all members of each of the populations. The populations in the second group can be conceptualized but never listed in their entirety, since some future members do not even exist at present.

A *sample* is a group made up of some of the members of a population. For those who have been exposed to the language of "sets," a sample is a subset of the population. Because statisticians wish to learn about a population by examining a sample from that population, they are concerned that the sample provide good rather than misleading information about the population. Obviously, if it could be done, samples would be chosen in such a way as to be exactly like the population, only with fewer members. But this goal is unrealistic, and we must instead be satisfied with choosing samples in such a way that there is no systematic bias involved. In attempting to avoid systematic bias, the statistician is concerned about both intentional and unintentional bias. A person with a particular point to prove could intentionally choose a sample likely to support his position. For example, a person proposing a new method of treating a disease could deliberately choose from among all persons with the disease a sample of young patients who would be expected to respond quickly to treatment.

On the other hand, an entirely unintentional systematic bias can exist. For example, suppose you wish to know whether a bond issue for construction of a new hospital is likely to pass. The population you are interested in is composed of all persons who will vote on the bond issue. You decide to take a sample of potential voters to see what percentage favor the bond issue. You choose your sample by calling registered voters, starting in alphabetical order, until you have reached 100 persons. There are numerous potential sources of bias in the sample you have chosen. First, only persons with telephones will be included. Second, if the calls are made during the day, only persons who are near home phones during the day will be contacted. Third, if the survey is made during the summer, only those not away from home on vacation will be contacted. Fourth, even contacting persons in alphabetical order may introduce a bias by influencing the numbers of certain ethnic groups. Perhaps you can think of other sources of bias. The point is that choosing a sample in such a way as to avoid systematic bias is extremely important and is not to be taken lightly. The statistician chooses an unbiased sample by insisting on some form of *random* sample.

RANDOM SAMPLING

When a random sample was mentioned in an earlier lesson, it was stated that a random sample is a sample chosen in such a way that every member of the population has an equal chance of being selected. It is easy to define a sample in this way, but not so easy to choose one that meets the definition. In many contests described as sweepstakes, winners are chosen on the basis of a random selection. Usually all entries are placed in a large container, the container is rotated so that there will be thorough mixing of all the cards, and then someone reaches into the container and chooses a card. Even this process is not guaranteed to give a random sample. Persons who regularly enter such contests have learned to send their entries in late and to wrinkle the cards somewhat in an effort to increase their chances of winning. The slightly wrinkled, late entries do not become bunched as easily with the rest of the entries and tend to remain on the top of the container when it is rotated.

Statisticians avoid the need for drawing cards from a barrel or numbered balls from a revolving cage by using a table of random numbers (see Table E-1 in Appendix E). A table of random numbers is one in which the numbers zero through nine appear in random sequence. If you were to select a single digit number by placing a pin anywhere on the table, each of the numbers zero through nine would have an equal probability of being selected. Since there are ten numbers, each would have a probability of 0.10. The number immediately next to this number in any direction—up, down, right, left, or diagonally—would be independent of the number chosen, and again each of the numbers zero through nine would have a probability equal to 0.10.

From this description of a table of random numbers, it follows that to choose a random number of any desired number of digits, all you need do is start anywhere on the table and go in any direction until you have the number of digits needed. To choose another number, you could either start at another place on the table or you could just continue in the same direction.

To use the table in choosing a sample of any designated size you will need to do the following:

1. List all members of the population. It does not matter what the order is—alphabetically, by age, or any convenient order.
2. Number all the members, beginning with 1, 2, 3, and so forth up to N, the total number in the population.
3. Determine how many digits there are in the number assigned to the last member of the population. This is how many digits you will need for each of the numbers you choose from the table of random numbers.
4. Decide in which direction you will read numbers on the table of random numbers. (Left to right a row at a time is as good as any.)
5. Place a pencil somewhere on the table and read the required number of digits. If this number corresponds to a number assigned to a member of your population, that member becomes a part of your sample. If not, move to the right and again read the required number of digits.
6. To get the next number, continue reading numbers in the same direction. Continue until you have a sample of the desired size.

Now, let us see how this process would work to choose a random sample of size 5 from the 48 babies whose weights are given in the exercise for Lesson 1 of Chapter 3. The first two steps have already been accomplished, since the weights already appear in a numbered order. As the third step, we decide that two-digit numbers are needed, since 48 has two digits. We then decide to read numbers left to right a row at a time. Now, using Table E-1, suppose we happen to place our pencil on the first digit on column three on the tenth line. That digit is a 5 and the next one is a 3. Our first possible member of the sample would be the baby who has number 53 in our listing. Of course, there is no number 53, so we will have to go to the next two-digit number. This is a 34, so the first member of the sample is baby number 34, whose weight in ounces is 130. The next two-digit number is 25 (ignore the spaces between columns), so the second member of the sample is baby number 25, with a weight of 107 ounces. Following the same procedure, we see that baby number 39 becomes the third member. To obtain the fourth member, we skip over numbers 88 and 53 and arrive at the number 06. The fourth member of the sample is therefore the baby listed sixth, with a weight of 122 ounces. The final member of the sample is the baby listed fifth, with a weight of 98 ounces.

Sampling without replacement

When the population of interest is of some finite size, a decision must be made as to whether a single member of the population can appear more than once in the sample. If the population is rather small and the sample to be chosen is a significant portion of the population, it is quite likely that, at least once, the same number will be selected from the table of random numbers. If that "repeat" number is skipped, you are said to be "sampling without replacement." However, in some instances when a repeat number occurs, that member of the population is included twice (or more times) in the sample. In this case you are said to be "sampling with replacement." Later, when we look at sampling distributions of statistics, you will learn that a modification of some formulas is required when sampling without replacement from a finite population.

Simple versus stratified random sampling

If *one* random sample is chosen from the *entire* population of interest, the sample is said to be a "simple random sample." In many studies the population is divided into several groups (strata) and then random samples of varying sizes are selected from each stratum. This "stratified sampling" is often more efficient, in terms of the amount of information obtained, than is simple random sampling. Sampling is a whole discipline by itself. If you wish to learn more about it, there are numerous books devoted to the topic. For our purposes it is sufficient to confine our examples to simple random samples.

SYMBOLS FOR POPULATIONS AND SAMPLES

A distinction needs to be made between the numbers that represent characteristics of a population and the numbers that represent characteristics of a sample. The first are referred to as *parameters* of the population and the second as *statistics* for the sample. In other words, the mean computed for the entire population is a parameter, and the mean computed for a sample is a statistic. As was indicated in an earlier lesson, different symbols are used to distinguish between the two. Unfortunately, there is not total agreement in statistics books as to the symbols used. Generally, Greek letters are used for parameter values, with lower case Arabic letters sometimes used for the corresponding statistics. Table 5-1 contains the symbols that will be used in this book.

The parameter symbol for the mean of a population is the Greek letter for M and, as indicated earlier, is pronounced "mew." A sample mean is designated by placing a dash over a letter used to represent the variable being measured. If X is used to represent the variable height and Y the variable weight, the sample mean for height is $\overline{X}$ and the sample mean for weight is $\overline{Y}$. The symbol is read "X bar" or "Y bar."

The Greek letter for S is σ, which is pronounced "sigma." Sigma refers to the standard deviation of a population. The corresponding standard deviation for a sample is indicated by a lower case s. Since the variance is the square of the standard deviation, the variance is designated by using the symbol for the standard deviation raised to the second power.

The Greek letter for P is the symbol π, which is no doubt familiar, since it appears as part of the formula for the circumference of a circle. It is pronounced "pie" and will be used to designate the proportion of some trait in a population; for example, the proportion

Table 5-1. Symbols used to represent parameters and statistics

Concept	Symbols Population parameter	Sample statistic
Mean	μ	$\bar{X}$
Standard deviation	σ	s
Variance	σ^2	s^2
Proportion	π	p
Number	N	n

of persons who survive for 5 or more years following a diagnosis of lung cancer. The proportion in a sample will be designated by a lower case p. In contrast, the upper case P will be used as an abbreviation for probability. As indicated earlier, the statistician's definition of probability is related to the concept of proportion.

MODIFICATION IN THE FORMULA FOR STANDARD DEVIATION

The exercise that follows requires you to choose two simple random samples by use of a table of random numbers and to compute statistics for the sample chosen. The results of the exercise will be used as illustrations for the next lesson, on sampling distributions.

Before doing the exercise, you should be aware of a modification in the formula for computing the standard deviation and variance when a sample is involved. Earlier you learned that the formula for the standard deviation is computed from a formula first given, in its definitional forms, as Formula 3.2-4 or, in its computational form, as 3.2-7.

$$\sigma = \sqrt{\frac{\Sigma x^2}{N}} \quad \text{or} \quad \sigma = \sqrt{\frac{\Sigma X^2 - \frac{(\Sigma X)^2}{N}}{N}}$$

In many instances it is not possible or it is too costly to compute the standard deviation of a population. An estimate must be made based on the standard deviation computed for a random sample from the population. In such cases the formula used to compute the standard deviation of the sample is as shown in Formulas 5.1-1 and 5.1-2.

$$s = \sqrt{\frac{\Sigma x^2}{n-1}} \qquad \textbf{5.1-1}$$

or

$$s = \sqrt{\frac{\Sigma X^2 - \frac{(\Sigma X)^2}{n}}{n-1}} \qquad \textbf{5.1-2}$$

An explanation for this modification will be given in a later chapter for confidence intervals. For now it is sufficient to say that when the standard deviation is computed with this modified formula, the statistic s, computed for the sample, will be a better estimate of the parameter σ. Use this modification when computing sample standard deviations for samples in the exercise that follows. Do not use $N-1$ when finding σ for the population, however. You will need to be careful in this regard if you use a calculator that has a built-in function for computing the standard deviation. Probably it will use $n - 1$ in its internal computations. Some calculators use N as the denominator for calculating variance and $n - 1$ in the denominator for calculating the standard deviation. If this is the case for your calculator, to find σ, the population standard deviation, first find σ^2 and then take the square root to get σ.

SUMMARY

This first lesson on sampling defined populations and samples. The advantages of a particular type of sample, called a random sample, were described. Choosing a sample in such a way as to allow every member of the population an equal chance of selection is not easy. The use of a table of random numbers, as described in this lesson, achieves the goal. Sampling with and without replacement was described, as were procedures for stratified versus simple random sampling.

The symbols for sample statistics and population parameters were listed in Table 5-1. The lesson ended by describing a modification of the formulas for standard deviation and variance. The modification was the use of $n - 1$ in the denominator of the formula.

LESSON 1 EXERCISE

A population consists of 100 persons. For each person, information is available as to age *(X)* and sex *(Y)*. The data are as follows:

X	Y	X	Y	X	Y	X	Y	X	Y
10	M	22	F	24	F	42	F	37	M
48	F	36	F	13	F	16	F	57	M
1	F	44	F	4	M	79	M	3	M
50	F	65	M	13	M	30	F	29	M
11	F	73	F	60	F	43	M	75	F
17	M	25	F	22	F	6	M	11	M
53	M	59	F	52	F	24	M	83	M
60	F	58	M	79	M	36	F	91	M
13	M	53	M	26	F	16	M	66	M
18	M	93	M	57	F	20	F	56	M
16	F	30	F	63	F	7	F	6	F
47	M	49	M	23	F	15	F	12	M
91	M	58	M	64	M	61	F	19	F
64	F	31	M	80	M	63	M	17	F
66	F	28	M	31	M	36	F	82	M
17	F	27	F	51	F	39	M	60	F
41	F	82	M	79	M	44	F	46	M
41	M	53	M	24	F	5	F	28	F
94	M	40	M	23	F	35	F	13	F
62	M	29	M	4	M	37	M	5	F

1. Find: π, the proportion of females in the population; μ, the mean age in the population; and σ, the standard deviation in the population.
2. Now use a table of random numbers and choose a sample of size 5 with replacement. Find p, the proportion of females in your sample, and $\overline{X}$, the mean age in your sample.
3. Now choose a sample of size 25 with replacement and find the same statistics.

Lesson 2 □ Sampling distribution of a proportion

In the exercise for the previous lesson you were given a population of 100 persons for whom π, the proportion of females, was 0.50. As part of the exercise, you were asked to choose a random sample of size 5 and to compute p, the proportion of females in the sample. Suppose each person in a large class of 400 students carries out this assignment, each choosing a random sample by use of a table of random numbers and then recording p for the sample. The students then wish to pool their results to see how the sample p's are distributed.

To illustrate the results of 400 such random samples, a computer was programmed to carry out the assignment, choosing 400 random samples of size 5 and another 400 samples of size 25 from the table in the exercise. Each sample was chosen with replacement. The statistics p and $\overline{X}$ were computed for each sample. The results of the computer's samples will be used as examples in both this lesson, in which sample p's are studied, and the following lesson, in which sample $\overline{X}$'s are studied.

Let us consider the values of p calculated for the 400 random samples of size 5. How can these results be summarized? First, what values of p were possible, either when you carried out the assignment, or when the computer carried out the assignment 400 times? Since the samples were of size 5, the number of females in any sample could range from 0 to 5. The proportion of females in any sample could, therefore, be 0.00, 0.20, 0.40, 0.60, 0.80, or 1.00. Table 5-2 shows the frequency of each of these values among the 400 sample p's calculated by the computer. It also shows the percentage of each result. Fig. 5-1 is a histogram based on the percentages shown in Table 5-2.

Fig. 5-1 represents the distribution of sample proportions. Since each sample proportion is a statistic, the distribution is called a *sampling distribution of the statistic p*. Specifically, Fig. 5-1 represents a sampling distribution of the statistic p for samples of size 5 taken from a population in which π equals 0.50.

If many classes of students all carry out the assignment at the end of the last lesson and the students in each class pool their results, each class will end up with sampling distributions similar to that of Fig. 5-1. The larger the number of students carrying out the

Table 5-2. Sample proportions for 400 random samples ($\pi = 0.5$, $n = 5$)

p	f	% f
0.00	7	1.75
0.20	63	15.75
0.40	133	33.25
0.60	134	33.50
0.80	57	14.25
1.00	6	1.5

assignment in each class, the more similar the various sampling distributions will be. Can you guess what the sampling distribution would look like if a class of infinite size were to carry out the assignment? You have seen it in an earlier lesson. To refresh your memory, turn to Fig. 4-1 and look at the middle picture.

You will recall that the middle picture of Fig. 4-1 is the result of applying the binomial theorem to find the probability of 0, 1, 2, 3, 4, or 5 positive outcomes for the situation in which the probability of a positive outcome is 0.5 for each of five events. It should be apparent that these conditions are met when random samples of size 5 are taken from a population in which π equals 0.5. It is not surprising, therefore, that the percentages in Table 5-2 are very close to those represented by the middle picture of Fig. 4-1.

What is the relationship between the picture in the middle of Fig. 4-1 and the sampling distribution of p resulting from the samples chosen by the computer? One is a sampling distribution derived on the basis of *theory,* and the other is a sampling distribution derived *empirically.* Both represent the same concept: the distribution of sample p's for random samples of size 5 taken from a population in which π equals 0.5. When many random samples are taken from a population and the statistics (one from each sample) are used to form a distribution, the sampling distribution of the statistic is said to be derived empirically. For many situations it is not necessary to derive the sample distribution of the statistic on an empirical basis. Rather, the sampling distribution of the statistic is derived on a theoretical basis using mathematical formulas.

In actual practice, it is unusual for a statistician to have to resort to empirically deriving a sampling distribution for a statistic. In the great majority of cases, the sampling distribution of the statistic has already been established on a theoretical basis. However, the student just learning about sampling distributions will find it much easier to understand

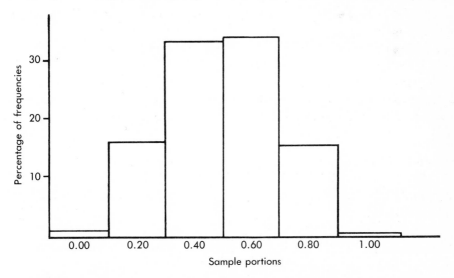

Fig. 5-1. Histogram of sample portions for 400 random samples; $\pi = 0.5$, $n = 5$.

the concept of a sampling distribution by thinking about how the sampling distribution can be empirically derived. Since the concept of a sampling distribution of a statistic is central to all of inferential statistics, the importance of this understanding cannot be overstated.

To test your understanding of the idea of a sampling distribution of the statistic p, think about the following situation. The 5-year survival rate for a specified disease is only one out of five. In a certain clinic, there are four patients who have been diagnosed as having the disease. Suppose that the patients are considered to be a random sample of size 4 and that p is the proportion surviving in the sample. Can you describe how the sampling distribution of p can be derived both theoretically and empirically? Think the problem through as best you can before proceeding.

Let us take the theoretical situation first. The solution is to use the binomial theorem. If we consider survival to be a positive outcome, $P(A)$, or π, = 0.2. The possible results are 0, 1, 2, 3, or 4 positive outcomes. Using Table E-4 in Appendix E for π = 0.2 and n = 4, we find the probabilities: 0.4096, 0.4096, 0.1536, 0.0256, and 0.0016. The theoretically derived sampling distribution is, therefore, as shown in Table 5-3.

Looking at this table, we see that its interpretation is straightforward. For example, if samples of size 4 are chosen from a population in which π is 0.2, only 16 out of every 10,000 samples can be expected to result in a sample p of 1.00; 256 out of 10,000 would result in a sample p of 0.75.

Now let us consider how the sampling distribution could be derived empirically. We could do as was done in the exercise at the end of the previous lesson. In other words, we could take many random samples of size 4 from among a population of patients with the disease. Then, we would have to observe each patient in each sample for a period of 5 years to determine p, the proportion surviving in each sample. We could then make a percentage frequency distribution of p values. This would take some length of time to carry out. However, stop to think for a minute.

Do we really have to sample from persons with the disease, who have a one in five chance of survival? Or could we sample from some other, more convenient population in which π for some characteristic equals 0.2? The answer is that the sampling distribution will be the same so long as π is 0.2 and n is 4. It does not matter whether π represents the proportion of surviving patients, the proportion of black balls in a barrel, or the proportion of zeros and ones in a table of random numbers. In each case, we are merely finding the

Table 5-3. Theoretical sampling distribution of a proportion (π = 0.2, n = 4)

Number of positive outcomes	Corresponding sample p	Probability
0	0.00	0.4096
1	0.25	0.4096
2	0.50	0.1536
3	0.75	0.0256
4	1.00	0.0016

sampling distribution of p when samples of size 4 are taken from a population in which π is 0.2.

Probably the simplest procedure for empirically deriving the desired sampling distribution would be to take samples of size 4 from a table of random numbers. Each sample will be composed of four single-digit numbers, and p will be the proportion of numbers that are either zeros or ones. Since there are ten digits, and two of them are either zero or one, π for the population of single-digit random numbers is 0.2. If we take many such samples, we would expect the p's computed for the various samples to occur with about the relative frequencies computed for the theoretically derived sampling distribution (Table 5-3). You may wish to verify this empirically by taking a set of samples and plotting the distribution of p. The more samples you take, the more closely the results will match the distribution of Table 5-3.

LARGE SAMPLES AND p

Now let us consider the results of the part of the previous exercise in which you were asked to take a random sample of size 25 with replacement. You were again asked to calculate p, the proportion of females in your sample. Suppose students in a class of 400 again wish to pool their results. What would the distribution of their sample p's look like? To find out, let us see what the distribution looked like for the 400 p values calculated for the 400 samples of size 25 chosen by the computer. A frequency distribution and a percentage distribution of the results are shown in Table 5-4. A histogram of the information in Table 5-4 appears as Fig. 5-2.

How would you describe the distribution represented by Fig. 5-2? In other words, what is its shape, its central tendency, and its variability? The shape should appear familiar. If you make a frequency polygon of the histogram, it will resemble the figure of the normal curve in Fig. 3-3. The central tendency of the distribution can be found by finding the mean of all 400 p values in the frequency distribution. Similarly, the variability can be found by finding the standard deviation of the p values. These two values are 0.4936 and 0.1015 respectively. To summarize, the distribution has a mean of 0.4936, a standard deviation of 0.1015, and a shape that is approximately normal.

Remember that the results shown in Table 5-4 and Fig. 5-2 represent an empirically derived sampling distribution for the statistic p for random samples of size 25 taken from a population in which π equals 0.5.

Once more it would also be possible to use the binomial theorem to theoretically derive the sampling distribution. There would be considerable work involved in the computations, however, since there are now a total of 26 possible results and the elements in Table E-4 in Appendix E do not go as high as $n = 25$. Fortunately, when the computations get tedious because the sample size is large, there is another theoretical solution. As we have observed, the shape of Fig. 5-2 is approximately that of the normal curve. As it turns out, the shape of the sampling distribution of p approaches that of a normal curve as the sample size increases, and as π gets nearer to 0.5. It is generally stated that the approximation is sufficiently good whenever both the values of $n\pi$ and $(n)(1 - \pi)$ are equal to or greater than five. Accordingly, if both $n\pi$ and $n(1 - \pi)$ are five or more, we will use the normal curve as an approximation to the sampling distribution that would result from the binomial theorem.

Once we decide that the shape of the distribution is normal, we can find the probability of various sample proportions by changing the sample p's into z scores. To do so requires that we know the mean and standard deviation of the sampling distribution. (Remember, z equals the observed value minus the mean, divided by the standard deviation.) You should have no trouble guessing the mean of many sample p's. The average of an infinite number of sample proportions will be equal to the population proportion, π. Statisticians, therefore, refer to p as an *unbiased* statistic.

It is not as readily apparent that the standard deviation of the many sample p's can be obtained from Formula 5.2-1.

$$\text{Standard deviation of sample } p\text{'s} = \sqrt{\frac{\pi(1 - \pi)}{n}} \qquad \textbf{5.2-1}$$

Proof of this formula is beyond the scope of this book. You should simply know what it means and how to use it. For our example of samples of size 25 taken from a population in which π is 0.5, the formula would result in a value equal to 0.10.

At this point, another notation needs to be introduced. When we are describing the amount of variability in a sampling distribution, we are referring to the standard deviation of the sample statistics, one statistic from each sample. We will refer to this standard deviation of a statistic as the *standard error* of the statistic. Later when we are considering quantitative data, this will make it easier to distinguish between the *standard deviation of*

Table 5-4. Sample proportions for 400 random samples ($\pi = 0.5$, $n = 25$)

p	f	% f
0.12	1	0.25
0.16	0	0.00
0.20	1	0.25
0.24	2	0.50
0.28	8	2.00
0.32	12	3.00
0.36	26	6.50
0.40	39	9.75
0.44	59	14.75
0.48	67	16.75
0.52	63	15.75
0.56	37	9.25
0.60	37	9.25
0.64	28	7.00
0.68	15	3.75
0.72	3	0.75
0.76	2	0.50
TOTAL	400	100.00

scores within an individual sample, and the *standard deviation of the statistics* computed from many samples. The term *standard error* of a statistic makes sense. If each sample were a perfect miniature of the population, the statistic computed for each sample would equal the parameter value. There would be no variability in the sample statistics. In a sense, then, the variability in sample statistics arises from sampling "error." It makes sense to call a measure of the amount of that error the standard error of the statistic rather than the standard deviation of the statistic. Virtually all statistics books use this terminology when referring to the variability of sample means. We will use it for sample proportions as well. The formula for the standard error of sample proportions can then be stated as follows.

$$SE_p = \sqrt{\frac{\pi(1 - \pi)}{n}}$$

5.2-2

where n is the sample size and
        SE_p is the standard error of sample proportions.

The reader should once more be reminded that this formula is used when the sampling distribution of p is being derived theoretically. As stated earlier, we could derive the sampling distribution empirically by taking many samples and computing p for each sample. We could then plot a percentage polygon based on the sample p's to determine the shape of the distribution; we could average all the p's to find the mean of the distribution; and finally we could find the standard deviation of all the p's to find the variability in the

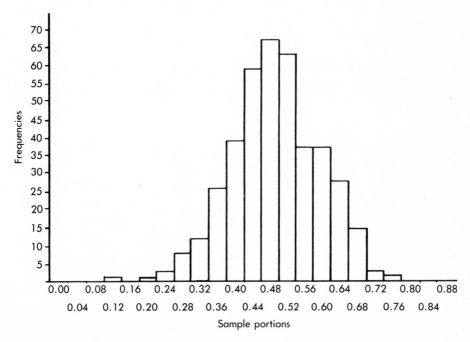

Fig. 5-2. Histogram of sample portions for 400 random samples; $\pi = 0.5$, $n = 25$.

distribution. This variability we would call SE_p. We have carried out this process earlier, choosing 400 samples and showing the results in Table 5-4 and Fig. 5-2.

Now let us check to see how closely our empirically derived sampling distribution matches that expected on the basis of theory. Since $n\pi$ and $n(1 - \pi)$ are both larger than five, we expect the sampling distribution of p to be shaped like the normal curve. We have already noted that this was the case for the distribution shown in Fig. 5-2. Our theoretical distribution is said to have an average equal to the population proportion 0.5. We found the average of the frequency distribution in Fig. 5-2 to be 0.4936, very close to the theoretical value. We call the standard deviation of the theoretical distribution SE_p, and we calculate it by taking the square root of π times $1 - \pi$, divided by the sample size. For our samples of size 25, from a population in which π equaled 0.5, we calculated the SE_p to be 0.1. The standard deviation of the 400 p's in Table 5-4 was 0.1015, again very close to the expected value.

DETERMINING THE PROBABILITY OF A SPECIFIC SAMPLE STATISTIC

Now consider the p you obtained for your sample of 25. Was your sample result a common or uncommon result? Change your p to a z score, using 0.5 as the mean and 0.10 as the standard deviation. Estimate the proportion of all sample p's as extreme as the one you obtained. Do this using the table of the normal curve (Table E-6) in Appendix E. The probability of a sample p as extreme as yours in either direction will be found in the "proportion beyond" column of the two-tail section of the table. For example, if your sample p was 0.60, you would calculate as follows.

$$z = \frac{p - \pi}{SE_p} = \frac{0.60 - 0.50}{0.10} = +1.00$$

For a z score of 1.00, Table E-6 lists a two-tail probability of 0.3173. The interpretation is that approximately 32% of all sample p values will be as far from π as the one you observed.

ANOTHER EXAMPLE

To be certain that you understand the idea of the sampling distribution of the statistic p, consider another example. Suppose 10,000 persons each choose a sample of size 25 from a barrel containing an infinite number of beads, of which 80% are red. How many people will choose a sample of 15 or fewer red beads? What is the statistic computed for each sample and how is it calculated? What are the shape, central tendency, and variability of the sampling distribution?

The statistic is, of course, p, the proportion of red beads in each sample. It is calculated by dividing the number of red beads in a sample by n, the sample size. Since the proportion π of red beads in the barrel is 0.80, we expect the average of all sample p's to be 0.80. Since both $n\pi$ and $n(1 - \pi) \geq 5$, we expect the sampling distribution to be approximately normal. Finally, we calculate SE_p, using Formula 5.2-2.

$$SE_p = \sqrt{\frac{\pi(1 - \pi)}{n}} = \sqrt{\frac{(0.8)(0.2)}{25}} = 0.08$$

We want to know how many people will get 15 or fewer red beads. Fifteen red beads would be a sample proportion equal to 0.60. Where would this sample p fall on the sampling distribution just described? To find out, we change 0.60 to a z score by subtracting the mean (0.8) and dividing by SE_p (0.08).

$$z = \frac{0.60 - 0.80}{0.08} = -2.5$$

Referring to Table E-6, we find a one-tail probability of 0.0062. We expect only 62 out of the 10,000 people to get 15 or fewer red beads in their sample.

SUMMARY

In this lesson you learned about the sampling distribution of a statistic, specifically the statistic p (proportion). The sample distribution of the statistic p refers to the distribution that would result if many random samples, all of equal size, were chosen from a population, and p, the proportion of some trait within each sample, were computed. It is possible to determine the characteristics of this sampling distribution empirically by taking many samples and then determining the characteristics of the resulting distribution of sample p's (shape, central tendency, and variability). It is much more usual to determine the characteristics of the sampling distribution on a theoretical basis. When the sample size is small, the binomial theorem will provide the required probabilities for each possible p. When sample sizes are large, the normal curve provides a good approximation of the probabilities that can be obtained from the binomial theorem. When this approximation is used, the mean of the distribution is π and the standard deviation, SE_p, is equal to the value of $\sqrt{\pi(1 - \pi)/n}$.

In the next lesson you will learn about the sampling distribution of another statistic, $\overline{X}$. Before going to that lesson, you are given some exercises to test your understanding of the sampling distribution of p.

LESSON 2 EXERCISES

1. Find the sampling distribution of p both empirically and theoretically for samples of size 2 taken from a population in which π equals 0.5. Choose 25 samples when deriving the distribution empirically. How well do the empirically and theoretically derived distributions match? How could you make them match better?

2. Find the sampling distribution of p theoretically for samples of size 64 taken from a population in which π equals 0.8. Describe the distribution by telling its shape, central tendency, and variability.

3. Describe in detail how you would empirically determine the shape, central tendency, and variability of the sampling distribution of p for samples of size 10 taken from a population in which π equals 0.5.

4. What is the probability that a person who is merely guessing at answers on a 100-item true-false test will guess 70 or more answers correctly?

Lesson 3 □ Sampling distribution of a mean

As part of the exercise at the end of the first lesson of this chapter you were asked to choose two random samples with replacement from a population in which μ was 40.17 and σ was 24.6. The first sample was to be of size 5 and the second of size 25. For each sample you found p, the proportion of females, and $\overline{X}$, the mean age. In the last lesson we considered what the distribution of sample p's would look like if 400 students each carried out the assignment. We called this the sampling distribution of the statistic p. In this lesson we will look at the sampling distribution of the statistic $\overline{X}$.

MEANS OF LARGE SAMPLES

Let us first consider the $\overline{X}$'s from the samples of size 25. Again, suppose 400 students each independently choose samples of size 25. Each student finds $\overline{X}$, the mean of his or her sample. The sample $\overline{X}$'s will not all be identical. Some will be slightly above μ, the population mean, and some will be slightly below; but in general they will be clustered about μ. The distribution of the sample $\overline{X}$'s will be called the sampling distribution of $\overline{X}$.

Let us see what happened when the computer carried out the assignment 400 times. For each sample of size 25, the statistic $\overline{X}$ was calculated. A frequency distribution of the 400 $\overline{X}$'s is shown in Table 5-5, and a histogram based on the frequency distribution appears as Fig. 5-3.

How would you describe the distribution represented by Fig. 5-3? Again, we need to know its shape, central tendency, and variability. The shape should look familiar. If you were to make a frequency polygon of the histogram, it would resemble a normal curve. The central tendency of the distribution can be found by finding the average of all 400 sample means represented by the frequency distribution in Table 5-5. Similarly, the variability of the distribution can be found by finding the standard deviation of the $\overline{X}$'s. We

Table 5-5. Sample means for 400 random samples ($\mu = 40.17$, $\sigma = 24.6$, $n = 25$)

Class limits	f	%f
24-26	1	0.25
27-29	4	1.00
30-32	13	3.25
33-35	41	10.25
36-38	75	18.75
39-41	91	22.75
42-44	76	19.00
45-47	61	15.25
48-50	25	6.25
51-53	9	2.25
54-56	3	0.75
57-59	1	0.25
TOTAL	400	100.00

will call this latter the $SE_{\bar{x}}$. For the data in Table 5-5, the mean and standard deviation are 40.85 and 5.231 respectively. To summarize, the sampling distribution represented in Table 5-5 has a mean of 40.85, an $SE_{\bar{x}}$ equal to 5.231, and a shape that is approximately normal. It should be remembered that this is an empirically derived sampling distribution for samples of size 25 taken from the population of scores given as an exercise at the end of Lesson 1 of this chapter.

As indicated in the preceding lesson, the statistician rarely has to resort to empirical methods to learn about the sampling distribution of a statistic. As was true for the sampling distribution of p, the sampling distribution of $\overline{X}$ can also be described on the basis of theory. First, let us consider shape of the distribution. Most statistics books agree that the sampling distribution closely approximates the normal curve whenever the sample size is 25 or more. Indeed, if the population from which the samples are drawn is normally distributed, the sampling distribution of $\overline{X}$ will be a normal distribution no matter what the sample size.

As was true with the statistic p, the statistic $\overline{X}$ is also an unbiased statistic. This indicates that as the number of samples increases, the mean of the sample $\overline{X}$'s approaches the population mean, μ. In other words, if it were possible to empirically choose an infinite number of samples and compute an infinite number of $\overline{X}$'s, the mean of all these $\overline{X}$'s would be μ, the parameter value. The central tendency of the sampling distribution of $\overline{X}$ is therefore equal to μ, the mean of the population. You would probably have guessed that this would be the case.

The only other thing we need to know is the variability of the sampling distribution of

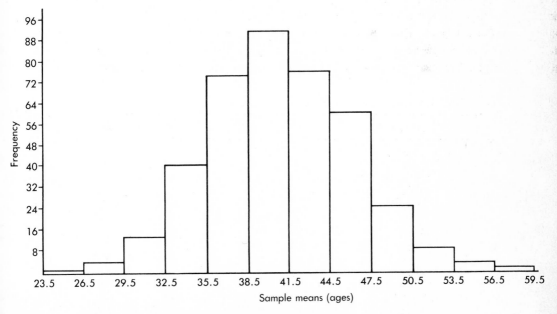

Fig. 5-3. Histogram of 400 sample means; $\mu = 40.17$, $\sigma = 24.6$, $n = 25$.

$\overline{X}$. In other words, we need to know $SE_{\bar{x}}$, the standard deviation of all the sample $\overline{X}$'s. The theoretical formula is not self-evident, but you should be able to see that it makes sense. What are the factors that you think will affect the variability of sample $\overline{X}$'s? In other words, what do you think will affect how tightly the sample $\overline{X}$'s will be clustered about the value of μ? There are two such factors. You would probably guess that sample size would be a factor, and indeed it is. Now what is the relationship? In other words, as the sample size increases, what happens to $SE_{\bar{x}}$? Intuitively, you would no doubt conclude that larger samples give closer estimates of μ. Therefore, as sample size increases, $SE_{\bar{x}}$ decreases. Similarly, as sample size decreases, $SE_{\bar{x}}$ increases.

There is another factor that affects $SE_{\bar{x}}$, but it is not quite as obvious. It is the variability of the scores in the population from which random samples are chosen. The less spread there is among scores in the population, the less spread there will be among sample means. If you carry this to an extreme, you will see that eliminating variability in the population will eliminate spread in sample $\overline{X}$'s. Scores in the population will all be identical, as will scores in all samples.

In Formula 5.3-1 you will see that there is a direct relationship between $SE_{\bar{x}}$ and the variability in the population, and an inverse relationship between $SE_{\bar{x}}$ and the size of the sample.

$$SE_{\bar{x}} = \frac{\sigma}{\sqrt{n}}$$ 5.3-1

where σ equals the standard deviation of the population and
          n equals sample size.

For our example:

$$SE_{\bar{x}} = \frac{24.6}{\sqrt{25}} = 4.92$$

We can now summarize the theoretical description of the sampling distribution of the mean: (1) the distribution has approximately a normal shape if $n \geq 25$ or if the population has a normal distribution; (2) the mean of the distribution is μ, the parameter value; and (3) the standard deviation of sample $\overline{X}$'s can be found by dividing σ by the square root of n. We will call this value $SE_{\bar{x}}$.

Now let us return to the 400 sample $\overline{X}$'s calculated for the random samples chosen by the computer from a population in which μ was 40.17 and σ was 24.6. First, since the samples were of size 25, we expect the distribution to have approximately a normal shape. In Fig. 5-5 you can see that this is true. We expect the mean of the $\overline{X}$'s to be approximately equal to μ (40.85 vs. 40.17), and we expect the standard deviation of the means to be approximately equal to $\sigma/\sqrt{n}$ (5.23 vs. 4.92). Once again, you can see that the empirically derived sampling distribution is very close to what would be expected based on the theoretical distribution.

We can also compare observed frequencies of the empirical distribution of Table 5-5 with expected values for the normal curve. We expect virtually all sample means to be within 3 $SE_{\bar{x}}$'s of μ (25.41 to 54.93). Approximately 95% should be within 2 $SE_{\bar{x}}$'s of μ (30.33 to 50.01), and approximately 68% should be within 1 $SE_{\bar{x}}$ of μ (35.25 to 45.09). An examination of the distribution will reveal that the frequencies conform quite well to these expectations.

Knowing that the sampling distribution of $\bar{X}$'s is approximately normal (with a mean of 40.17 and a standard error of 4.92) makes it possible to use the table of the normal curve to estimate the probability of various sample $\bar{X}$'s. Just as in the last lesson, you can find the probability of a sample $\bar{X}$ as extreme as the one you observed by converting your sample mean to a z score. For example, if you calculated a mean of 33.7, you would calculate as follows:

$$z = \frac{\bar{X} - \mu}{SE_{\bar{x}}} = \frac{33.7 - 40.17}{4.92} = -1.31$$

Looking at Table E-6 Appendix E, we find a two-tail probability of 0.1902.

MEANS OF SMALL SAMPLES

It has been stated that the sampling distribution of $\bar{X}$'s is approximately normal when the sample size is 25 or more or when the population from which samples are taken is normally distributed. What can be done when the sample is less than 25 and you do not believe the population to be normally distributed? It would be possible to derive the distribution by empirically choosing many samples and making a frequency distribution of sample $\bar{X}$'s. This is not as easy a task as it is in the case of sample p's, however. You remember that in the case of sample p's we can substitute any convenient population in which π is equal to π in our population. We can then take random samples from a convenient population rather than random samples from the population in which we are actually interested. This is very helpful in situations where it is not feasible to take repeated samples from the population of interest.

The sampling distribution of $\bar{X}$'s depends on the shape of the population from which samples are chosen. However, it also depends on μ and σ from that population. As a result, it is very difficult to find a convenient population that is appropriate for use in empirically deriving the sampling distribution of $\bar{X}$. The one chosen must have not only the same parameter value μ, but also the same σ and the same shape. From this it can be seen that attempting to empirically derive the distribution of sample means for small samples is often not feasible.

Unfortunately, if the population from which samples are drawn is a skewed distribution and sample size is small, neither is there a good theoretical solution to determining the sampling distribution of $\bar{X}$. The smaller the sample size, the more the sampling distribution of $\bar{X}$ tends to be shaped like the population from which the sample is drawn. At the other extreme, the larger the sample size, the more the sampling distribution looks like the normal curve. Our rule of thumb that the sampling distribution will be approximately normal when $n \geq 25$ is only a rule of thumb. If the population is severely skewed, a sample size of 25 will not completely eliminate the skewness in the sampling distribution. On the other hand, if the population is nearly a normal distribution, the distribution of means of samples of size 5 may be very nearly normal. The generalization is that if a population distribution is roughly symmetrical, with a clustering of scores near the mean, the sampling distribution of $\overline{X}$ rapidly approaches the normal curve as n increases.

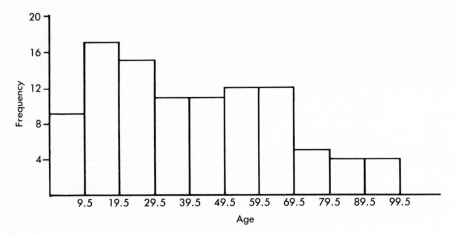

Fig. 5-4. Histogram of ages of 100 persons.

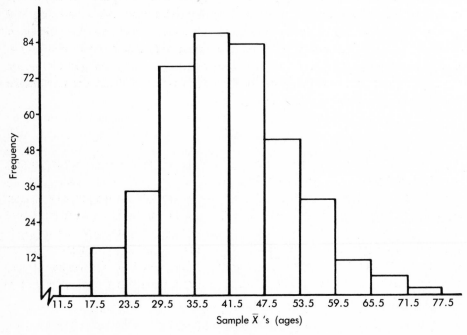

Fig. 5-5. Histogram of 400 sample means; $\mu = 40.17$, $\sigma = 24.6$, $n = 5$.

Going back to our example of 400 samples of size 5 from the population of scores at the end of Lesson 1 of this chapter provides an illustration. A histogram of the *scores* appears as Fig. 5-4. Notice that the distribution does not resemble a normal curve. The distribution is rather even, with some skewness to the right.

Now look at the histogram in Fig. 5-5. This histogram represents the sample *means* of the 400 samples of size 5 chosen by the computer. It is quite evident that the sampling distribution is already taking on the form of the normal curve with a sample of size 5. It appears that statistics books may be rather conservative when they suggest that a sample size of 25 is needed for the sampling distribution of the means to be approximately normal. Use of the normal curve with samples of size less than 25 from populations that are not normal will always result in some error in the reported percentages. The extent of these errors will depend on the extent to which the population from which the samples are drawn departs from normality. Particular concern needs to be paid to skewness in the population. If you know that the population is skewed, do not assume that the sampling distribution of sample means will be approximately normal unless $n \geq 25$. For severely skewed populations, the number should be even higher. If the population is nearly symmetrical, with a clustering near the mean, you can assume that the sampling distribution of $\overline{X}$ will be approximately normal even through sample size is considerably less than 25.

THE FINITE MULTIPLIER

Before we conclude this chapter on sampling distributions, it is necessary to indicate the effect of sampling without replacement. Sampling without replacement from finite populations reduces the variability of the sampling distribution. When sampling with replacement, you can select extreme scores more than once in a sample, thus making the mean for that sample more extreme. Sampling without replacement allows the use of an extreme score only once in any sample, thus curtailing the range of possible means. Allowance is made for this curtailment of the range of possible $\overline{X}$'s by multiplying the estimate of $SE_{\bar{x}}$ by what is called the finite multiplier:

$$\sqrt{\frac{N-n}{N-1}}$$

You can see that when the sample size is small in comparison to the number in the population, the finite multiplier is very near one. As the sample size increases relative to the size of the population, the finite multiplier will get smaller and smaller. Since, when sampling is without replacement, the finite multiplier is multiplied by the standard error of the mean, you can see that the effect of the finite multiplier is to reduce the estimate of the variability in the sampling distribution of $\overline{X}$. This is consistent with our earlier conclusions about the curtailment of the range of $\overline{X}$'s. The appropriate formula for the standard error of the mean modified for sampling from a finite population without replacement appears as Formula 5.3-2.

$$SE_{\bar{x}} = \left(\sqrt{\frac{N-n}{N-1}} \right) \left(\frac{\sigma}{\sqrt{n}} \right) \qquad \textbf{5.3-2}$$

ANOTHER EXAMPLE

To be certain you understand the idea of a sampling distribution of the statistic $\overline{X}$, consider another example. Suppose "average daily rates" for hospitals throughout the United States are normally distributed, with a mean of 180 and a standard deviation of 15. If you choose nine hospitals at random and figure their mean "average daily rate," what is the chance that the mean will be either above 190 or below 175?

Once again, we need to know about the sampling distribution, this time for the statistic $\overline{X}$. Will the sampling distribution be normally distributed with samples of size 9? Careful now! The answer is yes, because it was stated that the population of all "daily hospital rates" is normal. We do not need to worry about our sample size being less than 25. The mean of the sampling distribution will be the parameter value 180, and the $SE_{\overline{x}}$ will be $\sigma/\sqrt{n}$, or $^{15}/_3$, or 5.0.

We want to know how often a mean "daily hospital rate" greater than 190 or less than 175 will occur for random samples of size 9. We change each number to a z score: $+2.00$ and -1.00. Looking at the table, we find that 2.28% of the scores are above $+2.00$ and 15.87% of the scores are below -1.00. We conclude that the probability that we will get a sample mean greater than 190 or less than 175 is about 0.18 (0.1587 + 0.0228).

SUMMARY

The sampling distribution of $\overline{X}$ can be determined either empirically or theoretically. Empirically, we take hundreds of samples of the desired size from the population of interest. We then plot a frequency polygon of the sample $\overline{X}$'s to discover the shape of the sampling distribution; we find the mean of the $\overline{X}$'s to discover the central tendency of the sampling distribution; and we compute the standard deviation of sample means to discover variability of the sampling distribution. We call this $SE_{\overline{x}}$. On a theoretical basis, if the sample size is 25 or more, we know that the sampling distribution will be approximately normal in shape. No matter what the size of the sample, we know that the mean of the sample $\overline{X}$'s will be μ and that the standard deviation of sample $\overline{X}$'s $(SE_{\overline{x}})$ will be $\sigma/\sqrt{n}$.

For small sample size, the shape of the sampling distribution of $\overline{X}$ takes on the characteristics of the population from which samples are drawn. If the population is

Table 5-6. Summary and comparison of sampling distributions for statistics p and $\overline{X}$

Parameter	Statistic	Sampling distribution		
		Shape	Central tendency	Variability
π	p	Approximately normal if $n\pi$ and $n(1-\pi) \geqslant 5$	parameter π	$SE_p = \sqrt{\dfrac{\pi(1-\pi)}{n}}$
μ	$\overline{X}$	Approximately normal if $n \geqslant 25$ or normal if population is normal	parameter μ	$SE_{\overline{x}} = \dfrac{\sigma}{\sqrt{n}}$

normal, the sampling distribution of X will be normal regardless of sample size. The nearer the population is to normal, the more rapidly the sampling distribution of $\overline{X}$ approaches the normal curve as sample size increases.

Table 5-6 provides a summary and comparison of the sampling distributions for the statistics p and $\overline{X}$. We could add other statistics, such as σ and the median, but these two examples are sufficient to teach the basics of inferential statistics.

You will now have an exercise to test your knowledge of the sampling distribution of $\overline{X}$. Following that, you will finally begin to apply what you have learned by looking at one of the approaches to inferential statistics, confidence intervals.

LESSON 3 EXERCISES

Results of the Graduate Record Examination for the entire population of students taking the exam are said to have a μ of 500, a standard deviation of 100, and a distribution that is approximately normal.

1. Describe what the theoretical distribution of sample $\overline{X}$'s from this population is like when random samples of size 36 are chosen.
2. Describe how you could empirically derive the distribution referred to in the first question.
3. Describe what the effects would be if the sample size were 16 rather than 36. Careful now!
4. Suppose the average length of stay in a chronic disease hospital for a certain type of patient is 60 days, with a standard deviation of 15. It is reasonable to assume an approximately normal distribution of lengths of stay. Suppose many samples of 16 patients are selected at random with replacement from the hospital. For each sample the average length of hospital stay is computed. What two limits would include 99% of all sample means?

CHAPTER 6

Confidence intervals

Lesson 1 □ Confidence intervals for proportions

Congratulations! You have now made it to the point where you can apply some of the basic concepts you have learned. As was stated earlier, inferential statistics involves learning about populations by examining the characteristics of a sample. Specifically, the statistician makes an inference about a parameter, based on an observed statistic. You are now ready to learn how this is done.

As an example, a physician might wish to know what the likelihood is of an undesirable side effect when patients are treated with a therapeutic drug. The population he is interested in is made up of all the persons for whom the drug might be an appropriate treatment. His interest is in π, the proportion who will suffer the undesirable side effect. To learn about π, he will choose a random sample of persons from the population, treat them with the drug, and then compute p, the proportion who exhibit the side effect. He will then make an inference about π, based on the sample p.

In another situation, an educator might wonder what the effect would be of switching from larger lecture sessions to small discussion groups as a method for teaching biostatistics. He knows from past years what average level of performance can be expected using the large lecture sessions. The question is whether shifting to small discussion groups will result in any change in the level of performance.

The two situations just described are different in terms of the basic question being asked. In the first, the question is: "What is the parameter value for the population of interest?" Specifically, "What proportion (π) of patients treated with the drug will experience undesirable side effects?" In the second situation, the question is: "Is there a difference between the unknown parameter value for the test population and the known parameter value of some standard population?" Specifically, "Is there a difference in the average amount learned when students are taught in small discussion groups, compared with the average amount learned under the old method of large group sessions?" The population of all possible persons who are taught by the new method is referred to as the *test population*. The population of all persons taught by the old method for which results are known is called the *standard population*.

To answer each of the two different types of questions just described, the statistician uses two different approaches. When the question is: "What is the parameter value?" the approach used is called *setting up a confidence interval*. When the question is: "Is the unknown parameter of the test population different from the known parameter of the standard population?" the approach used is called *hypothesis testing*. One of these two approaches is used whenever inferential statistics is used. Which is to be used depends on which of the two questions is relevant to a particular situation. Since one of the two will always be used, a student who understands the rationale and methods of each of these two approaches will understand the approach used to solve any problem of inferential statistics. Details of computation for various statistics, and the sampling distributions associated with the statistics, may be complicated and unfamiliar, but the basic approach will always be either setting up a confidence interval or hypothesis testing.

This chapter, which includes two lessons, will describe how a confidence interval is established. The next chapter will describe hypothesis testing. The remainder of this first lesson on confidence intervals will be concerned with general procedures for setting up a confidence interval, followed by specific details of the process when the parameter of interest is π, the population proportion. The second lesson will be devoted to the process of setting up a confidence interval for the parameter μ.

CONFIDENCE INTERVALS: GENERAL RATIONALE

As indicated earlier, a confidence interval is established when an attempt is made to answer the question, "What is the parameter value for some defined population of interest?" The parameter might be π, μ, or any other parameter in which the researcher is interested. If it were feasible or possible, the answer to the question would be obtained by computing the parameter from data obtained for every member of the population. When this is either impossible or not feasible because of the costs involved, the researcher chooses a random sample from the population, computes the statistic, and then uses this statistic to make an inference about the unknown parameter.

Suppose you had just completed the process of choosing a random sample and computing the relevant statistic and someone asked you to make a single guess as to the value of the parameter. You do not need a course in statistics to tell you that, in the absence of any other information, your best guess is the value you just calculated for the statistic. The advantage you have now that you are studying statistics is that you will also be able to provide information as to how "good" that guess (estimate) is likely to be.

In the case where we are trying to determine the value of a parameter, the "goodness" of an estimate refers to how close the estimate is *likely to be* to the actual value of the parameter. Here it is important to make a distinction between the goodness of any *particular estimate* and the goodness of the *process* used in making the estimate. A person with no knowledge of statistics may resort to intuition, star charts, or any other method and guess, on a single occasion, the exact value of the parameter. That guess is as good as it can possibly be. The statistician, with all his disciplined methods of choosing random samples and computing relevant statistics, cannot make a better estimate. In fact, he almost certainly will do worse. The statistician will be the first to admit that this can happen for any individual instance. He is secure in the knowledge that over the long run his *method* is better and that over many instances his estimates will be better on the average.

It is because of the distinction just made that the words *likely to be* were emphasized earlier. The statistician never knows how good any particular estimate is; instead he knows how good, in general, estimates will be when his method is used. An analogy can be made with attempts to describe the accuracy of a particular rifle. The statement is made that a particular rifle is capable of grouping a certain percentage of its shots within a circle of a given diameter from a certain distance. The goodness of the rifle determines the diameter of the circle required. In a similar manner, the estimates of the statistician center on the parameter value. The goodness of the method used in arriving at the estimates can be described by stating how far you would have to go to each side of the parameter to include a given percentage of the estimates. It might be stated, for example, that 95% of the estimates of π will be within ten percentage points of the actual value, or that 90% of the estimates of μ will be within 3 inches of the actual value.

A researcher does not take many samples and make many estimates of the parameter. Instead, he takes one sample and states that his best estimate of the parameter is the value of the observed statistic. He then indicates how good that estimate is likely to be by setting up a band for his estimate, rather than stating it as a single point.

Let us return to our earlier analogy. When we describe the accuracy of a particular rifle, the diameter of the circle will depend on two things: (1) the percentage of shots to be included in the circle (obviously, a larger circle is needed to include 95% of the results than to include 80% of the shots) and (2) the precision of which the rifle is capable.

In a similar manner, the width of the band that the statistician gives when asked about a parameter depends on two things: (1) the percentage of times the band is expected to include the parameter value and (2) the precision of which his sampling procedure is capable. A larger band is needed if he intends for the band to include the parameter value 95% of the time than if he is satisfied with 80%. Accordingly, when a band (interval) is reported, it is called an 80% confidence interval, a 90% confidence interval, a 95% confidence interval, and so forth. The researcher arbitrarily makes the choice of the level of confidence.

CONFIDENCE INTERVALS: GENERAL COMPUTATIONS

From the description in the previous section, it follows that the confidence interval will be computed for any parameter by a formula that has the following form.

$$\text{Parameter} = \text{observed statistic} \pm (A)(B) \qquad \textbf{6.1-1}$$

The A in the first parentheses relates to the specific confidence level chosen and the B in the second parentheses relates to the precision of estimates resulting from the sampling procedures used.

Earlier, when learning about sampling distributions of statistics, you were told that this concept would be central to all of inferential statistics. Where does the sampling distribution enter into the setting of confidence intervals? The answer is that the variability of the sampling distribution is the basis for parentheses *B* and that the shape of the distribution determines the source of the number that goes into parentheses *A;* the fact that the mean of the sampling distribution equals the parameter value is the basis for starting with the observed statistic.

Let us begin with parentheses *A*. If we know that a sampling distribution is shaped like the normal curve and if we have decided to use a 95% confidence interval, we will look at the table of the normal curve (Table E-6, Appendix E) to see which two limits would include 95% of all sample statistics. Using the two-tail column of the table of the normal curve, we find that we must go 1.96 standard deviations above and below the mean. Accordingly, the number 1.96 will go into parentheses *A*. If we wanted a 90% confidence interval, what value would be entered? You should respond 1.645. Right?

Notice that it was only because we knew that the sampling distribution had a normal shape that we were able to make use of the table of the normal curve. If the sampling distribution were not shaped like the normal distribution, we would have to find the value from some other appropriate table, assuming that one were available.

As stated, going 1.96 standard deviations in either direction includes 95% of the scores in a normal distribution. We still need to know the size of the standard deviation. Where will that come from? Since we are talking about sample statistics, it will be the variability of the sampling distribution of the statistic. We have learned to call this the standard error of the statistic. The formula for a confidence interval that fills in the parentheses for *A* and *B* is therefore as follows.

Parameter = observed statistic ± (appropriate tabled value)(standard error of the statistic)

$$\text{6.1-2}$$

where   the appropriate tabled value depends on the confidence level chosen and the shape of the sampling distribution of the statistic.

CONFIDENCE INTERVALS FOR π

Now let us be specific and describe the process for finding a confidence interval for the unknown proportion of some trait or characteristic within a population. Let us return to the example given at the start of the lesson, in which the question was: "What proportion of patients treated with a certain drug will experience an undesirable side effect?" Since the question is "What is π?" the approach used will be to set up a confidence interval.

To make use of the normal curve, we will want to use a sample size large enough for p to be normally distributed. The difficulty is that the shape of the sampling distribution depends not only on n, but also on π, which is unknown. We will have to make some preliminary guess as to the value of π. Usually we will have some prior information on which this guess may be based. For our purposes, let us guess that π is approximately 0.10. If so, we will need a sample of size 50 for both $n\pi$ and $n(1 - \pi)$ to be equal to or greater than 5. To be on the safe side, let us double that and use a random sample where $n = 100$.

After choosing a random sample of 100 patients and treating them with the drug, we will compute the statistic p, the proportion with the side effect. Suppose it turns out to be 0.20. Our best single estimate as to π, the proportion of all patients treated with the drug who will suffer side effects, is therefore 0.20. However, to indicate how good that estimate is, we will set up a confidence interval using Formula 6.1-3.

$$\pi = p \pm [z \text{ value}]\left[\sqrt{p(1-p)/n}\right] \qquad \textbf{6.1-3}$$

The value to be substituted in each set of parentheses depends on the value of π. Since π is unknown, we have to substitute p in its place. First, to determine whether we can use the table of the normal curve as we had planned, we have to check to see if both np and $n(1-p)$ are equal to or greater than 5. In this example, where p is 0.20 and n equals 100, we conclude that the table of the normal curve is appropriate.

For the second set of parentheses, we know that the *SE* of the statistic p is equal to $\sqrt{\pi(1-\pi)/n}$. Not knowing π, we again have to substitute p. We therefore estimate as follows:

$$SE_p = \sqrt{\frac{p(1-p)}{n}} \text{ or } \sqrt{\frac{0.20(0.80)}{100}} \text{ or } 0.04$$

The only remaining decision is to choose the level of confidence to use. As was stated earlier, this decision is entirely arbitrary. For this example let us set up a 95% confidence interval. For the first parentheses, we will need to find the score on the table of the normal curve that leaves a total of 5% in the two tails. Using Table E-6 in Appendix E, look under the "π beyond" column of the two-tail part of the table until you find 0.05. Alternatively, you could look under the "π remaining" column for 0.95. The score listed for this row is 1.96. This is the value that goes into the first parentheses. As we have already stated, the estimate of the standard error that goes in the second parentheses is 0.04. Our 95% confidence interval for the proportion of patients who will suffer the side effect can therefore be stated as follows.

$$\pi = 0.20 \pm (1.96)(0.04)$$
$$\pi = 0.20 \pm 0.0784$$

The lower limit of the interval is 0.20 minus 0.0784, or 0.1216, and the upper limit is 0.20 plus 0.0784, or 0.2784. It should be stated again that there is no way of knowing for sure that π is between these two limits. If you choose other random samples you will get different values for p. The 95% confidence limits you set will therefore differ from sample to sample. The proper interpretation is that 95% of these confidence intervals will include the unknown parameter, π.

EMPIRICAL DEMONSTRATION

In actual practice, when confidence intervals are set one never knows the value of the parameter. There is no way to test whether the process really works. We can, however, make an empirical test by starting with a population with a known parameter value. Suppose we take all the two-digit numbers in a table of random numbers and circle those below 15. These would be the numbers 00 to 14. Since 100 different two-digit numbers are possible, all with an equal probability of occurrence, 15% of all the two-digit numbers would be circled. The proportion (π) of two-digit numbers $\leq$ 14 is 0.15.

Now let us imagine taking a random sample of 100 two-digit numbers and finding p, the proportion of numbers $\leq$14. We then will use the procedure just described to set a confidence interval. This time let us choose a 90% level of confidence. The parameter value π(0.15) may or may not be within these limits. However, if we repeat the process over and over, 90% of the time the limits should include the value 0.15.

A computer was programmed to choose 400 random samples of size 100 from a population in which π was known to be 0.15. In each case the computer calculated p and set a 90% confidence interval. Table 6-1 describes the results of the samples. The first three columns are familiar from our discussions of frequency distributions. Note that with samples of size 100, a proportion of 0.08 indicates that eight numbers in the sample were $\leq$14; a proportion of 0.12 means that 12 were $\leq$14. As might be expected, the largest number of samples had 15% positive outcomes (numbers $\leq$ 14). One sample had only five numbers below 15, and one sample had 26 numbers below 15. The fourth and fifth

Table 6-1. Confidence intervals of 90% for π based on 400 sample proportions ($n = 100$, $\pi = 0.15$)

Sample proportion	f	cf	Confidence interval limits		
			Lower	Upper	Includes π?
0.05	1	1	0.0141	0.0858	No
0.06	0	1	0.0210	0.0990	No
0.07	2	3	0.0280	0.1119	No
0.08	4	7	0.0353	0.1246	No
0.09	8	15	0.0429	0.1370	No
0.10	25	40	0.0506	0.1493	No
0.11	22	62	0.0585	0.1614	Yes
0.12	31	93	0.0665	0.1734	Yes
0.13	42	135	0.0746	0.1853	Yes
0.14	35	170	0.0829	0.1970	Yes
0.15	50	220	0.0912	0.2087	Yes
0.16	41	261	0.0996	0.2203	Yes
0.17	39	300	0.1082	0.2317	Yes
0.18	25	325	0.1168	0.2431	Yes
0.19	29	354	0.1254	0.2545	Yes
0.20	20	374	0.1342	0.2658	Yes
0.21	14	388	0.1429	0.2770	Yes
0.22	2	390	0.1518	0.2881	No
0.23	5	395	0.1607	0.2992	No
0.24	2	397	0.1697	0.3102	No
0.25	2	399	0.1787	0.3212	No
0.26	1	400	0.1878	0.3321	No

columns list the upper and lower boundaries of the 90% confidence interval that results from each sample p. For example, for the sample in which there were five positive outcomes, p would be 0.05 and the estimate of SE_p would be $\sqrt{(0.05)(0.95)/100}$, or 0.0218. The value from the normal curve that leaves a combined 10% in the two tails is 1.645. We would therefore compute the 90% confidence interval as follows:

$$0.05 \pm (1.645)(0.0218) \quad \text{or} \quad 0.0141 \text{ to } 0.0858$$

Of course, this would be one of those times when the interval fails to include π, as indicated by a "No" in the last column.

Since we have used a 90% confidence interval, we expect only 10% of the 400 samples to have a "No" in the last column. The sum of frequencies for confidence intervals with a 'No" in the last column of Table 6-1 is 52, or 13% of the 400 samples. The sum of the frequencies with a "Yes" is 348, or 87%. In other words, 87% of the confidence intervals included π. But our method was supposed to result in 90% of the confidence intervals including π. What is the explanation for this discrepancy? One possible explanation is that we have not taken enough samples. Perhaps we should take 1000 or 10,000 samples rather than just 400. Indeed, increasing the number of samples will give us a clearer picture of the true proportion of misses.

Rather than taking more samples, let us consider another solution, which is equivalent to choosing an *infinite* number of samples. We know from Table 6-1 that when the number of positive outcomes is anywhere between 11 and 21 inclusive, the confidence interval computed will include π. All we need to do is figure the probabilities for each of these outcomes and then add these probabilities. We can use the binomial expansion for figuring the probability of each outcome. For 11 positive outcomes, we figure as follows:

$$P(11) = {}_{100}C_{11} (0.15)^{11} (0.85)^{89}$$

Although it would be a lot of work to carry out the computations for the number of positive outcomes 11 through 21, the computer can do it very quickly. The probabilities for the results 11 through 21 add up to 0.861. In others words, if we take an extremely large number of samples, we can expect 86.1% of the resulting 90% confidence intervals to include π. What is wrong? The answer is that our confidence intervals are supposed to be based on values of π, but, not knowing π, we had to substitute sample values of p. We can see from this example that this substitution causes problems, in that our 90% confidence interval is really an 86.1% confidence interval. The seriousness of the problem will vary for differing values of π and n. We should make some modification either in our procedures or in the interpretation of our resulting confidence intervals.

EFFECTS OF SUBSTITUTING p FOR π

Whenever we want to set confidence intervals for π we face a dilemma, in that π is needed for our computations. We are forced to use p, the sample proportion, in place of π. As a result, we can consider the resulting confidence intervals as approximations only. If we happen to get a sample p value nearer to 0.5 than the actual value of π, our estimate of SE_p will be an overestimate. If we happen to get a sample p value further from 0.5 than the actual value of π, our estimate of SE_p will be an underestimate. Since the width of the interval for any confidence level will be directly related to SE_p, the estimated interval may be either slightly too narrow or slightly too broad. In a real situation, since we never know π, we never know in which direction we have erred in setting any single confidence interval. Unfortunately, we will generally have more "misses" than expected. There are methods for improving the procedure of setting confidence intervals for π that involve complicated calculations. Most statistics books simply recommend using p in place of π and recognizing that the result is only an approximation. If the sample size is fairly large and π is near 0.5, the approximation will be good. If sample size is small or π is near 1.00 or 0.00, the approximation can be quite far off, with "misses" occurring two or three times as often as expected. A conservative approach is to use 0.5 as the estimate of π when calculating the standard error of p. The formula becomes:

$$SE_p = \sqrt{\frac{(0.5)(0.5)}{n}} \text{ or } \frac{0.5}{\sqrt{n}}$$

Usually π will not turn out to be 0.5. As a result, the magnitude of SE_p and the width of confidence intervals calculated using this standard error will be overestimated. A reported 95% confidence interval will be *at least* a 95% confidence interval.

SUMMARY

When the goal of a research project is to answer the question, "What is the parameter value?" the approach used is to set up a confidence interval. The level of the confidence interval is arbitrary, usually 80%, 90%, 95%, or 99%. The procedure involves taking a random sample from the population of interest and computing the statistic corresponding to the parameter of interest. A confidence interval is then established by starting with the observed statistic and adding and subtracting a constant to get an upper and lower boundary. The constant to be added and subtracted is based on two factors. The first is obtained from a table appropriate for the shape of the sampling distribution of the statistic. The second is the standard error of the sampling distribution of the statistic.

When the goal is to answer the question, "What is π?" the sample statistic is p, the proportion of positive outcomes in the sample. If both np and $n(1 - p)$ are equal to or greater than 5, the sampling distribution is *assumed* to be approximately normal, and the confidence interval can be computed from Formula 6.1-3.

The resulting confidence interval must be considered an approximation, since p has been substituted for π in estimating the SE_p for the second set of brackets. When sample size is large and π is near 0.5, the approximation will be good. Caution is needed when π is near 0 or 1 or when the sample size is small, because the process results in a significantly larger number of confidence intervals that fail to include π than is expected. A conservative approach is to substitute 0.5 in place of π or p when calculating SE_p. The result is Formula 6.1-4.

$$\pi = p \pm [z \text{ value}] \left[\sqrt{\frac{(0.5)(0.5)}{n}} \right] \qquad \textbf{6.1-4}$$

LESSON 1 EXERCISE

1. A television survey team wishes to estimate the number of persons in a certain high school who watched a TV special dealing with the issue of abortion. A random sample of 80 students is asked about the program. Of this number, 60 indicate that they viewed the program. Set an approximate 99% confidence interval for π, the proportion within the school who viewed the program.
   a. What is your best single guess as to π?
   b. Can you use the table of the standard normal curve? Why or why not?
   c. What value will go into the first set of parentheses?
   d. What is the estimated standard error of p?
   e. What are the upper and lower boundaries of the approximate 99% confidence interval?
   f. What is the meaning of this interval?
   g. What will happen if you change to a 90% confidence interval?
2. Answer items *d* through *f* using the conservative approach to calculate SE_p.

Lesson 2 □ Confidence intervals for means

In the last lesson you learned that when a researcher wishes to ask the question, "What is the parameter value?" the approach is to establish a confidence interval. The first step is to arbitrarily choose a level of confidence; for example, 90%. A confidence interval is then computed by starting with the observed statistic and then adding and subtracting a constant. The constant results from multiplying the standard error of the statistic by a value obtained from the appropriate table.

For the parameter π, the appropriate table is the table of the normal curve when both $n\pi$ and $n(1 - \pi)$ are equal to or greater than five. The standard error of the statistic is:

$$SE_p = \sqrt{\frac{\pi(1 - \pi)}{n}}$$

Since we do not know π, we substitute the observed value of p in place of π in both of these statements. This substitution does not cause serious problems as long as n is large and π is near 0.5.

CONFIDENCE INTERVAL FOR μ: LARGE SAMPLES

When the question is, "What is the parameter value μ?" the same general procedure that was used in the preceding lesson is appropriate. The only differences are: (1) the statistic observed will be $\overline{X}$, (2) a different criterion is needed to determine whether the table of the normal curve is appropriate, and (3) the standard error of the statistic is $SE_{\overline{x}}$.

To illustrate, we will begin with an example in which a large sample is taken from a population that is assumed to be normally distributed. Suppose the goal is to learn the average starting salaries of graduates of baccalaureate nursing programs throughout the state of Illinois. The cost of obtaining data from all students who graduate in a given year may be prohibitive. We might, however, have sufficient funds to allow us to obtain the information about starting salary from a random sample of 100 new graduates. After tracking down all 100 persons, we find $\overline{X}$ and s to be 11,200 and 400 respectively. Our best single guess as to μ is 11,200. We will want to provide some information as to how good that estimate is, however, by setting up a confidence interval. Once more, the level of confidence will be chosen arbitrarily.

Our confidence interval will again be arrived at by starting with the observed $\overline{X}$ (11,200) and adding and subtracting a constant. As before, the constant will be determined by multiplying the values in two parentheses: the first containing an appropriate tabled value and the second the standard error of the statistic. Since we are assuming that the population is normal, we know that sample means will be normally distributed. To find the value for the first parentheses, we need only choose a level of confidence and then consult the two-tail column of our table of the normal curve. Let us choose an 80% confidence interval, in which case the tabled value is 1.282.

To find the $SE_{\bar{x}}$ to go in the second parentheses, we should use the formula $\sigma/\sqrt{n}$. The difficulty is that we do not know σ. We will have to sutstitute s, the standard deviation of our sample, which we earlier stated to be 400. (We will discuss the effect of this later.) Carrying out this computation for our example, we estimate the $SE_{\bar{x}}$ to be $400/\sqrt{100}$, or 40.

Our confidence interval for the average starting salary is, therefore:

$$11,200 \pm (1.282)(40) \qquad \text{or} \qquad 11,200 \pm 51.28$$

Our lower limit is 11,148.72, and our upper limit is 11,251.28.

SUBSTITUTION OF s FOR σ

When the purpose of choosing a random sample is to learn about the unknown population mean, μ, usually the population standard deviation is also unknown. Since we will need to know the standard error of $\bar{X}$ $(\sigma/\sqrt{n})$, this will cause a problem. Our solution will be to use the sample standard deviation as a substitute for σ. What implications does this have? First, we will want to use the best possible estimate of σ. Second, even when the best possible estimate is used, we will need to be aware of the effects of the substitution and make allowances for these effects. Let us think about each of these two considerations.

An unbiased statistic: s^2

You remember that it was explained earlier that modifications of the usual formulas are appropriate when the standard deviation and variance of a sample are computed. This is in contrast to other statistics, such as the mean or the proportion, where the same computations are used to find the statistic for a sample and the parameter for the population. The reason given for using the modification ($n - 1$ rather than n in the denominator) was that the results would be better estimates of σ and σ^2.

Now that we have come to the point where we are actually going to substitute s for σ, we should be a bit more specific. We have already indicated that $\bar{X}$ and p are *unbiased* estimates of μ and π. By this, we mean that the average of these statistics computed for an infinite number of samples would equal the corresponding population parameter values.

In contrast, neither s^2 nor s is an unbiased statistic when the formulas with n in the denominator are used to calculate them. Although proof is beyond the scope of this book, it can be shown that using $n - 1$ in the denominator will make s^2 an unbiased estimate of σ^2. Unfortunately, even when $n - 1$ is used in the denominator, s does not become an unbiased estimate of σ. The best that can be said is that s becomes less biased. At first it may not seem possible for s^2 to be an unbiased statistic while s, its square root, remains biased. It is possible, however, since the square root of an average of the squares of a set of numbers is not necessarily equal to the average of the numbers.

Happily, the bias remaining when $n - 1$ is used in computing s is rather inconsequential when the population from which the sample is taken is approximately normal and sample size is relatively large. For example, when the population has a normal distribution and the sample size is 25, an unbiased estimate of σ can be obtained by multiplying s, computed with $n - 1$ in the denominator, by 1.0104. With larger samples the adjustment is even smaller. Because the amount of bias is so small, most statistics books simply advise using $n - 1$ in the denominator when computing s and then using s as the substitute for σ.

Effects of substituting s for σ: population normal

Even if s were a completely unbiased estimate of σ, we would still have to consider the effects of substituting this estimate for the real value of σ in the formula for $SE_{\bar{x}}$. Each sample taken will result not only in a different estimate of μ but also in a different estimate for $SE_{\bar{x}}$. We learned earlier that if samples are taken from a normal population, the means of samples will be normally distributed and the z scores computed for individual sample means will be standard normal scores with a mean of zero and a standard deviation of one. However, the z scores for sample means are computed as indicated in Formula 6.2-1.

$$z_{\bar{x}} = \frac{\overline{X} - \mu}{\dfrac{\sigma}{\sqrt{n}}} \qquad\qquad \textbf{6.2-1}$$

What happens if s is substituted for σ in this formula? Even though sample means will still be normally distributed, we are no longer dividing by the true standard error of the mean. Our resulting values will still have a mean of zero, but the standard deviation will no longer be one, and use of the table of the standard normal curve will no longer be appropriate.

To distinguish between the z score for a sample mean computed when σ is known and the similar value computed for a sample mean when σ is unknown, let us call the latter $t_{\bar{x}}$. The formula is given as Formula 6.2-2.

$$t_{\bar{x}} = \frac{\overline{X} - \mu}{\dfrac{s}{\sqrt{n}}} \qquad\qquad \textbf{6.2-2}$$

where $\quad s = \sqrt{\dfrac{\Sigma x^2}{n - 1}}$

How closely will the distribution of sample $t_{\bar{x}}$'s be to a standard normal distribution? The answer is that it depends on how close the various estimates of $SE_{\bar{x}}$ are to the actual value of $SE_{\bar{x}}$. This in turn depends on how close the sample s's are to σ, which in turn depends on the size of the samples. Obviously, large samples give close estimates of σ and very small samples give imprecise estimates of σ. It follows then that the effect of substituting s in place of σ depends on the sample size. If the sample is very large, there will be little effect, and the distribution of sample $t_{\bar{x}}$'s will be very close to a standard normal distribution. As sample size gets smaller and smaller, the estimate of $SE_{\bar{x}}$ gets worse and worse, and use of the table of the standard normal curve to evaluate the probability of any particular $t_{\bar{x}}$ will result in bigger and bigger errors.

What is needed is a table of probabilities for $t_{\bar{x}}$'s that takes into account sample size. As you might guess, such a set of probabilities has been worked out. A table of t distributions is included in Appendix E as Table E-7.

The t distribution

Since we will need to use the t distribution when setting up a confidence interval for μ when σ is unknown, let us pause to discuss the t distributions in Table E-7. Notice that since there is a different t distribution for every sample size, it has not been feasible to provide a table that includes all of the possible t values from 0 to 4, as was done for the z scores in the table of the standard normal curve. Instead t values are given only for selected points on the t distribution for different sample sizes. The points given are for probabilities listed for the table. There is a row for one-tail probabilities at the top and a row for two-tail probabilities at the bottom of the table.

The numbers under the heading *df* on the far left-hand side of the table relate to sample size. The *df* stands for *degrees of freedom,* which is a relatively complicated concept. For our purposes, when computing confidence intervals, all you need to know is that *df* is equal to the sample size minus one $(n - 1)$.

Let us begin by looking at the bottom row of the table, where the symbol under *df* stands for infinity. Of course, if it were possible to take a sample of infinite size, s would be equal to σ and the estimate of $SE_{\bar{x}}$ would be perfect. In effect, we would be back to the situation in which σ is known and the use of the standard normal curve would be appropriate. Therefore, based on what we learned earlier about the normal curve, we would expect 95% of sample means to be within 1.96 standard errors of μ. You can see that the number 1.96 appears in the column labeled ".05" among the two-tail probabilities listed at the bottom of the table.

Now look at what happens as the *df* gets smaller. Instead of going 1.96 standard errors of the mean in either direction to include 95% of all sample means, you must go an increasingly larger number of *estimated* standard errors of the mean in either direction. The meaning of this statement is a bit tricky to grasp. Suppose we have a normally distributed population with μ equal to 50 and σ equal to 10. Now suppose we take samples of size 4 from this population. The means of all these samples will be normally distributed, since the population is normally distributed. The standard deviation *(SE$_{\bar{x}}$)* of the sample means will be $\sigma/\sqrt{n}$, or 5. We can expect 95% of all sample means to be within 1.96 standard errors of the mean $(50 \pm 1.96 \times 5)$.

Suppose, however, that the σ were not known to be 10 and instead had to be estimated from each sample s. With samples of size 4, s would vary greatly from sample to sample. If you were to take many samples and compute $\bar{X}$, s, and the estimated $SE_{\bar{x}}$ for each one, 95% of the sample $\bar{X}$'s, would not be within 1.96 estimated standard errors of the population mean. Looking at Table E-7 for the row corresponding to 3 *df* (4 minus 1), we find that 95% of all sample means can be expected to be within 3.182 estimated standard errors of the population mean.

Note that in the example just given, the problem is not that the sample means are not normally distributed. Neither is it that going 1.96 standard errors either side of μ will fail to include 95% of all sample means. The problem is that the standard error is not known and must be estimated from s. The smaller the sample size, the poorer this estimate will be and the larger the tabled value of t will be in comparison to the corresponding tabled value for z.

Returning to our original goal of setting a confidence interval for μ, we conclude that when the population is normal but σ is unknown and must be estimated from s, our confidence interval will be:

$$\mu = \bar{X} \pm (\text{tabled } t \text{ value})(s/\sqrt{n})$$

The value for the first parentheses will depend on the level of confidence chosen and the *df* ($n - 1$). The value in the second parentheses will be computed using the standard deviation computed for the sample, that is, using $n - 1$ in the denominator of the formula for s.

Effects of substituting s for σ: population not normal

When the population from which a sample is chosen is not normal, s is not a good estimate of σ unless samples are very large (100 or more). It is a common misconception that this problem can be solved by using the t distribution. Unfortunately, such is not the case. When the population is not normal, σ is unknown, and sample size is not large, there is really no good solution to the problem of setting a confidence interval for μ. However, if the sample can be made large enough (100 or more) so that s becomes a good estimate of σ, we can set a confidence interval. The t values and the values from the standard normal curve will be nearly the same, and use of the t distribution as opposed to the z distribution will have little effect. To keep things simple, we might as well use the standard normal curve. Notice that in this case sample means can be expected to be approximately normally distributed, since n will be well over 25.

SUMMARY

In general, the confidence interval for the mean of a population is computed by starting with the mean of the sample and then adding and subtracting a constant. The constant is arrived at by multiplying the standard error of the mean by an appropriate tabled value. Since the standard deviation (σ) of the population is generally unknown, it is necessary to substitute the sample standard deviation *(s)* in place of the population standard deviation in the formula for the standard error of the mean *(SE$_{\bar{X}}$)*. If the population from which the sample is chosen is normal, allowance can be made for this substitution by using a table of t as the appropriate table rather than a table of the standard normal curve.

If the population from which the sample is chosen is not normal, you will have to use a large sample (100 or more) for s to be a good estimate of σ. In this case, there will be little difference whether you use the value found in the table of t or in the table of the standard normal curve.

LESSON 2 EXERCISES

1. Suppose that for a population of 1000 army recruits, height is normally distributed, with σ equal to 6 inches. For a sample of nine randomly selected recruits, $\overline{X}$ equals 69 inches. Give a 99% confidence interval for μ.

2. Suppose that for a population of 1000 hospitals the number of beds is a normally distributed variable. You choose nine hospitals at random and find the average number of beds to be 380. The standard deviation of the number of beds for the nine hospitals in the samples is 60. Compute an 80% confidence interval for μ.

3. Suppose that the number of days until discharge from the hospital for patients with childhood rheumatic fever is a positively skewed distribution, with σ known to be 12. A randomly selected sample of 36 patients results in an average time until discharge of 22 days. Compute a 90% confidence interval for μ.

4. Suppose that the distribution for batting averages of all major league players is negatively skewed. For 25 randomly chosen players the average batting average is found to be .242 and the standard deviation of the 25 averages is found to be .50. Compute a 99% confidence interval for μ.

CHAPTER 7

Hypothesis testing

Lesson 1 □ Basic concepts and tests of proportions

You learned earlier that there are two approaches to inferential statistics, depending on the question being asked. You then learned how to set a confidence interval when the question is, "What is the parameter value?" In this chapter you will learn the rationale and methods for a process called hypothesis testing. This approach is used when you want to answer the question, "Is the parameter value of some test population different from the parameter value of some comparison (standard) population?" An example might be as follows. "Does adding a vitamin supplement to the diet of pregnant women result in a lower probability of low birth weight babies?" The test population is a group made up of women who receive the vitamin supplement. What is the standard population? This is the population of all pregnant women who do not receive the food supplement. The parameter you are interested in is π, the proportion of women who give birth to low birth weight babies. You wish to know whether π for the test population (π_T) is lower than π for the standard population (π_S).

RATIONALE OF HYPOTHESIS TESTING

In hypothesis testing, the parameter value is unknown for the test population but known or assumed to be some specified value in the standard population. Let us, for this example, suppose that the proportion of low birth weight babies in the general population is known to be approximately 0.10. To see whether the vitamin supplement reduces this proportion, we will choose a random sample of pregnant women, provide the supplement, and then compute the proportion of low birth weight babies in the sample. Naturally, we are looking to see if p is less than π_S, or less than 0.10. The difficulty is that even if we took random samples from the standard population, we would not expect every sample p to be 0.10. About one half of all sample p values will be less than 0.10, even when sampling from the standard population. Getting a sample p value less than 0.10 when sampling from the test population would therefore not be very convincing. What we need is a sample p value that is so far below 0.10 that it would be unlikely to occur when sampling from the standard population. To determine whether an observed p from the test population is so low as to be a rare event when sampling from the standard population, we will need to know what is likely to happen when samples are taken from the standard population.

You guessed it! We need to know the sampling distribution of the statistic for samples chosen from the *standard* population. We can then determine where our observed sample statistic from the *test* population would fall on this distribution. If there would be a very small probability of a sample p this low occurring among samples from the standard population, we will conclude that there is some evidence that $\pi_T < \pi_S$. All that is left to be determined is how small the probability must be before we decide that $\pi_T < \pi_S$. As when we were setting the level of a confidence interval, this choice will be arbitrary. We will examine some of the factors that go into that choice a little later. For now, let us just say that the probability must be small, and the smaller the probability, the more sure we will be that π_T is indeed less than π_S.

Fig. 7-1 is a symbolic diagram of the process we will use for hypothesis testing when the parameter of interest is π.

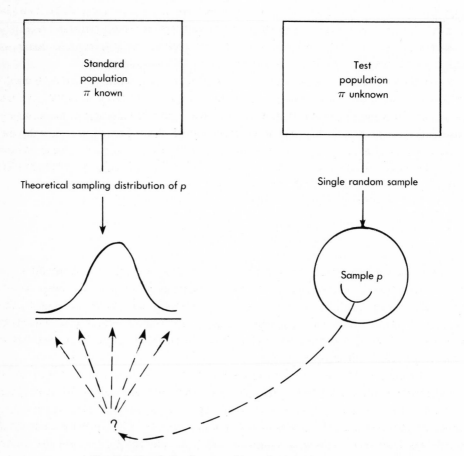

Fig. 7-1. Symbolic diagram of hypothesis testing process.

On the left side of the diagram is the standard population, in which the parameter is known. For our example, this population is all pregnant women without the diet supplement, and π_S is known to be 0.10. On the right is the test population, in which π_T is unknown. For our example, this is a hypothetical population of all pregnant women to whom the supplement might be given. Although π_T is unknown, it is hoped that the diet supplement will result in a lowering of π_T to below 0.10.

Notice that only a single random sample of some chosen size n is taken from the test population. For this sample, the statistic p is computed. In contrast, for the standard population we must be able to describe the *sampling distribution* of the statistic p for many samples of size n. To do this empirically we would need to take many samples and then describe the resulting distribution of the individual sample p's. Usually this is unnecessary, because the sampling distribution is known on a theoretical basis. As a result, there is no need to take any samples from the standard population.

Let us now be specific. Suppose that you have tried the vitamin supplement with a randomly chosen sample of 100 pregnant women and that only six of these women have had low birth weight babies. The observed sample proportion is $^6/_{100}$, or 0.06. You are pleased to see that for this sample the proportion is below the proportion generally expected (0.10). The question is, "Does a sample proportion of 0.06 indicate that π_T is really below 0.10?"

To answer the question, we will have to determine where a sample p of 0.06 would fall on the sampling distribution of p for samples of size 100 taken from the standard population, in which π_S equals 0.10. First, we determine that the shape of the sampling distribution can be expected to be approximately normal. Why? Second, we know that the mean of the sampling distribution will be the parameter π_S, or 0.10. Finally, we know that the standard error of the sampling distribution will be $\sqrt{\pi(1-\pi)/n}$, or $\sqrt{(0.1)(0.9)/100}$, or 0.03. Where will our observed sample proportion of 0.06 fall on this sampling distribution? To find out, we can change our sample value to a z score.

$$z = \frac{p - \pi_S}{SE_p} = \frac{0.06 - 0.10}{0.03} = -1.33$$

Now we can use the table of the standard normal curve (Table E-6, Appendix E) to find what portion of the standard normal curve is below a z score of -1.33. Since we are interested only in scores below -1.33, we will use the one-tail column and look for the proportion beyond 1.33. The table value is 0.0918. This tells us that if we take random samples of size 100 from any population in which π equals 0.10, about 9% of the time the sample proportion will be 0.06 or less.

What shall we conclude? Our sample p would not be a common result. On the other hand, neither would it be extremely rare. We would certainly have been more comfortable with the decision that the diet supplement is having an effect if our sample result could be expected only one time out of 100 or one time out of 1000. Yet we must reach some conclusion based on the information available. Shall we announce to the world that adding a vitamin supplement to the diet of pregnant women reduces the risk of low birth weight babies? If our sample statistic happens to be one of those 9% that can be expected to happen by chance when π is really 0.10, we will have made an error. We will end up recommending vitamin supplements that have no effect. On the other hand, if we conclude that this result is merely a chance happening and that π remains 0.10 even with the vitamin supplement, we may also be making an error, this time resulting in the failure to recommend vitamin supplements that could reduce the risk of low birth weight babies. Either way, we risk making an error.

VOCABULARY OF HYPOTHESIS TESTING

Although it will not help solve our dilemma, it will make it easier to talk about the problem if we introduce some vocabulary used by statisticians and researchers. First, the statistician defines two hypotheses: the *null hypothesis* and the *alternative hypothesis*. The null hypothesis is that there is no difference between the parameter of the test population and that of the standard population. In most situations this is equivalent to saying that a treatment has no effect. Obviously, the researcher will be hoping that his treatment does have an effect and that the null hypothesis can be proved wrong. If so, he can then conclude that the alternative is true; that there is a difference between the parameter values of the test and standard populations.

The alternative hypothesis may be either a *directional* or a *nondirectional* hypothesis, depending on the logic of the experiment. In most situations the treatment is expected to result in a difference in a given direction; for example, lowering of the proportion of low birth weight babies. In such a case, the alternate hypothesis is said to be a *directional,* or *one-tail,* hypothesis. In the symbols often used, the null and the alternate hypotheses for the experiment we have described can be stated as follows:

$$H_o: \qquad \pi_T \geq \pi_S \quad \text{or} \quad \pi_T \geq 0.10$$

where H_o stands for null hypothesis.

$$Alt.: \qquad \pi_T < \pi_S \quad \text{or} \quad \pi_T < 0.10$$

If we were conducting the experiment to learn whether the diet supplement has *any* effect on the proportion of low birth weight babies without regard to direction, the hypothesis would be a *nondirectional,* or *two-tail, hypothesis* and would be stated:

$$H_o: \qquad \pi_T = \pi_S \quad \text{or} \quad \pi_T = 0.10$$

$$Alt.: \qquad \pi_T \neq \pi_S \quad \text{or} \quad \pi_T \neq 0.10$$

As was stated, the researcher usually is hoping to find evidence against the null hypothesis. What constitutes such evidence? In general, if the sample taken from the *test* population results in a statistic that would be a rare occurrence for the sampling distribution of the statistic from the *standard* population, we will reject the null hypothesis and conclude that the alternative is correct. If the statistic would be a common occurrence, we will conclude that there is no evidence against the null hypothesis.

We said earlier that the goal of hypothesis testing is to answer the question, ''Is there

Table 7-1. Types of errors in hypothesis testing

Decision		Reality	
		Null false	*Null true*
Reject null		Correct decision	Type I error
Accept null		Type II error	Correct decision

a difference between the parameter of the test population and the parameter of the standard population?'' If you reject the null hypothesis, you are saying, ''Yes, there is a difference; my treatment had an effect.'' If you fail to reject the null hypothesis, you are saying, ''I have not found convincing evidence of a difference; my treatment remains unproven.'' In either case, there is always the possibility that you have made an error. Statisticians distinguish between the error made in each of these two cases. The error of wrongly rejecting the null hypothesis (deciding an ineffective treatment is effective) is called a *Type I error*. The error of failing to reject a false null hypothesis (failing to identify that an effective treatment is indeed effective) is called a *Type II error*. Table 7-1 describes these two types of error.

In most situations, the researcher wants to be cautious about making false claims for a treatment, and is therefore chiefly concerned with guarding against a Type I error. He accomplishes this by only rejecting the null hypothesis when the sample statistic from the test population falls in the extreme part of one of the tails of the sampling distribution from the standard population. The more extreme he requires a sample statistic to be before he will reject the null hypothesis, the less are his chances of making a Type I error. The process of deciding how extreme a sample statistic must be in order to reject the null hypothesis is referred to as *choosing a level of significance* or *setting an alpha level*. A person might decide to reject the null hypothesis only if the sample statistic falls in the area of the sampling distribution that includes only those 5% that are the most extreme values. If so, he would be choosing a 0.05 alpha level. If a directional, or one-tail, test is used, only one tail of the sampling distribution is considered. If a nondirectional test is used, extreme scores in either tail will cause rejection of the null hypothesis.

Before looking at specific examples of hypothesis testing, we should examine more closely the implications of the choice of a level of significance. Suppose that in spite of all the care with which we do our background research, the treatment we are testing does not have the anticipated effect and in fact has no effect at all. In this case the null hypothesis is true. The parameter values of the test population and of the standard population are equal. Samples taken from either the test population or the standard population will have statistics that cluster about this parameter value. In fact, sampling distributions of the statistic, taken from either population, will be the same. To avoid making an error, we should accept the null hypothesis.

Suppose now that we choose a significance level of 0.05. What will happen when we take a sample from the test population? Will we get an extreme sample value and erroneously reject the null hypothesis? What is the chance of this occurring? We will reject the null hypothesis anytime our sample statistic falls in the extreme 5% of the sampling distribution of the standard population. However, since the sampling distribution from both populations coincide, we will erroneously reject the null hypothesis 5% of the time.

The implication of all of this is that when you choose a level of significance you are establishing the level of the risk that you are prepared to take of saying your treatment is effective in those instances in which it is not. Notice that this is a conditional probability. What is the probability of rejecting a null hypothesis when it is true? This is the definition often given to *level of significance* or *alpha level*. The alpha level is the probability of rejecting a *true* null hypothesis, or alternatively, it is the probability of making a Type I error. Since the researcher gets to choose the signifiance level, he can set the risk of making a Type I error at any level he chooses.

The next obvious question is: "Why doesn't the researcher always guard very diligently against a Type I error by making the alpha level very low, such as 0.01 or even 0.001?" The answer will be discussed more thoroughly in the next lesson. For now, it is sufficient to state that, in general, being more careful not to make Type I errors by setting extreme alpha levels increases the risk of making Type II errors. We could reduce the probability of making a Type I error virtually to zero, but to do so would increase the probability of a Type II error to an unacceptable level.

Traditionally, the chosen level of significance has been 0.05, 0.01, or 0.001. Many statistics books have recommended that the level of significance be chosen at the outset of the experiment. In recent years, however, there has been an indication of a change of thinking with regard to hypothesis testing and the choosing of a level of significance. We will delay discussion of that development until the next lesson. For now, in the examples in this lesson and for the exercises at the end of this lesson, the level of significance will always be indicated for you.

AN EXAMPLE OF HYPOTHESIS TESTING WITH π

Let us now look at a specific example of hypothesis testing for the parameter π. Suppose an examination of psychological theories leads you to the hypothesis that living in crowded social conditions will reduce ability to learn tasks involving higher mental functions. To test this hypothesis, you decide to conduct an experiment in an animal psychology laboratory. For years the laboratory has used a certain strain of white rats in learning experiments with a maze. It has been found that only 20% of all rats fed and housed under normal conditions are unable to escape the maze within a 1-hour time period on the first trial. You decide to test the hypothesis regarding the effect of crowding by increasing the number of rats per cage from four to eight and then testing them in the maze. Each rat is still provided with all the food and water it wants. Sixty-four rats are housed under these conditions. After 1 month, each of the 64 rats is placed in the maze.

Seventeen fail to escape the maze in the 1 hour alloted. What do you conclude? Is there a difference between π_T, the proportion of rats failing to escape the maze following 1 month of crowding housing, and π_S, the proportion failing under normal housing conditions? To make the process of hypothesis testing clear, let us examine a set of questions that must always be answered when hypothesis testing is carried out.

1. *What is the parameter of interest?* The answer in this case is π, the proportion of rats failing to escape the maze.
2. *What is the test population?* Here it is all rats of this strain who are housed under the crowded conditions.
3. *What is the standard population?* Here it is all rats of this strain who are housed under normal conditions.
4. *What is the known parameter for the standard population?* Past history has shown π_S to be equal to 0.20.
5. *Is this a two-tail or one-tail test?* Since the hypothesis is that crowding will inhibit learning ability, we will make a one-tail test.
6. *What is the alternative hypothesis?* Alt: $\pi_T > \pi_S$, or $\pi_T > 0.20$.
7. *What is the null hypothesis? H_o: $\pi_T \leq \pi_S$.*
8. *What level of significance will be used?* Let us arbitrarily choose 0.05.
9. *What size sample is used?* In this experiment $n = 64$.
10. *What is the sampling distribution of the statistic for samples of the designated size taken from the standard population?*
    a. *Shape:* In this case, the shape is normal, because $n\pi_S$ and $n(1 - \pi_S)$ are both greater than 5.
    b. *Central tendency:* This is the parameter value π_S, or 0.20.
    c. *Variability:* $\sqrt{\pi_S(1 - \pi_S)/n}$, or $\sqrt{(0.2)(0.80)/64}$, or 0.05.
11. *What is the sample statistic from the test population?* In this case, there were 17 out of 64 rats that failed to escape the maze. Therefore, $p = 0.266$.
12. *Is the probability of the observed statistic less than the chosen level of significance?* There are two ways to find the answer to this question. Let us call one approach A and the other approach B.

    For approach A, you identify the *critical region* of the sampling distribution. Any sample statistic that falls in this area will lead to rejection of the null hypothesis. If a two-tail test is used, a 0.05 significance level will result in two critical regions, each containing $2^1/_2\%$ of the sampling distribution. For a one-tail test, there will be a single critical region containing 5% of the sampling distribution. For our example we want to identify the part of the sampling distribution that includes only 5% of the highest sample p's. We look for 0.05 in the one-tail section of the table of the normal curve under the "proportion beyond" column. The corresponding z score is 1.645. We therefore need to identify the part of our sampling distribution that is 1.645 standard deviations above the mean. With a mean of 0.20 and SE_p equal to 0.05, the sample proportion that is 1.645 standard deviations above the mean is 0.20 + (1.645)(0.05), or 0.2822. Our critical region is made up of all sample p's equal to or greater than 0.2822.

For approach B, you find the probability of a sample statistic as extreme as or more extreme than the one you have observed. When the sampling distribution is normal, you will change the sample statistic into a z score and then use the table of the standard normal curve to find the probability of a z score this extreme or more extreme. If a nondirectional test is used, you will look at the two-tail section of the table; for a directional test, you will use the one-tail section. In both cases the probability is read from the "proportion beyond" column. For our problem:

$$z = \frac{0.266 - 0.20}{0.05} = +1.32$$

Looking at the one-tail section of the standard normal curve, we find that the probability of a z score equal to or greater than $+1.32$ is 0.0934.

13. *What do you conclude concerning the null hypothesis?* Once again, we may use either approach A or approach B to find the answer.

Using approach A, we found the critical region to include all sample p's equal to or greater than 0.2822. Since our sample p is 0.266, it does not fall in the critical region, and we conclude that there is insufficient evidence to reject the null hypothesis.

Using approach B, we found the probability of a sample p as great as or greater than the one we observed to be 0.0934. We had decided to reject the null hypothesis only if we observed a sample statistic so extreme as to occur only 5% of the time or less. Sample p's as high as ours would occur 9.34% of the time. We therefore conclude that there is insufficient evidence to reject the null hypothesis.

14. *What type of error might you be making, and what is the implication of this error?* For our example, we may be making a Type II error, failing to reject the null hypothesis when it is false. The implication is that we would fail to get a confirmation of our theory that crowding inhibits the learning of a maze by white rats when we should have gotten such confirmation.

Notice that in this example both approach A and approach B result in the same conclusion with regard to the null hypothesis. This will always be the case. The choice of which approach to use is therefore, to some extent, a matter of personal preference. In the next lesson you will learn that in some situations approach B is not possible and approach A must be used. There is some advantage to approach B, in that you learn not only whether the observed statistic falls in the critical region, but where it falls. If it is not in the critical region, is it close? If it is in the critical region, did it barely make it, or would it be significant not only at the chosen alpha of 0.05, but even at alpha 0.001? When it is possible to do so, the use of approach *B* is recommended. It should be noted that all of the previous terms and concepts that have been introduced in this lesson are standard in most statistic books. Although the ideas of approach *A* and approach *B* are discussed in some form in other statistics books, often they are not clearly delineated. Calling one approach *A* and the other approach *B* is not standard terminology; it is specific to this book.

SUMMARY

In this lesson you learned about an approach to inferential statistics called hypothesis testing. This approach is used when you want to compare the unknown parameter of some test population with the known parameter of a standard population. In general, the approach is to choose a *single* random sample from the *test* population and to see where the relevant statistic from this sample would fall on a *sampling distribution* of the statistic for samples of this same size taken from the *standard* population. If the observed sample statistic would be a rare event on this sampling distribution, we conclude that there is a difference in the parameter values for the test population and for the standard population. The decision as to what constitutes a rare event is somewhat arbitrary and will be discussed more fully in the next lesson.

The following concepts relating to hypothesis testing were introduced. Be sure that you understand each of these:
Test population
Standard population
Null hypothesis
Alternate hypothesis
Directional, or one-tail, test
Nondirectional, or two-tail, test
Significance, or alpha level
Type I error
Type II error
Critical region

An example of hypothesis testing was then presented. The procedure for the analysis was illustrated by listing 14 questions, which were answered as part of the analysis. You need not go to the trouble of writing down your answer to each of these questions each time you use hypothesis testing, but it is probably a good idea to do so while you are learning the process. Near the end of the process, the question can be answered by either of two approaches. You will reach the same conclusion about the null hypothesis in either case.

The exercise that follows will give you a chance to test your understanding of the material in this lesson. Then, the following lesson will discuss the use of hypothesis testing when the parameter of interest is μ. It will also contain a more detailed discussion regarding the choice of a level of significance.

LESSON 1 EXERCISE

For several years, records have been kept of incidents of violence among prisoners at a large federal correctional institution. The data show that over any 30-day period, the proportion of prisoners involved in violent incidents averages 0.07. A prison psychologist claims that involving the prisoners in transcendental meditation (TM) will reduce violent behavior. To test this theory, prison officials agree to provide instruction in TM to 200 randomly selected prisoners for a period of 60 days. During the 30 days following this period, 9 of the 200 prisoners who have been provided with TM instruction are involved in violent incidents. Test the appropriate hypothesis, using an alpha level of 0.10.

Write out your answers to all 14 questions listed at the end of this lesson and use both approach *A* and approach *B*.

Lesson 2 □ Tests of the mean

In the first lesson of this chapter you learned the general rationale of hypothesis testing. You also learned how this rationale is applied when the parameter of interest is π, the population proportion. This lesson will explain how this same rationale is applied to hypotheses about μ, the population mean. If you understood the general rationale of hypothesis testing and the lesson dealing with confidence intervals for μ, you will not have difficulty with this material. In fact, it is so straightforward that you will probably find you can make the application yourself. The remainder of the lesson will be more challenging. It will contain a discussion of factors to be considered when choosing a level of significance.

HYPOTHESIS TESTS INVOLVING μ

Suppose that in the example of the last lesson, instead of keeping records of the proportion of rats failing to escape the maze in 1 hour, the *length of time* required by each animal to escape had been recorded. Based on extensive experience, it is known that rats housed under usual conditions take an average of 41 minutes to escape. The distribution of individual times is approximately normal, with a standard deviation of 16. An examination of psychological theory still leads one to believe that crowded conditions will inhibit learning, resulting in an increase in the average length of time required for escape. There is no reason to expect any change in the *variability* of the times of individual animals or in the shape of the distribution of the times. You again house a certain number of rats under crowded conditions. Suppose that the number this time is only 16, rather than 64. You then find that the average length of time the rats need to escape the maze after a period of crowded housing is 44.6 minutes.

Can you now answer the questions from the last lesson, and test the appropriate hypothesis? Use the same 0.05 level of significance. Before proceeding, take time to see how well you can do.

1. *What is the parameter of interest?* Now the answer is μ, the average time required for a rat to escape from the maze.
2. *What is the test population?* As before, it is all rats of this strain who are housed under the crowded conditions.
3. *What is the standard population?* Again, it is all rats of this strain who are housed under normal conditions.
4. *What is the known parameter for the standard population?* Extensive experience has shown μ_S to be equal to 41 minutes.
5. *Is this a two-tail or one-tail test?* Again, since the hypothesis is that crowding will inhibit learning ability, we will make a one-tail test.
6. *What is the alternative hypothesis?* Alt: $\mu_T > \mu_S$, or $\mu_T > 41$ minutes.
7. *What is the null hypothesis?* H_o: $\mu_T \leq \mu_S$, or $\mu_T \leq 41$ minutes.

8. *What level of significance will be used?* As before, let us arbitrarily choose 0.05.
9. *What sample size is used?* This time $n = 16$.
10. *What is the sampling distribution of the statistic for samples of the designated size taken from the standard population?*
    a. *Shape:* The shape is normal, since the distribution of individual observations is approximately normal.
    b. *Central tendency:* This is the parameter value μ_S, 41 minutes.
    c. *Variability:* $SE_{\bar{x}} = \sigma/\sqrt{n} = 16/\sqrt{16} = 4$.
11. *What is the sample statistic from the test population?* For the sample of size 16, the average escape time was 44.6 minutes.
12. *Is the probability of the observed statistic less than the chosen level of significance?* Using approach A, we want to identify the part of the sampling distribution that includes only 5% of the highest sample means. As before, we look at the one-tail section of the standard normal curve and find 0.05 in the "proportion beyond" column. The corresponding z score is 1.645. Thus, we need to determine the part of our sampling distribution that is 1.645 standard deviations above the mean. With a mean escape time of 41 minutes and $SE_{\bar{x}} = 4$, the value that is 1.645 standard deviations above the mean is $41 + (1.645)(4)$, or 47.58 minutes. Our critical region is made up of all sample means equal to or greater than 47.58 minutes.

    Using approach B, we again convert the sample statistic to a z score (since the sampling distribution is normal), and then use the table of the standard normal curve to find the probability of a z score this extreme. As before, since this is a directional test, the one-tail section of the table is used, and the probability is read from the "proportion beyond" column. For this problem:

$$z = \frac{44.6 - 41}{4} = \frac{3.6}{4} = 0.9$$

    Looking at the one-tail section of the standard normal curve table, we find that the probability of a z score equal to or greater than 0.9 is 0.1841.
13. *What do you conclude concerning the null hypothesis?* Using approach A, we find the critical region to include all sample means equal to or greater than 47.58 minutes. Since our sample mean is 44.6 minutes, it does not fall in the critical region. We therefore conclude that there is not sufficient evidence to reject the null hypothesis.

    Using aproach B, we find the probability of a sample mean as great as or greater than the one we observed to be 0.1841. We had decided to reject the null hypothesis only if we observed a statistic so extreme as to occur only 5% of the time or less. Sample means as high as 44.6 would occur 18.41% of the time. Hence, we conclude there is not sufficient evidence to reject the null hypothesis.
14. *What type of error might you be making, and what is the implication of this error?* As before, we may be making a Type II error, failing to reject the null

Table 7-2. Factors affecting hypothesis tests of the mean

	σ *known*	σ *unknown*
Population normal	Cell 1 1. Sample size not an issue 2. $SE_{\bar{x}} = \sigma/\sqrt{n}$ 3. z distribution	Cell 2 1. Sample size reflected in *df* 2. $SE_{\bar{x}} = s/\sqrt{n}$ 3. t distribution
Population not normal	Cell 3 1. Requires sample size of at least 25 2. $SE_{\bar{x}} = \sigma/\sqrt{n}$ 3. z distribution	Cell 4 1. Requires sample size of at least 100 2. $SE_{\bar{x}} = s/\sqrt{n}$ 3. z or t distribution

hypothesis when it is false. The implication is that we would fail to confirm our theory that crowding of rats inhibits their learning of a maze when, in fact, we should have.

You can see that when σ is known and the sampling distribution is normal, there is no problem in applying the rationale of hypothesis testing to population means. Can you identify the changes that would have to be made if σ were unknown, or if the population were not normally distributed, or if we confronted a combination of these conditions? In general, we know that the sampling distribution of means will be a normal distribution if the population is normal or if sample size is 25 or more. This means that if the population is not normal, it will be necessary to take samples of size 25 or more to make use of the z values of the standard normal distribution. We also know that when σ is not known but must be estimated from the sample s, we will need to use the t distribution rather than the normal curve. When the population is not normal and σ is not known, s is not a good estimate of σ unless samples are very large. We will have to use a sample of size 100 or more, in which case the use of the t distribution will give about the same result as the use of the standard normal distribution. Table 7-2 summarizes the effects of these conditions on hypothesis tests involving μ.

To test your understanding of the information in Table 7-2, suppose that σ had not been known from past experience to be 16. Rather, suppose that the sample standard deviation, s, had been found to be 16. How would your previous analysis have changed? The relevant cell from Table 7-2 is cell 2. The information in this cell tells us to use the t distribution. Since n is 16, we will use 15 *df*. Try the analysis, using approach A.

Since 16 is now the sample standard deviation, s, rather than the population standard deviation, σ, the variability of the sampling distribution will now be the *estimated $SE_{\bar{x}}$*, or $s/\sqrt{n}$. Our standard error is still $16/\sqrt{16}$, or 4, but now it is an estimate. With a significance level of 0.05, a one-tail test, and 15 *df*, you find the critical region is 1.75 estimated $SE_{\bar{x}}$'s above the population mean of 41. The critical region, therefore, includes all means equal to or greater than $41 + (1.75)(4)$, or 48. What has happened to the critical region as a result of having to use the sample standard deviation as an estimate of σ? Is this consistent with our earlier observations?

Now try the analysis, using approach *B*. What problem do you encounter? When you try to use the *t* table to find the probability of your observed *t* value, you find that the table is not complete. There is no way of finding the probability of the specific *t* value. This is what was meant by the statement in the last lesson that in some situations approach *B* cannot be used. As a generalization, you can use either approach *A* or approach *B* when you will be consulting the table of the standard normal curve, but approach *B* is not possible when an incomplete table of the *t* distribution will be used.

We started by discussing the analysis of this problem with a situation in which the information in cell 1 was relevant. We then changed the situation so that the information in cell 2 was relevant. We will not go into detail regarding cells 3 and 4, because the application is very straightforward. The only caution that needs to be made is that neither cell 3 nor cell 4 is relevant to small samples. Cell 3 requires a minimum sample size of 25, and cell 4 requires a minimum sample size of 100. Both of these restrictions are stated for the case where the population departs rather radically from normality, particularly for skewed distributions. If the populations are nearly symmetrical with some clustering about the mean, the restrictions are conservative. You would probably not make serious errors by treating the situation for cell 3 as though it were for cell 1 and the situation for cell 4 as though it were for cell 2.

CHOICE OF LEVEL OF SIGNIFICANCE

In all of the examples included until now you have been informed that a specified level of significance was to be used. Generally, this has been the rather traditional level, 0.05. As stated earlier, choosing to use a 0.05 significance level means that in 5% of those experiments in which the treatment has no effect, you will make the mistake of saying that it does. The consequences of this type of error could range from extremely serious to almost trivial. Suppose you claim that a proposed new method of treating obesity is more effective than previous methods, when in fact it is no better. The seriousness of this error depends on the cost of the new method compared with the cost of the old. If the costs are roughly comparable, claiming that the new is better will not be a serious error. On the other hand, suppose the new method involves major surgery for the removal of a segment of the small intestine. You would want to be very certain that the new method is indeed superior to other, less costly methods of treatment.

Just as the relative seriousness of a Type I error will vary from one experiment to another, so will the cost of making a Type II error. Failing to recognize an improved treatment that would increase a patient's chances for survival would be a serious error indeed! On the other hand, failing to confirm one's pet psychological theory as the result of an experiment that should result in a confirmation will probably not have serious lasting consequences. The theory's development will be delayed somewhat, but subsequent investigations are likely to reveal the error.

Since the seriousness of both Type I and Type II errors can vary from one experiment to another, and since the probabilities for the two types of errors are related, it appears that a rational approach would be to assess the cost of each before choosing a level of significance. If the cost of a Type I error is great and the cost of a Type II error is small, it would make sense to set the alpha level at 0.01 or even 0.001. On the other hand, there are situations in which it is the cost of a Type II error that is great, whereas the cost of a Type I error is relatively small. Suppose, for example, that a new method of treating a form of skin cancer is proposed. The new method involves the traditional treatment plus an infrared light exposure for 10 minutes, once weekly. In this situation, what is the cost of saying the treatment improves a patient's chance for survival when in fact it is of no benefit? On the other hand, what is the cost of failing to recognize that the treatment improves a patient's chance for survival when in fact it does? Here it would not make sense to set alpha at 0.01, 0.05, or, for that matter, 0.20. Instead, if there is any indication at all that the treatment works, most patients given the option would accept the inconvenience of the 10 minutes of infrared light treatment weekly.

Assuming that it makes sense to compare the costs of Type I and Type II errors before choosing a level of significance, who should make this comparison? Suppose a new treatment has the potential for reducing the crippling effects of arthritis, but in a certain percentage of patients it increases the risk of strokes. Who is to decide what constitutes a proper level of significance? Ideally, the *patient* would be given all the information and then allowed to make the decision whether to undergo the treatment. What type of information should be provided? He should be told what increase there will be in his risk of suffering strokes if he undergoes the treatment. The patient generally will not have difficulty understanding this information. The other information he needs will be harder to provide. He will want to know whether the treatment reduces crippling. He wants a simple yes or no. Unfortunately, an experiment seldom provides unqualified evidence for the effectiveness of a treatment. Instead, a sample statistic is observed from the new treatment group that would occur some of the time even when sampling from patients provided with the usual treatments. It might be a rather common result or it might be a very extreme result. If traditional hypothesis testing procedures are followed, the researcher will choose a level of significance and decide whether to accept or reject the null hypothesis.

In recent years there has been some movement away from the use of a stated significance level in favor of reporting the probability of the observed statistic. In this case, a modification of approach B is used. Questions 8, 13, and 14 are not answered by the researcher. Instead, the reader of the research report is told that if there were no differences between the parameter values of the standard and test populations, a sample statistic as extreme as the one observed would occur 9%, 12%, 1%, or some other percentage of the time. When this procedure is followed, each reader must decide what the relative cost of each type of error is and choose a level of significance. Although this seems to be the ideal arrangement, it works only when the research report is intended for rather sophisticated readers. Since the rationale of hypothesis testing is not understood even by many readers of the scientific literature, traditional hypothesis testing procedures will probably continue to be widely used.

It should be noted that when the newer method is followed, the terminology used will cause confusion with the symbols we have been using. The wording commonly used is as follows: "The *p* value of the observed statistic is 0.023, or 0.09, or whatever." However, we have used the lower case *p* to stand for the proportion within a sample. The lower case *p* in the preceding statement is intended to imply probability. Since we earlier used a capital *P*, as in *P(A)*, for this purpose we will be consistent and state "the *P* value of the observed statistic is 0.023." Using this modified procedure for the example given at the outset of this lesson would have resulted in the following statement. The *P* value of the observed statistic, 44.6, is 0.1841.

SUMMARY

In this lesson you saw how the process of hypothesis testing applies to tests of the population parameter μ. If the population is normally distributed and the population standard deviation is known, the application is straightforward, with the table of the standard normal curve being used just as when the test involved the parameter π. When the population standard deviation is unknown but the population is normal, the only change in the procedure is the use of the *t* distribution in place of the *z* distribution. When the population is not normal but the population standard deviation is known, the only caution is that the use of the standard normal curve is appropriate only if the sample size is 25 or more. When the population is not normal and the population standard deviation is unknown, the advice is to use a very large sample (100 or more), in which case it makes little difference whether you use the *z* or *t* distribution.

In the second half of this lesson you were advised that blindly following tradition by guarding against a Type I error at a 0.05 level of significance for all hypothesis tests does not make sense. Some consideration must be given to the relative costs of making a Type I error as opposed to the costs of making a Type II error. You saw that your choice of a significance level sets the probability of your rejecting true null hypotheses (Type I error) and that it affects the probability of your failing to reject false null hypotheses (Type II error). In the next lesson you will learn more about the relationship between choice of a level of significance and the probability of making a Type II error. First, complete the following exercises to test your understanding of the process of hypothesis testing as it relates to the population mean.

LESSON 2 EXERCISES

1. For years a manufacturing process has been producing an average of 14.8 units per hour, with a standard deviation of 2.4. The distribution has been approximately normal. After careful study, a consulting engineer suggests some machinery modifications that he believes will increase production. Before making the changes in all of the machines throughout the plant, a test is made by modifying 16 machines and recording the results. After a week of observation, the average rate per hour is calculated for each of the 16 modified machines. The average of these 16 hourly averages turns out to be 16.4, with a standard deviation of 2.7. Test the appropriate hypothesis at the 0.01 level. Assume that the modification does not affect the shape or variability of results.

2. Suppose that for a certain medical school the average performance of students on the anatomy section of the National Boards has been 82, with a distribution that is approximately normal. Twelve students are chosen at random from an entering class and are taught anatomy through computerized instruction with only limited laboratory experience. The mean and standard deviation on the National Board for this group of students are respectively 79.4 and 9.1. There does not appear to be any skewness in the distribution of scores, and there is some clustering about the mean of 79.4. Use a 0.05 level to test whether the switch can be expected to result in a change in the level of performance.

3. The average radiation dose level in the body of residents of a county is known to be about 0.20 rem per year. There is concern that residents living near an atomic power plant may experience higher levels because of radiation leaks. You plan to take a random sample of 200 residents living within a mile of the plant. What null hypothesis will you test? What would be the consequences of a Type I error? What would be the consequences of a Type II error? What level of confidence will you recommend?

Lesson 3 □ Type II errors and power of a test

Just to be different, let us start this lesson with a trick question. Suppose a scientist decides early in his career to use the 0.05 level of significance to analyze the data from all of his experiments. What proportion of all his experiments will result in Type I errors? Careful now. Do not make up your mind too quickly.

With that kind of introduction, you know that the obvious answer—5 out of every 100—is wrong. Can you figure out why? The answer lies in a statement in the first lesson of this chapter: "When you set the probability of a Type I error at 0.05, or 0.01, or whatever, *it is a conditional probability.*" A significance level of 0.05 means that the probability of a Type I error is 0.05 *when the null hypothesis is true.* To make a Type I error, the null hypothesis must be true. If the treatment is ineffective in half of the experiments our young scientist conducts, then in 5% of these he will make a Type I error. In the other half of his experiments, in which the treatment does have the expected effect, a Type I error is not possible.

If we call the occurrence of a true null hypothesis A, and the rejecting of this true null hypothesis B; then the probability of both occurring will be found through the multiplicative rule.

$$P(\text{Type I error}) = P(A) \times P(B|A)$$

We know that $P(B|A)$ will equal 0.05. If one half of all the experiments conducted have ineffective treatments, then $P(A) = 0.50$. The probability of Type I errors throughout our young scientist's career will be $(0.5)(0.05)$, or 0.025. Of course, if more than half of the experiments he conducts result in effective treatments, the probability of a Type I error will be even less.

Under what condition will the young scientist make a Type I error in 5% of all the experiments he conducts? It is only if none of the experiments he conducts results in effective treatments, so that $P(A)$ equals 1.00. From this it can be seen that setting an 0.05 level of significance fixes an upper limit for the overall percentage of Type I errors.

PROBABILITY OF TYPE II ERRORS

Now suppose we change the question to, "What proportion of all the experiments conducted by our young scientist will result in Type II errors?" The last question was more complicated than you may have initially thought. Brace yourself, this one is worse! The probability of a Type I error depends on only two factors: the chosen level of significance and whether the treatment was effective. The probability of a Type II error (sometimes called beta), depends on four factors.

Remember, a Type II error occurs when an effective treatment goes unrecognized. You were already told earlier that guarding carefully against a Type I error by setting a very extreme level of significance increases the probability of a Type II error. The chosen level of significance is, therefore, one of the four factors affecting beta. Before trying to show why this is so, let us identify the others. Can you think of what they might be?

Two of the factors relate to the precision of the sample statistic, as measured by the standard error of the statistic. If you think back to the formula for the standard error of the mean, you will be able to name two more factors that affect the probability of Type II errors. They are the variability in the populations and the size of the sample. In general, the more variability there is in the population being compared, the higher the probability of a Type II error. On the other hand, the larger the sample from the test population, the lower the probability of a Type II error.

The fourth factor that affects the probability of a Type II error is the most important of all. It is the effectiveness of the treatment. The effectiveness of a treatment can range from being completely ineffective, in which case a Type II error is not possible (why?), to extremely effective. Between these two extremes, the probability of a Type II error goes from very high, when the treatment causes only a small difference, to extremely low, when the treatment results in large differences between the parameters of the test and the standard populations.

Let us examine an example to clarify the role of each of these four factors in influencing the probability of a Type II error. Suppose a statistics textbook is to contain a special table of 5000 random three-digit numbers with a mean of 500 and a standard deviation of 20. Each of the numbers 000 through 999 does not occur with equal frequency. Instead, the numbers are normally distributed. The table is therefore a table of random, normal, three-digit numbers. In one of the early drafts of the manuscript, the mean for the table was 510 rather than 500. The presses are all ready to roll when a question arises as to which version of the table is included in the appendix. You doubt that the early version has been included but you are not certain. As a quick check you find the mean of the first 100 numbers and use hypothesis testing with a 0.05 level of significance to decide whether to let the presses roll. The null hypothesis is that the correct table is included. What is the probability of mistakenly including the wrong table in the book (making a Type II error)?

First, let us examine the hypothesis testing procedure you will use. The standard population has a normal distribution with a mean of 500 and a standard deviation of 20. Following approach A, you will use a one-tail test (why?) and identify the critical region as 1.645 standard errors above 500. The standard error will be $20/\sqrt{100}$, or 2.00. The critical region will then be composed of all means larger than $500 + (1.645)(2.00)$, or 503.29. In other words, if the mean of the first 100 numbers is equal to or greater than 503.29, you will hold up the presses. If the mean is less than 503.29, you will let the presses roll.

Now what is the chance that *when the wrong table is included,* the first 100 numbers will have a mean that *fails* to fall in the critical region? In other words, what is the probability that the mean of the 100 numbers will be less than 503.29? To answer this, we need to know what the sampling distribution from the test population looks like when the wrong table has been used. Like the standard population, such a sampling distribution

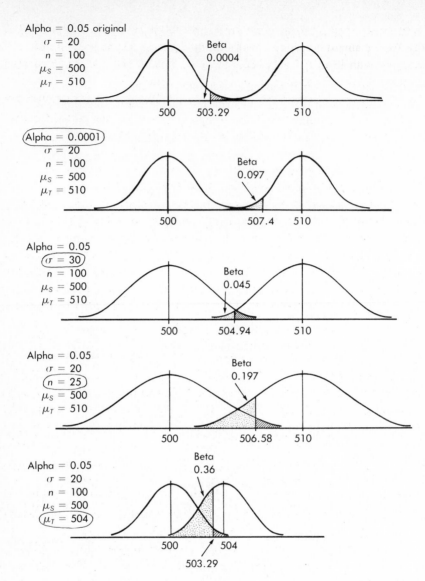

Fig. 7-2. Effects of factors influencing the probability of a Type II error: the sampling distribution from the standard population and the test population.

from the test population has a normal distribution and a standard error of the mean equal to 2.0, but this time the mean is 510. What is the chance of drawing a random sample of size 100 from this test population and getting a mean less than 503.29? To find out, we change 503.29 into a z score.

$$z = \frac{503.29 - 510.0}{2} = \frac{-6.71}{2} = -3.355$$

The chance of getting a sample with a mean this far below 510 is only about 4 out of 10,000. We are almost certain to identify the wrong table if it has been included. Note, however, that with an alpha level of 0.05 there is a 5 out of 100 chance of holding up the presses *when the correct table is included* (a Type I error).

Many statistics books refer to the probability of a Type II error as beta, in contrast to the alpha level of a Type I error. It is also rather common to speak of the *power of the test*. This is 1.00 minus beta. In our example, beta would be 0.0004 and the power of the test would be 0.9996.

Now let us see why we are so unlikely to make a Type II error; specifically, how does each of the four factors we have identified contribute to this result?

The effects of the four factors in influencing the probability of a Type II error can be seen by a careful examination of Fig. 7-2. Each of the drawings shows the sampling distributions from the standard and the test populations for the case when the wrong table has been used. The critical region of the sampling distribution from the standard population has been identified by hatching. That portion of the sampling distribution from the test population which fails to fall in the critical region is identified by dots. The dotted area therefore represents the probabilty of a Type II error; the remaining portion represents the power of the test.

The first drawing at the top represents the situation as originally described. Notice that the critical region is composed of all sample means above 503.29. Almost all of the sampling distribution from the test population falls in the critical region. That part which is to the left of 503.29 is so small that it cannot be identified with dots. It represents only four ten-thousandths of the total area of the sampling distribution from the test population. Each of the other drawings shows what happens as one of the four factors affecting beta is independently changed. The change, which has always been made in a direction that increases the probability of a Type II error, has been identified by circling. Let us look at the drawings one at a time, starting with the second drawing.

Drawing 2: choice of a level of significance. We originally chose a 0.05 level of significance and established a critical region 1.645 standard errors above 500. This means that even if the correct table has been included, there is a 0.05 probability that we will hold up the presses while we check out the table. What if we had chosen an alpha level of 0.0001? The critical region would have been that part of the curve 3.719 standard errors above 500. The area of the normal curve 3.719 standard deviations above the mean is too small to be able to mark with hatching, but we can identify the point at which it starts. Our critical region would include all means above $500 + (3.719)(2)$, or 507.4. Notice that the result is a critical region closer to the mean of the test population, increasing the probability of a sample mean from the test population that is not in the critical region. As in the original situation, we can change 507.4 to a z score to find the probability of means as low as this when samples are taken from the test population.

$$z = \frac{507.4 - 510}{2} = -1.3$$

Since about 9.7% of the area of the standard normal curve is below -1.3, beta would now equal 0.097 and the power of the test would be $1.0 - 0.097$, or 0.903. The interpretation is that if we decide to stop the presses only if we get a sample mean so extreme as to occur only one time in 10,000 when the correct table in included, we have almost a 10% chance of letting the presses roll even if the incorrect table is included.

Drawing 3: variability of the populations. Our populations had standard deviations of 20, resulting in a $SE_{\bar{x}}$ of 2 for a sample size of 100. What if the standard deviation had been 30, rather than 20? The result would be a standard error of 3 instead of 2. Returning to the original 0.05 significance level, our critical region would be composed of all means greater than $500 + (1.645)(3)$, or 504.94. The area to the right of 504.94 for the sampling distribution from the *standard population* represents 5% of the total area of the curve and has been identified with hatching. The area to the left of 504.94 for the sampling distribution of the *test population* represents those sample means that will *not* fall in the critical region, therefore resulting in a Type II error. To find this probability, we again find the z score.

$$z = \frac{504.94 - 510}{3} = -1.69$$

Since about 4.5% of the area of the standard normal curve is below -1.69, beta would now equal 0.045 and the power of the test would equal 0.955. Notice that an increase in the population standard deviation from 20 to 30 has increased the probability of a Type II error from the original 0.0004 to 0.045. It is interesting to note that in this instance the probabilities for the two types of error are about equal, 0.05 and 0.045.

Drawing 4: size of sample. Our original sample of size 100 resulted in a standard error of 2.0 when the standard deviation was 20. What if the first 25 numbers from the table were used instead of the first 100? The standard error would be $20/\sqrt{25}$, or 4.0, resulting in a critical region made up of all means above $500 + (1.645)(4)$, or 506.58. Reducing sample size has moved the critical region toward 510, increasing the chances of a Type II error. Beta is now 0.197 and the power of the test is 0.813. Notice that reducing the sample size to 25 has changed the probability of a Type II error from 0.0004 to almost 0.20. In other words, taking a sample size of 25 rather than 100, while holding alpha at 0.05, means that if the wrong table has been included, instead of being almost certain to discover the error, there is a one-in-five chance that it will go undetected.

Drawing 5: effectiveness of the treatment. For the example given, the early version of the table had a mean of 510. If the mean had been closer to 500—for example, 504—it would be harder to tell whether the wrong table had been included. Notice in this drawing that the critical region is made up of all means over 503.29, the same figure as in the top drawing, representing the original situation. However, the sampling distribution of the test population centers on 504 rather than 510. As a result, there is a much greater chance of a Type II error.

$$z = \frac{503.29 - 504.0}{2} = -0.35$$

Since approximately 36% of the standard normal curve falls below a z score of -0.35, beta equals 0.36 and the power of the test equals 0.64. The interpretation is that if the wrong table, with a mean of 504, has been included, the probability of discovering the error is only 0.64 when we use a sample of 100 and an alpha level of 0.05.

Reflecting on what has been said about the four factors that affect the probability of a Type II error, what would you do to improve the chances of discovering the error if the wrong table has been used? Since you cannot do much about the variability in the tables or the actual amount of the difference between μ_S and μ_T, you have two choices. You must either be willing to increase the risk of a Type I error (stopping the presses unnecessarily) or to increase the sample size.

Let us now summarize our conclusions from this example.

1. *Level of significance.* The more extreme the chosen significance level, the greater the probability of a Type II error, or the less the power of a test.
2. *Variability in the populations.* The more variability in the populations, the greater the probability of a Type II error, or the less the power of a test.
3. *Size of the sample.* The smaller the sample size, the greater the probability of a Type II error, or the less the power of the test.
4. *Effectiveness of the treatment.* The smaller the difference between the parameters of the test population and of the standard population, the greater the probability of a Type II error, or the less the power of the test.

POWER OF A TEST AND SAMPLE SIZE

In reality, the amount of the difference between the parameters of the test population and of the standard population is never known. Without this knowledge, the probability of a Type II error and, hence, the power of a test, cannot be determined. Of what use, then, is the whole foregoing analysis? The answer is that knowing the relationship of each of the factors to beta provides us with information as to how we can make our tests more powerful. We have four options: (1) make alpha less extreme, (2) reduce population variability, (3) choose a larger sample, or (4) make the treatment more effective. Often, the only one of these that is attainable is to choose a larger sample.

As a matter of fact, it is possible to figure out just what size sample is necessary for beta to be held to some given level for any specified difference between the parameter values of the test and standard populations. To do so requires an estimate of σ and a designation of the level of significance to be used. The computations are a bit complicated, and you will probably have to seek the help of a statistician when the time comes to ask the question, "What size sample do I need?" You might give some thought ahead of time to the questions the statistician will ask you.

1. What difference do you expect as the result of the treatment, or, alternatively, how small of a difference do you wish to be able to identify?
2. What risk of a Type I error are you prepared to accept?
3. What risk of a Type II error are you prepared to accept?
4. What is the variability of the population?

These are tough questions to answer, and the usual reaction is to respond, "I have no idea." In such a case, the statistician cannot give you advice regarding the size of sample required. The statistician will only tell you to take the largest sample you can. If you can provide some approximate answers to the questions, the statistician will be able to calculate an approximate minimum sample size. To be on the safe side, you will probably want to increase this if it is feasible.

Suppose the study is then conducted and you fail to get a result that causes you to reject the null hypothesis. What is your conclusion? The answer, again, is not that the null hypothesis is correct. It is simply that this experiment has not provided you with strong enough evidence to convince you that the null hypothesis is false. You may be making a Type II error, but there is no way of knowing this for any experiment.

SUMMARY

The three lessons you have just finished are difficult. Although the logic of hypothesis testing is rather straightforward, the possible results and the probabilities for the two types of errors can be confusing and difficult to grasp. The exercise that follows is designed to test your understanding of the concepts from this lesson involving the probability of a Type II error and the power of a test.

Experience has shown that students have difficulty integrating the many concepts of hypothesis testing. Appendix C contains a problem that includes all of the complexities and subtleties of hypothesis testing. It will serve as a review exercise for this entire chapter on hypothesis testing. It contains a description of a problem and then a set of questions relating to the problem. A detailed set of answers is also provided. If you understand this reference problem in Appendix C and can answer the questions, you have a thorough grasp of hypothesis testing. Try to answer all of the questions before looking at the detailed explanations provided. Good luck!

LESSON 3 EXERCISE

Suppose that is it discovered that infection with rabies results in an elevation in the amount of a certain chemical present in blood samples from the infected patient. In patients not infected, the reading is normally distributed, with a mean of 50 and a standard deviation of 10. Among all persons infected with rabies, the readings are also normally distributed, with a standard deviation of 10, but the average reading is 70. Accordingly, the diagnosis of rabies infection is made whenever there is a high reading. The problem is: "How high must the reading be?" Setting the level too low results in many false positive findings. Setting the level too high results in many false negative findings. A certain clinic decides to make a diagnosis of rabies whenever a reading is so high as to occur only 5% of the time or less among persons without the disease. What is the probability of false positive and false negative findings? Analyze the problem by using the logic of hypothesis testing. Each patient represents a sample of size 1.

1. What are the standard and test populations?
2. What are the parameter values of the standard and test populations?
3. Describe the sampling distributions from the standard and test populations for samples of size 1.
4. What readings will result in the diagnosis of rabies in the clinic?
5. Draw the overlapping distributions from question 3 on a line like the line below. Use your

knowledge of the normal curve. Inflection points are at plus one and minus one standard deviation from the mean, and almost all scores are within three standard deviations of the mean.

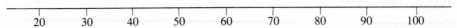

6. Mark the point where the clinic begins to diagnose rabies infection. What part of the distribution of the standard population is to the *right* of this point (false positive finding, or Type I error)?

7. What part of the sampling distribution from the test population is to the *left* of the point marked (false negative finding, or Type II error)?

8. What part of the sampling distribution from the test population is to the *right* of the point marked (correct identification of rabies, or power of the test)?

9. The treatment for rabies involves a series of painful injections. Untreated rabies infection results in death. Comparing the costs of Type I and Type II errors and your responses to questions 6 and 7, would you recommend a different significance level? Why?

10. How would the problem change if four children had all been bitten by the same wild dog, which could not be located but is suspected of being rabid?

PART THREE

Applications of inferential statistics

CHAPTER 8

Two-sample experiments

Lesson 1 □ Sampling distribution of a difference between statistics

In the last two chapters you learned the two approaches to inferential statistics, using as examples the proportion and the mean. Have you noticed any difference between the examples and exercises that have been given, and the experiments you read about in scientific journals? Think back to the last research report you read. In all probability, the study involved a comparison of two or more treatments, both of which had unknown parameters.

To keep things as uncomplicated as possible, all examples given in this book until now have been *one-sample* experiments. In Chapter 6, there was a *single* population with an unknown parameter for which you learned to set a confidence interval. In Chapter 7, there was a single population with an unknown parameter, which you learned to compare with the known parameter of a standard population.

It is far more usual in scientific studies to make a comparison of two or more populations, both with unknown parameters. As an example, suppose that two different methods of treating damaged knee ligaments have been advocated. For each patient, the results of treatment can be evaluated on a scale from 0 to 12, considering such factors as flexibility, strength, and absence of pain. The method that results in the highest average score will become the treatment of choice in the future.

If we call the first treatment method A, and the second treatment method B, it is the difference between μ_A and μ_B that interests us. Our question might be, "What is the difference between the average scale score for method A and the average scale score for method B?" Or the question might be, "Is the difference between the average scale score for method A and the average scale score for method B equal to zero?" Do you recognize these two questions? If you make the difference between μ_A and μ_B the parameter of interest, the first question indicates that a confidence interval is called for. The second question calls for hypothesis testing, with the standard population being a population with a parameter value of zero.

Once we decide that it is the difference between μ_A and μ_B that is the parameter of

139

interest, everything you have learned in the two previous chapters becomes directly applicable. In the case of confidence intervals, we will attempt to answer the question, "What is the parameter value $\pi_A - \pi_B$ or $\mu_A - \mu_B$?" In the case of hypothesis testing, we will try to answer the question, "Is the parameter value $\pi_A - \pi_B$ or $\mu_A - \mu_B$ different from that of the standard population?" Usually, the standard population has a parameter value of zero, but that is not necessary.

Whether setting confidence intervals or testing hypotheses, you will need to know the sampling distribution for the observed statistic. Because the sampling distribution of a difference between statistics can get a bit complicated, this first lesson will be devoted to that topic. Once you know how the sampling distribution of a difference between statistics (either p's or $\overline{X}$'s) is computed, you can then apply this knowledge to inferences involving proportions and means.

CHARACTERISTICS OF A SAMPLING DISTRIBUTION OF A DIFFERENCE BETWEEN STATISTICS

Can you visualize what is meant by the sampling distribution of the difference between statistics? Let us return to our example of two proposed methods of treating knee injuries. Suppose there really is no difference in the results of the two treatments, A and B, and that both treatments result in scale scores that are normally distributed, with μ equal to 6. Suppose now that you empirically take 1000 *pairs* of random samples, one sample from the population using treatment A and the other from the population using treatment B. Each sample is of size 10. For each pair of samples you compute the statistic $\overline{X}_A - \overline{X}_B$. You then make a frequency distribution of the 1000 observed differences between sample means. This represents an empirically derived sampling distribution of the difference between means for two populations in which μ_A equals μ_B. To find its central tendency, you find the average of the 1000 differences; to find its variability, you find the standard deviation of the 1000 differences; to find its shape, you examine a histogram of the 1000 differences.

If our interest had been in the difference between proportions rather than in the difference between means, we could still take 1000 pairs of samples. The only difference is that we would now record the difference between p_A and p_B instead of the difference between $\overline{X}_A$ and $\overline{X}_B$. Everything else would be the same. We would then have an empirically derived sampling distribution of the difference between sample proportions.

As you learned earlier, the statistician does not often find it necessary to empirically discover what a sampling distribution is like. Instead, the sampling distribution is known on a theoretical basis. This applies as well to the sampling distribution of the difference between statistics. As always, we will have to know shape, central tendency, and variability.

Shape. Determining the shape of a sampling distribution of a difference between sample statistics is easy. You need only to apply the same criteria you learned for the single-sample case. For sample proportions, you learned that the sampling distribution is approximately normal if both $n\pi$ and $n(1 - \pi)$ are equal to or greater than five. This same criterion can be applied to determine whether *differences between proportions* will be normally distributed. The criterion must be met for each population.

For sample means, you learned that the sampling distribution is normal if the population is normal or if samples were large (25 or more). Again, the sampling distribution of the difference between sample means will be normal if one of these two criteria is met in relation to each sample.

Central tendency. Determining the central tendency of the sampling distribution of the difference between sample statistics is also easy. It is the parameter value: the actual difference between π_A and π_B or between μ_A and μ_B. In many instances of hypothesis testing, the null hypothesis will be that there is no difference, and the central tendency of the sampling distribution will be zero.

Variability. Once again, the standard deviation of the sampling distribution will be called the standard error of the sampling distribution ($SE_{p_A - p_B}$ or $SE_{\bar{X}_A - \bar{X}_B}$). Understanding what is meant by $SE_{p_A - p_B}$ or $SE_{\bar{X}_A - \bar{X}_B}$ is not particularly difficult. It represents the standard deviation of all the observed differences between pairs of statistics $p_A - p_B$ or $\bar{X}_A - \bar{X}_B$. Neither is the formula particularly difficult to compute. It is Formula 8.1-1.

$$SE_{\text{stat }A - \text{stat }B} = \sqrt{(SE_{\text{stat }A})^2 + (SE_{\text{stat }B})^2} \qquad \textbf{8.1-1}$$

This formula indicates that to find the standard error of a difference between two statistics, you need to find the standard error for each statistic independently, square these values, add the results, and then find the square root of the sum. Knowing that $SE_p = \sqrt{\dfrac{\pi(1 - \pi)}{n}}$ and $SE_{\bar{X}} = \dfrac{\sigma}{\sqrt{n}}$ allows us to derive the two specific formulas for the standard error of a difference between proportions and the standard error of a difference between means. In the case of sample proportions, the formula appears as Formula 8.1-2.

$$SE_{p_A - p_B} = \sqrt{\frac{\pi_A(1 - \pi_A)}{n_A} + \frac{\pi_B(1 - \pi_B)}{n_B}} \qquad \textbf{8.1-2}$$

In the case of sample means, the formula used is Formula 8.1-3.

$$SE_{\bar{X}_A - \bar{X}_B} = \sqrt{\frac{\sigma_A^2}{n_A} + \frac{\sigma_B^2}{n_B}} \qquad \textbf{8.1-3}$$

It should not come as a surprise to learn that the variability in the difference between statistics for pairs of random samples depends on the variability of each of the statistics. A more detailed explanation of the formula for the standard error of the difference between statistics appears as a technical note at the end of this lesson. This explanation has been placed in a technical note because it is a bit complicated and is not essential to an understanding of the meaning of the concept or its application.

SUBSTITUTION OF STATISTICS FOR UNKNOWN PARAMETER VALUES

An examination of the specific formulas for the standard errors of differences between statistics reveals that the statistician is again faced with the situation where the standard error depends on parameter values that are often unknown. In the case of $SE_{p_A - p_B}$, π_A and π_B appear in the formula, but these are usually the precise values that we are attempting to learn about. In the case of $SE_{\bar{X}_A - \bar{X}_B}$, we need to know σ_A and σ_B, which are often unknown. Let us examine the problem and its solution for each of these examples.

Substitution of p for π

In the one-sample case you were told to substitute p for π in the formula for the standard error of a proportion. You were then cautioned that, as a result, confidence intervals established using this estimated SE_p would have to be regarded as only approximations. The approximation will be best when π is near 0.5 and n is large. This same procedure and warning are applicable regarding the standard error of a difference between proportions.

There is one additional problem that occurs for those situations in which π_A is assumed to be equal to π_B (often the case in hypothesis testing). In this instance, we will not want to substitute p_A for π_A and p_B for π_B in estimating $SE_{p_A - p_B}$. Instead, we will want to use, in each case, the best estimate possible of the common proportion π. Averaging p_A and p_B will work if samples are the same size. However, if this is not the case, this procedure would be giving equal weight to the information from two unequal-sized samples.

The best procedure *when you are assuming π_A equals π_B* is to pool the information from both samples. For example, suppose that in a sample of 10 patients treated by method A, 6 survive, and that in a sample of 100 patients treated with method B, 82 survive. The proportion surviving in sample A is 0.6, and in sample B, it is 0.82. Averaging would give a value of 0.71, but this would give equal weight to the sample proportions taken from two different-sized samples. Instead, we will pool the two samples and find that 88 survived out of a total of 110. Our best estimate of the common proportion π_A and π_B is therefore $^{88}/_{110}$, or 0.80. Since this is an estimate based on the proportion in the pooled samples, we will call it p. The estimated standard error of difference between p_A and p_B would therefore be:

$$SE_{p_A - p_B} = \sqrt{\frac{p(1-p)}{n_A} + \frac{p(1-p)}{n_B}} \text{ or } \sqrt{\frac{(0.8)(0.2)}{10} + \frac{(0.8)(0.2)}{100}} \text{ or } 0.133$$

Can you now conceptualize the meaning of the figure 0.133? Relate it back to the case of an empirically derived sampling distribution. The meaning is as follows. Suppose you have two populations, A and B, both having the same proportion (0.80) for some attribute. You repeatedly choose pairs of random samples ($n_A = 10$, $n_B = 100$) from these two populations, compute p_A and p_B, and subtract to find the difference. If you carry out this procedure an infinite number of times and then find the standard deviation of all these differences, you expect it to be 0.133.

Note that for this example you would not expect the distribution of differences to be shaped like the normal curve? Why?

Substitution of s for σ

The standard error of the difference between sample means also depends on parameter values. This time σ_A and σ_B are often unknown in Formula 8.1-3.

$$SE_{\bar{x}_A - \bar{x}_B} = \sqrt{\frac{\sigma_A{}^2}{n_A} + \frac{\sigma_B{}^2}{n_B}}$$

In the *one-sample case,* you learned that when σ is unknown and the population is normally distributed, substituting s for σ would not cause serious problems. You make use of the t distribution instead of the z distribution when setting a confidence interval or when testing a hypothesis. If the population is not normally distributed, you will need to make use of large samples (100 or more).

In the *two-sample case,* we will resort to the same solution of using the t distribution when σ_A and σ_B are unknown. Everything you have learned about $SE_{\bar{x}}$ in making inferences about μ is directly applicable to the use of $SE_{\bar{x}_A - \bar{x}_B}$ when making inferences about $\mu_A - \mu_B$. There is, however, one complication. It occurs when you assume that σ_A equals σ_B. You would not want to substitute s_A for σ_A and s_B for σ_B in the formula for $SE_{\bar{x}_A - \bar{x}_B}$. Instead you want to substitute the best estimate of the common sigma in both places in the formula. Again, if samples are of equal size, you would just take an average of $s_A{}^2$ and $s_B{}^2$ as the best estimate of σ^2. When samples are of unequal size, you will want to pool the information from the two samples and weight the information according to sample size. The appropriate formula is Formula 8.1-4.

$$s = \sqrt{\frac{\Sigma x_A{}^2 + \Sigma x_B{}^2}{n_A + n_B - 2}} \qquad \textbf{8.1-4}$$

Note: Remember that $\Sigma x^2 = \Sigma X^2 - \dfrac{(\Sigma X)^2}{n}$

Suppose that you had the following information:

	n	ΣX	ΣX^2	Σx^2	**s**	$SE_{\bar{x}}$
Sample A	10	40	170	10	1.054	0.333
Sample B	20	100	546	46	1.556	0.348

For these data, the standard error of the difference between the means of pairs of random samples is computed as follows. First, you find the best estimate of the common σ.

$$s = \sqrt{\frac{10 + 46}{10 + 20 - 2}} = \sqrt{\frac{56}{28}} = \sqrt{2} = 1.414$$

You then substitute s in place of σ_A and σ_B in the formula for $SE_{\bar{X}_A - \bar{X}_B}$.

$$SE_{\bar{X}_A - \bar{X}_B} = \sqrt{\frac{s^2}{n_A} + \frac{s^2}{n_B}} = \sqrt{\frac{2}{10} + \frac{2}{20}} = 0.548$$

Can you once more conceptualize the meaning of the figure 0.548? Again, think of empirically taking pairs of samples ($n_A = 10$, $n_B = 20$) from populations having, in this case, the same σ (1.414). For each pair of samples, the difference $(\bar{X}_A - \bar{X}_B)$ is calculated. The figure 0.548 represents the standard deviation of the differences that would result from an infinite number of such pairs of samples.

SUMMARY

In the previous chapters the concepts of inferential statistics were presented for the case of a single unknown parameter. This current chapter expands the situation to include inferences for two unknown parameters. We will either wish to estimate the difference between the two parameters (set a confidence interval) or decide whether the difference between two parameters is zero or some other specified value (test hypotheses). In either case, we wish to learn about the differences between two parameters ($\pi_A - \pi_B$ or $\mu_A - \mu_B$). To learn about this difference, we will take a sample from each population and find the difference between the sample statistics ($p_A - p_B$ or $\bar{X}_A - \bar{X}_B$). We will consider this difference to be the observed statistic and then apply all of the concepts learned for the one-sample case. To begin with, we will need to know the sampling distribution of the observed difference between statistics. We need to know shape, central tendency, and variablity. The criteria for determining when the distribution is normal in shape are the same as for the one-sample case, but the criteria must apply to both samples. The central tendency of the sampling distribution of differences between statistics will be equal to the difference between the parameters. The variability of the sampling distribution will be represented by the standard error ($SE_{p_A - p_B}$ or $SE_{\bar{X}_A - \bar{X}_B}$). It is obtained from Formula 8.1-1.

$$SE_{\text{stat }A - \text{stat }B} = \sqrt{(SE_{\text{stat }A})^2 + (SE_{\text{stat }B})^2}$$

The information that goes into the two sets of parentheses in this formula is the standard error of the statistic for each population. There is a complication in the case of proportions when it is assumed that π_A equals π_B, and there is a complication in the case of means when it is assumed that σ_A equals σ_B. In both cases, the solution involves pooling the information from both samples to come up with the best estimate of the common parameter.

The exercises that follow give you a chance to test your understanding of the concept of a standard error of a difference between statistics. Following the exercises is the technical note, which explains the basis for the formula for the standard error of the difference between proportions. The next lesson describes the process of setting confidence intervals and testing hypotheses for the two-sample case.

LESSON 1 EXERCISES

1. Suppose one takes pairs of random samples of size 25 of single-digit numbers from two different tables of random numbers. For each pair of samples, the proportions of even numbers are
computed (p_A and p_B). What will the sampling distribution of the difference between p_A and p_B
look like?
What will be the shape? Why?
What will be the central tendency? Why?
What will be the variability? Explain.

2. Now, suppose that for each pair of samples, we compute $\overline{X}_A - \overline{X}_B$ instead of $p_A - p_B$. What will
the sampling distribution look like? (Hint: Since all the numbers occur with equal frequency for
each population, σ^2 can be obtained by finding the variance of the numbers 0 through 9; and μ
can be obtained by finding the average of the numbers 0 through 9).

3. Suppose you wish to learn if the proportion of persons favoring a bond issue to build a new
county hospital differs depending on whether the person is 40 or more years old or under 40. You
do not know what π is for either age group. You choose a random sample of size 100 from the
registered voters of the county and obtain the following data.

	≥40	<40
Number favoring	32	48
Number opposed	4	16

If you assume π_A equals π_B, what would be your best estimate of this common proportion? What
is your estimate of $SE_{p_A - p_B}$?
If you do not assume π_A equals π_B, what is your estimate of $SE_{p_A - p_B}$?

4. In problem 3, suppose that instead of recording the persons' ages as over or under 40, you had
their actual ages. The summary statistics for age for the two groups are as follows:

	n	ΣX	ΣX²
Favoring	80	2800	101,665
Opposed	20	600	18,475.5

If you assume that σ_A equals σ_B what would be your best estimate of this common σ? What is
your estimate of $SE_{\overline{X}_A - \overline{X}_B}$?
If you do not assume σ_A equals σ_B what is your estimate of $SE_{\overline{X}_A - \overline{X}_B}$?

LESSON 1 TECHNICAL NOTE: VARIANCE OF A COMPOSITE

The formula for the standard error of a difference between statistics is actually a special case of
a more general formula for the variance of any composite, where a composite is any score resulting
from adding or subtracting two variables. If the two variables are designated A and B, the formulas
are:

$$(\sigma_{A+B})^2 = \sigma_A{}^2 + \sigma_B{}^2 + 2r_{AB}\sigma_A\sigma_B \qquad \textbf{8.1-5}$$

and

$$(\sigma_{A-B})^2 = \sigma_A{}^2 + \sigma_B{}^2 - 2r_{AB}\sigma_A\sigma_B \qquad \textbf{8.1-6}$$

where r_{AB} = correlation between variables A and B.

For example, suppose the scores of a midsemester exam are designated A and scores on the final
exam, B. If $\sigma_A = 4$, $\sigma_B = 8$ and r_{AB} equals 0.8, what will be the variance of the composite score
that results when the midsemester score is added to the final score?

$$(\sigma_{A+B})^2 = (4)^2 + (8)^2 + 2(0.8)(4)(8) = 131.2$$

The variance of the composite scores will be 131.2, and the standard deviation will be the square root of 131.2, or 11.45.

Although it is hard to imagine why one would do so, if one subtracted the midsemester score from the final score, the last term in the equation would be subtracted rather than added, and the variance and standard deviation would be 28.8 and 5.36 respectively.

To gain insight about the formula for the variance of a composite, try visualizing what happens if A equals B. Compare the formula under this circumstance with the information in Table 3-2. If A equals B, then adding A to B would be the same as coding the scores for A by multiplying by 2. What happens to the variance if you code by multiplying every score by 2? The variance would be multiplied by the square of 2, or 4. Notice that this is what happens according to Formula 8.1-5.

$$(\sigma_{A+B})^2 = \sigma_A^2 + \sigma_B^2 + 2r_{AB}\sigma_A\sigma_B$$

but if A equals B

$$(\sigma_{A+A})^2 = \sigma_A^2 + \sigma_A^2 + 2(1.00)\sigma_A\sigma_A$$

or

$$\sigma_{2A}^2 = \sigma_A^2 + \sigma_A^2 + 2\sigma_A^2$$
$$\sigma_{2A}^2 = 4\sigma_A^2$$

Even if you fail to see the relationship between the formula for the variance of a composite and coding, you should be aware of what happens in the case where r_{AB} equals zero. In this case, the formula is simplified and is the same whether you add or subtract to form the composite. When r_{AB} equals zero, the formula becomes Formula 8.1-7.

$$(\sigma_{A+B})^2 \text{ or } (\sigma_{A-B})^2 = \sigma_A^2 + \sigma_B^2 \qquad \textbf{8.1-7}$$

when $\quad r_{AB} = 0$

Notice that if we take the square root of both sides we have:

$$\sigma_{A-B} = \sqrt{\sigma_A^2 + \sigma_B^2}$$

Does the last formula suggest anything to you? What would happen if variable A were the *means* of samples from population A, and variable B the *means* of samples from population B? The variance of A would be $SE_{\bar{X}_A}$ squared, and the variance of B would be $SE_{\bar{X}_B}$ squared. We would then have:

$$SE_{\bar{X}_A - \bar{X}_B} = \sqrt{(SE_{\bar{X}_A})^2 + (SE_{\bar{X}_B})^2}$$

This, of course, is the formula given in the lesson for the standard error of the difference between means. Can you see why it is appropriate to assume that r equals zero and, therefore, to drop the last term of the general formula for the variance of a composite?

Lesson 2 □ Two-sample inferences involving proportions and means

In the last lesson you learned about the sampling distribution of a difference between statistics. Information was provided about differences between sample proportions and differences between sample means. In this lesson you will see how to apply these concepts while making inferences involving π and μ. Both confidence intervals and hypothesis tests will be discussed.

CONFIDENCE INTERVALS FOR THE DIFFERENCE BETWEEN PROPORTIONS

Students attempting to gain entrance to medical school often study from manuals that, it is claimed, will improve their performance on the Medical College Admission Test (MCAT). Suppose there are two manuals in common usage. You want to compare the two by estimating the difference between the proportion of students scoring above 600 using manual A and the proportion who score above 600 using manual B. The question is, "What is the difference between π_A and π_B?" You have 200 students planning to apply to medical school, and you randomly assign 100 to use study manual A and 100 to use study manual B. When they take the MCAT, the number scoring above 600 using manual A is 20 ($p_A = 0.2$) and the number scoring above 600 using manual B is 10 ($p_B = 0.1$). What is your best estimate as to the difference in the proportions scoring above 600 ($\pi_A - \pi_B$) using the two different study manuals? Once again, the best single estimate is the observed statistic $p_A - p_B$, or 0.10. Can you set a 95% confidence interval for $\pi_A - \pi_B$? Before reading on, stop to see if you can apply what you learned in the last lesson, along with your earlier knowledge of confidence intervals, to solve the problem.

You remember that a confidence interval is formed as follows:

Parameter = observed statistic $\pm$ (appropriate tabled value)($SE_{\text{observed statistic}}$)

For the first parentheses, we need to know whether it is appropriate to use the table of the standard normal curve. In the last lesson, it was stated that if $n\pi$ and $n(1 - \pi)$ are equal to or greater than 5 for both samples, the difference between sample proportions will be normally distributed. In this exmple, p_A and p_B are 0.2 and 0.1 respectively. These are our best estimates of π_A and π_B. With samples of size 100, np and $n(1 - p)$ exceed 5 for both samples. We therefore use the normal curve and find that the appropriate figure for the first parentheses for a 95% confidence interval is 1.96.

To find the standard error of the statistic $p_A - p_B$, we use Formula 8.1-2, from the last lesson.

$$SE_{p_A - p_B} = \sqrt{\frac{\pi_A(1 - \pi_A)}{n_A} + \frac{\pi_B(1 - \pi_B)}{n_B}}$$

Since we have no reason to assume π_A is equal to π_B, we will substitute p_A and p_B in the formula.

$$SE_{p_A-p_B} = \sqrt{\frac{p_A(1-p_A)}{n_A} + \frac{p_B(1-p_B)}{n_B}} = \sqrt{\frac{(0.2)(0.8)}{100} + \frac{(0.1)(0.9)}{100}} = 0.05$$

An approximate 95% confidence interval for $\pi_A - \pi_B$ is, therefore:

$$0.10 \pm (1.96)(0.05) \quad \text{or} \quad 0.10 \pm 0.098$$

Note that this is an approximate confidence interval, because it was necessary to substitute p_A and p_B for π_A and π_B in the formula for the standard error. As in the one-sample case, it is possible to use a conservative approach and calculate $SE_{p_A-p_B}$ by the formula:

$$SE_{p_A-p_B} = \sqrt{\frac{(0.5)(0.5)}{n_A} + \frac{(0.5)(0.5)}{n_B}}$$

If this is done, a reported 95% confidence interval will be *at least* a 95% confidence interval.

CONFIDENCE INTERVALS FOR THE DIFFERENCE BETWEEN MEANS

Suppose that instead of determining whether each student scored above or below 600 we had recorded the actual scores. The question might be, "What is the difference between the average score obtained by students studying with the aid of manual A and the average score of those using manual B ($\mu_A - \mu_B$)?" Let us assume that it is reasonable to conclude that the variability of students' scores would be about the same using either manual and that it is only the average level of the scores that differs. We also might expect the scores to be normally distributed in both instances. For the purpose of this example, let us suppose that there are only a total of 27 students and that for some reason they were not divided as evenly as possible. There were 15 students using manual A and 12 using manual B. When the 27 students took the exam, the results were as follows.

Group	n	ΣX	ΣX^2	Σx^2	$\overline{X}$	s
A	15	7,950	4,423,500	210,000	530	122.5
B	12	6,120	3,271,200	150,000	510	116.8

Can you now set a 95% confidence interval for $\mu_A - \mu_B$? Remember that you are to assume that σ_A equals σ_B. Before proceeding, try to find the solution to see what problems arise.

We will figure the confidence interval, using the same approach as before:

Parameter = observed statistic $\pm$ (appropriate tabled value)($SE_{\text{observed statistic}}$)

This time the statistic we observe is $\overline{X}_A - \overline{X}_B$, or $530 - 510$, or 20 points.

Where will the value to go into the first set of parentheses come from? Since you were told that the scores would be normally distributed, you expect the sample means and, therefore, the difference between sample means to be normally distributed. Still, you will not be able to make use of the table of the standard normal curve to find the value for the first parentheses. Why? The reason is that you are unable to get an exact z score, because you are unable to figure the true standard error for the denominator.

The true standard error of the sampling distribution requires σ. Since σ is unknown and must be estimated from the information in the samples, we will need to refer to the table of the t distribution (Table E-7, Appendix E). To do so, we will need to use the proper number of degrees of freedom when entering the table. For the one-sample case, df was $n - 1$. For the two-sample case, df will be $(n_A - 1) + (n_B - 1)$, or $(n_A + n_B - 2)$. For this example, we will look at the table of the t distribution with 25 df. The number appropriate for the 95% confidence interval is found by looking at the column labeled ".05" in the two-tail listings of proportion beyond t. Accordingly, we decide that the number 2.06 will go in the first parentheses.

For the second parentheses, we need the standard error of the observed statistic, or $SE_{\bar{x}_A - \bar{x}_B}$. The general formula is Formula 8.1-3.

$$SE_{\bar{x}_A - \bar{x}_B} = \sqrt{\frac{\sigma_A^2}{n_A} + \frac{\sigma_B^2}{n_B}}$$

Since we do not know σ_A or σ_B but are assuming that they are equal, we will need to substitute s as our best estimate of σ. We learned earlier that when we assume σ_A equals σ_B, we combine information from the two samples and compute s as follows:

$$s = \sqrt{\frac{\Sigma x_A^2 + \Sigma x_B^2}{n_A + n_B - 2}} \text{ or } \sqrt{\frac{210{,}000 + 150{,}000}{15 + 12 - 2}} \text{ or } 120$$

We then estimate the standard error of the difference between means as follows:

$$SE_{\bar{x}_A - \bar{x}_B} = \sqrt{\frac{(120)^2}{15} + \frac{(120)^2}{12}} = 46.5$$

The 95% confidence interval for the difference between the average score of students using manual A and those using manual B is, therefore:

$$20 \pm (2.06)(46.5) \quad \text{or} \quad 20 \pm 95.8$$

Notice that although in this case $\bar{X}_A$ was 20 points higher than $\bar{X}_B$, the confidence band is so wide that one would not be at all confident that μ_A is higher than μ_B. The 95% confidence band extends all the way from a difference of 115.8 in favor of A to a difference of 75.8 points in favor of B.

HYPOTHESIS TESTS INVOLVING THE DIFFERENCE BETWEEN PROPORTIONS

Let us now return to proportions as our example and see how hypothesis testing is applied in the two-sample case. Suppose that instead of asking what the difference is between π_A and π_B, we ask whether there is a difference. In this case, instead of setting a confidence interval for $\pi_A - \pi_B$, we will test the null hypothesis that $\pi_A - \pi_B$ equals zero against the alternate hypothesis that $\pi_A - \pi_B$ does not equal zero. If we wish, we can make this a directional test just as in the one-sample case. In fact, all of the concepts and the rationales used for the one-sample case are directly applicable.

Let us go back to the earlier example in which we were interested in the proportion of students scoring above 600 on the MCAT after reviewing using two different study manuals. Let us use the same data, but this time let us ask the question, "Is there a difference between π_A and π_B?'' We use a two-tail test and a 0.05 level of significance to test the null hypothesis that $\pi_A - \pi_B$ equals zero.

To illustrate a point, this time let us make the samples different sizes ($n_A = 100$, $n_B = 50$), while keeping the sample proportions unchanged ($p_A = 0.2, p_B = 0.1$). What will the sampling distribution of differences between proportions be like when the null hypothesis is true? First, can we again assume a normal shape? To decide this, we need to know whether $n\pi$ and $n(1 - \pi)$ are equal to or greater than 5 for each sample. Since π is unknown, we will have to substitute p when making this determination. However, should we use each sample p separately, or should we combine the data to find a single p? The sampling distribution is being determined under the assumption that the null hypothesis is true, or that π_A equals π_B. Accordingly, we should pool the information from both samples to get the best estimate of this common proportion π. Doing so, we find that 25 out of the 150 (0.167) received scores above 600. Our best estimate of the common π is, therefore, 0.167. Since this is a sample proportion, we will call it p. We then determine that np and $n(1 - p)$ are greater than 5 for both samples and that the sampling distribution of the difference between sample proportions will be normal. Notice that if we had simply averaged p_A and p_B, we would have estimated π to be 0.15. This would have given equal weight to the information from two unequal-sized samples.

Having determined that the shape of the sampling distribution is normal, we next consider its central tendency. Assuming that the null hypothesis is true, we expect the average of all observed differences to equal zero. So far, we have determined that differences will be normally distributed about a mean of zero. The only remaining problem is to determine the variability of these differences. The variablity of the sampling distribution will be represented by the standard error of the difference between proportions as expressed in Formula 8.1-2:

$$SE_{p_A - p_B} = \sqrt{\frac{\pi_A(1 - \pi_A)}{n_A} + \frac{\pi_B(1 - \pi_B)}{n_B}}$$

We will have to substitute our best estimate p (0.167) for both π_A and π_B. We get:

$$SE_{p_A - p_B} = \sqrt{\frac{(0.167)(0.833)}{100} + \frac{(0.167)(0.833)}{50}} = 0.065$$

Now, to see where our observed statistic falls on the sampling distribution based on the null hypothesis, we will change the observed statistic to a z score.

$$z = \frac{(p_A - p_B) - 0}{SE_{p_A - p_B}} = \frac{0.2 - 0.1}{0.065} = 1.54$$

Since we have decided to use an alpha level of 0.05, we refer to a table of normal curve and find that the z value 1.96 leaves a total of 5% in the two tails of the normal curve. Since our value is neither bigger than $+1.96$ nor less than -1.96, we will not reject the null hypothesis. We conclude that our data fail to provide evidence of a difference between π_A and π_B.

Once again, it should be noted that the z score computed is only approximate, since p was substituted for π in the computations for $SE_{p_A - p_B}$.

HYPOTHESIS TESTS INVOLVING THE DIFFERENCE BETWEEN MEANS: INDEPENDENT SAMPLES

The logic of hypothesis tests involving the difference between means is the same as for the difference between proportions. You will have to decide, however, whether it is reasonable to assume σ_A equals σ_B. If so, you will need to find the best estimate of this common sigma, as discussed in the previous lesson. The estimate s is then substituted for both σ_A and σ_B in the formula for the standard error. The t distribution will be used with $n_A + n_B - 2 \; df$.

In those instances in which it is not assumed that σ_A equals σ_B, s_A and s_B are computed and substituted for σ_A and σ_B in the formula for the standard error. The t distribution will again be used, but the appropriate number of degrees of freedom can no longer be calculated simply by $n_A + n_B - 2$. Some allowance must be made for the differences in sample sizes and σ's. The formula is rather involved. It is shown as Formula 8.2-1.

$$df = \frac{[(SE_{\bar{X}_A})^2 + (SE_{\bar{X}_B})^2]^2}{\dfrac{(SE_{\bar{X}_A})^4}{n_A} + \dfrac{(SE_{\bar{X}_B})^4}{n_B}} \qquad \text{8.2-1}$$

where $SE_{\bar{X}_A} = \dfrac{s_A}{\sqrt{n_A}}$ and

$SE_{\bar{X}_B} = \dfrac{s_B}{\sqrt{n_B}}$

The calculations from this formula will not generally result in an even number. You will have to round the result off to the nearest whole number to obtain the appropriate value.

To test your understanding of hypothesis tests involving means, go back to the earlier example, in which average MCAT scores resulting from the use of the two manuals, A and B, were compared. The data were:

	n	ΣX	ΣX^2	Σx^2	$\bar{X}$	s	SE
A	15	7950	4,423,500	210,000	530	122.5	31.6
B	12	6120	3,271,200	150,000	510	116.8	33.7

Perform a two-tail test of the null hypothesis that μ_A equals μ_B, using a 0.05 level of significance. Solve the problem twice, once assuming σ_A equals σ_B and once without making this assumption.

The first time you should get:

$$t = \frac{20}{46.5} \text{ with } 25 \; df$$

The second time you should get:

$$t = \frac{20}{46.2} \text{ with 26 } df$$

Notice that these two results are very nearly the same. This might be expected, since neither the sample sizes nor the sample standard deviations are far apart.

The examples that have been given for making inferences concerning the difference between two population means have deliberately been chosen to illustrate the complications that arise when sigma is unknown. If sigma is known for each population, the z distribution is used rather than the t distribution, and there are no complications involving pooled estimates of sigma and computations of degrees of freedom. In this case, the solution is straightforward. Calculate z using Formula 8.2-2.

$$z = \frac{(\bar{X}_A - \bar{X}_B) - 0}{\sqrt{\dfrac{\sigma_A^2}{n_A} + \dfrac{\sigma_B^2}{n_B}}} \qquad \text{8.2-2}$$

Although the problem is very straightforward when σ is known for each population, this is a condition that, unfortunately, is seldom met in actual practice. As a result, the t distribution is usually used when two population means are compared. As a result, it has become commonplace to speak of the "t test for differences between means." In fact, statisticians often speak of "doing a t test." You may have heard or read these phrases.

The formula for the t score for the difference between means when sigma is unknown but assumed to be *equal* in the two populations is shown as Formula 8.2-3.

$$t = \frac{(\bar{X}_A - \bar{X}_B) - 0}{\sqrt{\dfrac{s^2}{n_A} + \dfrac{s^2}{n_B}}} \qquad \text{8.2-3}$$

where s is computed using Formula 8.1-4 and
         df equals $n_A + n_B - 2$.

The formula for the t score for the difference between means when sigmas are unknown but assumed to be *unequal* in the two populations is shown as Formula 8.2-4.

$$t = \frac{(\bar{X}_A - \bar{X}_B) - 0}{\sqrt{\dfrac{s_A^2}{n_A} + \dfrac{s_B^2}{n_B}}} \qquad \text{8.2-4}$$

with df calculated using Formula 8.2-1.

On occasion you may also have heard of a *matched pair t test,* or a *correlated t test,* as opposed to an *independent sample t test.* What has been described to this point is referred to as an independent sample t test, in that there is no attempt to match or pair the members of the two samples.

HYPOTHESIS TESTS INVOLVING THE DIFFERENCE BETWEEN MEANS:
PAIRED SAMPLES

In the situation described in the preceding section, it was assumed that an indepen-dent sample was taken from each of the two populations. There are many studies in which matched pairs are chosen, with one member of each pair assigned to each of the treatment groups. In other instances, a subject serves as his own control by being measured before and after treatment. In such designs, the formula for the $SE_{\bar{x}_A - \bar{x}_B}$ is not appropriate, since there will be a correlation between sample means. The analysis of data from such experi-ments is usually referred to as the matched pair, or correlated, t test. The approach is clever and simple.

Suppose you want to know whether a therapeutic drug treatment will reduce blood pressure. You choose a random sample of 16 persons with high blood pressure and match them into pairs based on current blood pressure level: the two highest, the next two, and so forth. You then randomly assign one member of each pair to a treatment group and the other to a control group. The treatment group receives a therapeutic drug for 6 months. You then measure blood pressure levels, keeping the data in pairs. The data might look as follows where D is obtained by subtracting the reading for the member of the control group from the reading for the member of the treatment group.

Pair number	Control	Treatment	D
1	190	170	−20
2	180	150	−30
3	180	160	−20
4	170	170	0
5	190	170	−20
6	160	170	+10
7	200	150	−50
8	190	160	−30
			−160

The matched pair t test involves testing the null hypothesis that the average of all difference scores (D) is zero. This is, of course, what would be expected if the treatment has no effect. If the question is whether the treament has *any* effect, a two-tail test is used. In the majority of cases, the treatment will be expected to have an effect in a given direction, and a one-tail test will be used.

By changing the problem to one involving difference scores, you are back to the procedures for a one-sample test. You will use Formula 8.2-5.

$$t = \frac{\bar{D} - 0}{\frac{s_D}{\sqrt{n}}}$$

8.2-5

where n is the number of pairs of scores and
   df equals $n - 1$.

For this example we calculate as follows:

$$s_D = \sqrt{\frac{\Sigma D^2 - \frac{(\Sigma D)^2}{n}}{n-1}} = \sqrt{\frac{5600 - \frac{(-160)^2}{8}}{7}} = 18.52$$

$$\overline{D} = \frac{-160}{8} = -20$$

$$t = \frac{(-20) - 0}{\frac{18.52}{\sqrt{8}}} = \frac{-20}{6.55} = -3.05 \text{ with } 7 \ df$$

Using a one-tail test, the difference would be found to be significant at the 0.01 level. You will probably reject the null hypothesis that $\overline{D} \geq 0$ in favor of the alternate hypothesis $\overline{D} < 0$. When doing so, you are saying that the results show that the drug treatment has resulted in lower blood pressures for the treatment group.

INFERENCES BASED ON MORE THAN TWO SAMPLES

There will be situations in which you will be interested in testing for differences in the parameters of three or more populations. The analysis of data from such experiments will be discussed in lessons to come.

When you wish to compare π among more than two groups, the technique used is called a chi-square analysis. The analysis results in a test of the null hypothesis that $\pi_1 = \pi_2 = \pi_3 = \pi_4$, and so forth. The statistic computed is X_p^2, and this statistic is compared with values in a table of chi-square.

When the parameter of interest is μ, the procedure is called analysis of variance. The analysis results in a test of the null hypothesis that $\mu_1 = \mu_2 = \mu_3 = \mu_4$, and so forth. Instead of computing z or t, a statistic called F is computed and compared with values in an F table.

SUMMARY

At this point, congratulations are in order if you have comprehended the major concepts of these chapters on inferential statistics. You will be able to read the scientific literature with new insight. You may not be familiar with a specific statistical technique reported, but when it comes to the final conclusions, you will know the basis upon which they are made. You will know the meaning of a confidence interval that was set for the difference between the parameter values of two treatment groups. Or, alternatively, you will know that when an observed statistic is found to be significant at the 0.05 level, it is one that is so extreme as to happen 5% of the time or less when the null hypothesis is true. The volume of concepts covered has been large, and you were no doubt confused at many points along the way. Some confusion is inevitable when complex concepts are first introduced. It is hoped that as these concepts receive repeated use in subsequent lessons, they will become more meaningful to you. You may still be a bit "shaky" about some material from the most recent lessons. This material will be reinforced as you proceed to the remaining lessons.

LESSON 2 EXERCISES

1. Two different manufacturing processes (A and B) are used by two different companies to produce components for a dental x-ray machine. Occasionally, each process results in a faulty part. There is a merger of the two companies, and a decision is made to use a single manufacturing process. Many factors are included in the decision, but one important consideration is any difference in the proportion of faulty parts resulting from processes A and B. It is generally felt that process B produces a lower proportion of bad parts, but the amount of the difference is unknown. A random sample of 400 parts is selected from each manufacturing process, and the number of faulty parts is determined by careful testing. For process A, there were 36 bad parts; for process B there were 20. Find a 99% confidence interval for the difference in the proportion of bad parts between the two processes.

2. Two school systems that are in similar communities differ in that system A has comprehensive sex education programs starting in the first grade and continuing all the way through high school. System B has no formal program, but is considering adopting the same program as that of system A. Those who are opposed say that the program will lead to earlier sexual experience. To provide data to support this argument, random samples of 100 tenth-grade students are chosen from each school population. They are asked to provide answers to questions about their sexual experiences in a setting that ensures anonymity. Based on their responses, 22 students from school system A were reported to be sexually active, whereas the number from system B was 18. Use a 0.05 alpha level to test the appropriate hypothesis.

3. The same 100 students in each group were also asked to indicate the number of times they had dated during the preceding 30-day period. The results were as follows:

	n	s	$\overline{X}$
System A	100	2.1	5.1
System B	100	1.8	4.8

Test the appropriate hypothesis at an alpha level equal to 0.05. Do not assume $\sigma_A = \sigma_B$.

4. Ten of the students from each group were randomly chosen and tested for their knowledge of contraceptive practices. Scores on the test ordinarily have a normal distribution, with the number of correct responses ranging from 40 to 120. Two students from school system A failed to show up for the test. The results were as follows:

	n	ΣX	ΣX^2
System A	8	722	66168
System B	10	808	67090

Test the appropriate hypothesis at an alpha level equal to 0.05. Assume $\sigma_A = \sigma_B$.

5. After much debate, the program is adopted by school system B. After 1 year, the same ten students from B are retested. The scores are as follows:

Student	Original score	Score 1 year later
1	72	82
2	84	98
3	90	94
4	106	110
5	60	92
6	90	98
7	63	86
8	72	79
9	81	92
10	90	96

Test the appropriate hypothesis at the 0.05 level of significance.

CHAPTER 9

Comparisons of population proportions using chi-square

Lesson 1 □ Tests of goodness of fit

As your first introduction to hypothesis testing, you learned how to determine whether an unknown proportion in a test population (π_T) differs from a known proportion in a standard population (π_S). The procedure can be called a one-sample test of a proportion. In one-sample hypothesis testing, a sample result from a test population is compared with a known parameter value of a standard population. The parameter value of the standard population is known from some external source before the start of the experiment. For example, we might know that 20% (π_S) of patients with a given disease will die when traditional treatment methods are used. We then run an experiment in which the traditional treatment is modified in some manner. We would like to know whether the proportion dying (π_T) when this modified treatment is used differs from π_S. The sample proportion of patients who have died is computed and compared with the sampling distribution expected from the standard population. For small samples, the sampling distribution is calculated by using the binomial formula. For larger samples, the normal distribution can be used as an approximation to the binomial distribution.

Once you know the sampling distribution from the standard population, you determine where the sample proportion from the test population would fall on this distribution. For example, if 10% of the sample patients die, we ask, "What is the probability of as few as 10% dying, given that this sample was taken from the standard population, where it is known that 20% die?" A sample proportion that falls at one of the extremes of the sampling distribution provides evidence that the parameter value of the test population differs from that of the standard population. In this event, we conclude that π_T does not equal π_S.

Consider now the problem we would face if there were several outcomes of treatment that could be identified. For example, 6 months following the beginning of treatment we might be able to classify all patients as falling into one of the following categories:

1. Death
2. Permanent crippling

3. Minor residual effects
4. Complete recovery

Suppose we know that for the standard method of treatment the proportion of patients in each category is, respectively, 0.20, 0.30, 0.40, and 0.10. We can once more treat a sample of patients with the modified treatment and compare the sample results with the traditional proportions. We could ask questions such as, "Is the proportion who die the same?" or "Is the proportion who are crippled the same?" In each case, we would have a one-sample test of a proportion.

What is the matter with this procedure? As the number of classes for the characteristic increases, so does the number of one-sample tests to be made. Obviously, the amount of work is greater, but that is not the critical problem. What is it? The answer is that making multiple comparisons increases the probability that one or more of the comparisons will result in a Type I error. This statement should make sense intuitively. If four tests are run and each has a 0.05 probability of resulting in a Type I error, then the probability that at least one will result in a Type I error will exceed 0.05. We need not be concerned here about how much the probability increases. (Since the tests are not all independent, computing the probability is complicated.) We need only be aware that the more comparisons we make, the greater the probability will be that at least one of the comparisons will show a falsely significant difference.

What is required is a method of simultaneously comparing all of the proportions in the sample with all of the proportions from the standard population. We need a summary statistic that measures the differences between all of the observed values and the values expected when there is actually no difference resulting from the modified treatment method. The question we need to ask is, "Do our sample proportions differ from the standard proportions more than can reasonably be attributed to chance?" The procedure for answering this question is one that uses frequencies within each sample, rather than proportions. This causes no particular problems, since we can change an expected proportion into an expected frequency by multiplying the expected proportion by the sample size.

Suppose that we have 200 patients treated with the modified method and that the results are as follows:

Outcome	n
Death	30
Permanent crippling	40
Minor residual effects	90
Complete recovery	40
TOTAL	200

We need to know whether these observed frequencies are so different from those expected as to cause us to believe that the differences are not due merely to chance. You can calculate expected frequencies by multiplying the respective proportions from the standard population by the sample size of 200. It is convenient to identify these expected

Table 9-1. Observed and expected outcomes of treatment

	Number of outcomes
Death	30 (40)
Permanent crippling	40 (60)
Minor residual effects	90 (80)
Complete recovery	40 (20)
TOTAL	200

frequencies by placing them in parentheses in the lower right-hand corner of each cell of a table. Our table, then, appears as Table 9-1. Are the differences between sample frequencies and expected frequencies too great to be due solely to chance?

We need to compute some kind of statistic that summarizes the differences between these observed values and the expected values. Then we need to know the sampling distribution for this summary statistic when there is, indeed, no effect caused by the modification. We can then compare our observed statistic with the sampling distribution to see if it is an extreme value for that distribution. If it is extreme, we will reject the null hypothesis and conclude that the population proportions for the modified treatment differ from those of the standard treatment.

The required statistic is called Pearson's chi-square statistic, named after its inventor. (The Greek letter *chi* is pronounced kī.) It is often designated by the symbol X_p^2. When sample size is sufficiently large, this statistic has a sampling distribution that approximates a known *theoretical* distribution called the chi-square distribution. Unfortunately, statistics books frequently fail to distinguish between Pearson's chi-square *statistic* computed for a sample and the *theoretical* chi-square distribution. The Greek letter for chi with a power of 2 (χ^2) is sometimes used to refer to Pearson's chi-square statistic and at other times to the theoretical chi-square distribution. To avoid ambiguity, in this text the symbol χ^2 will always refer to the chi-square distribution. In contrast, X_p^2 will be used for Pearson's chi-square statistic. Keep in mind that, like a normal distribution or a *t* distribution, the chi-square (χ^2) distribution is a theoretical distribution with an exact formula for its shape. Under certain conditions, Pearson's chi-square statistic (X_p^2) has a sampling distribution that approximates the χ^2 distribution.

Let us delay for now an examination of a table giving probabilities for the distribution of χ^2. Instead let us first see how the statistic X_p^2 is computed. If you follow the discussion carefully, the logic of the computations should make sense.

How shall we summarize the differences between observed values and expected values? Your first inclination is probably to subtract the expected from the observed and add all these differences. The greater this sum of differences, the greater the evidence against the null hypothesis. Good thinking. The problem is, it does not work. If you try this procedure, you face a problem similar to what happened when we tried to find the sum of deviations from the mean when computing the variance: the sum of the differences equals zero. Our solution to this problem will be the same as before. We will *square* the differences before adding. It should be readily apparent that the magnitude of the sum of the squared differences will be influenced by whether the null hypothesis is true. If the null hypothesis is true, any differences are due to chance, and the sum of squared differences can be expected to be of limited magnitude. On the other hand, if the modified treatment actually results in different proportions for the various outcomes, the differences between observed frequencies and expected frequencies will be greater, and the sum of squared differences can be expected to be quite large.

It may not be as apparent that the sum we get is influenced by two other factors in addition to whether the null hypothesis is true. Can you identify these? What do you think would happen to the sum of the squared differences if the sample size chosen for the modified treatment were larger? If your answer is "a larger sum of squared differences might be expected," your intuition is correct. To standardize the statistic X_p^2 so that it is comparable for various sizes of samples, we need to somehow adjust the value for sample size. The total sample size for the modified treatment is divided among the various outcomes of treatment. The expected values indicate how the total sample size is expected to be divided among the total numbers of cells. Accordingly, the adjustment procedure that accomplishes the purpose of equating for various sizes of samples is to divide each squared difference by the expected value for that cell. The formula can then be expressed as Formula 9.-1.1

$$X_p^2 = \sum \frac{(f_o - f_e)^2}{f_e}$$

9.1-1

where f_o stands for observed frequency and
f_e stands for expected frequency.

For the data of our example, we would have the following:

$$X_p^2 = \frac{(30 - 40)^2}{40} + \frac{(40 - 60)^2}{60} + \frac{(90 - 80)^2}{80} + \frac{(40 - 20)^2}{20}$$

$$X_p^2 = 2.5 + 6.67 + 1.25 + 20.0$$
$$X_p^2 = 30.42$$

We now have a statistic that is an indication of how much our sample data deviate from the values to be expected under the assumption that the null hypothesis is correct. We refer to this statistic as $X_p{}^2$. Under conditions to be explained later, we expect $X_p{}^2$ to have a sampling distribution approximately the same as the theoretical χ^2 distribution.

At this point you may be thinking: "Wait a minute, you said that there were two extraneous factors affecting $X_p{}^2$ and we have only adjusted for one, sample size." You're right. The other extraneous variable is the number of squared differences entering into the computations. If another class for outcomes had been included, this would have added another squared difference to the computations. Somehow, we must take into account the number of classes we have used in reporting the data. The solution is once more familiar. It is to use the concept of degrees of freedom when comparing our computed $X_p{}^2$ with the tabled values of χ^2.

When figuring degrees of freedom for a *t* test of a sample mean, you learned to use $n - 1$ as the figure for *df*. The usual explanation given is that the mean for any group of scores is dependent on the sum of the scores. Then, if the sum of the scores is some fixed value, only $n - 1$ of the scores are free to vary. Once $n - 1$ of the scores are determined, the last score has to be a certain value. In this sense, *df* equals the numbers of things observed that are free to vary for any specified value of the statistic.

Now let us examine how this concept of degrees of freedom can be applied in the case of computations of $X_p{}^2$. In our present example, what restrictions have we placed on data? Only one. The column total must add up to 200. How many of the cell frequencies are therefore free to vary with this restriction in mind? The answer, of course, is three. Once three of the cell frequencies are known, the fouth has to be a certain value.

Now let us assume that, for our example, $X_p{}^2$ can be expected to be distributed as chi-square (χ^2). We therefore use the χ^2 distribution to determine whether our observed $X_p{}^2$ with 3 *df* is greater than can be expected to happen by chance when sampling from the standard population. A table of probabilities for χ^2 appears in Appendix E as Table E-8. A discussion of the theoretical basis for the χ^2 distribution will follow in the next chapter. For now, you need only know how to use a table containing the probabilities representing the distribution.

As with *t* distributions, there is a different χ^2 distribution for each value of *df*. As a result, only selected probabilities are included and listed at the top of the table. These are the probabilities of values as great as or greater than the χ^2 values in the body of the table. Suppose that we have decided to use an alpha level of 0.05 to test the null hypothesis that there is no difference in the proportions of outcomes as a result of the modified treatment. The value listed for 3 *df* in the column headed ".05" is 7.81. Are we looking for a value larger or smaller than 7.81 as evidence against the null hypothesis? Thinking back to our computation of $X_p{}^2$, we recall that we expect $X_p{}^2$ to be large when the modification causes real differences in proportions. This means that we will reject the null hypothesis if the computed $X_p{}^2$ value is equal to or greater than 7.81. Our value of 30.42 exceeds 7.8. In fact, our computed value is so large that it exceeds the value of 16.27, which would happen by chance only one time in 1000 when the null hypothesis is true. We therefore reject the null hypothesis and conclude that the proportions of outcome are affected by the modification of treatment. It should be noted that the rejection of the null hypothesis does not mean that all of the proportions differ from the proportions for the corresponding classes using the traditional treatment. Our test is an overall test that says that, among the various proportions, some proportions are different between the two treatment methods.

Having found that differences exist somewhere, we will often find it meaningful to further examine the data to see which observed frequencies are significantly different from those expected. Usually, this amounts to a post hoc examination of the data rather than a test of an experimental hypothesis. As a result, any finding will have to be regarded as tentative.

Remember the figures that went into the computations for X_p^2. They were:

$$X_p^2 = 2.5 + 6.67 + 1.25 + 20.0$$

The major contributor to the large value of X_p^2 is the fourth class. However, the second class—"permanent crippling"—also contributes a substantial amount to the total value of X_p^2. It is possible to combine classes in varying ways to determine which classes have frequencies that appear to be significantly different from those expected. For example, we could look at the category "permanent crippling" and combine all the remaining classes. We would have the following:

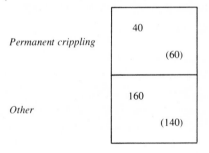

We again calculate X_p^2:

$$X_p^2 = \frac{(-20)^2}{60} + \frac{20^2}{140}$$

$$X_p^2 = 6.67 + 2.86$$
$$X_p^2 = 9.53 \text{ with 1 } df$$

This value turns out to be significant at the 0.005 level, and we conclude that the proportion of permanently crippling injuries differs for the two treatment methods. Once again, remember that the conclusion is tentative, based on this post hoc examination of the data.

In certain cases, individual comparisons may have been planned from the start. In this instance, the classes can be combined in such a way as to make each of the planned comparisons. The results are then interpreted as in any other hypothesis-testing situation. In such a case, the computation of the overall $X_p{}^2$ is unnecessary.

SOURCES OF EXPECTED FREQUENCIES

The procedure for comparing the observed frequencies in a sample, with expected frequencies based on some external factor, is generally referred to as a *goodness of fit test*. Using this procedure, you can determine how well sample data fit with expected values based on some external set of proportions. These external proportions can be based on (1) empirical data, as in the case of comparison with a traditional method of treatment; (2) theoretical values based on some existing theory; or (3) any hypothetical proportions that are meaningful to the investigator. The example just given illustrates the first situation. Let us now look at an example for each of the other cases. This will also provide additional practice in calculating and interpreting $X_p{}^2$.

An example in which expected proportions are based on *theory* can be found in discussions of mendelian theories of heredity, in which offspring are expected to exhibit traits according to certain ratios. For example, suppose we wish to study the hereditary distribution of a trait, such as color of coat, for an animal, such as the guinea pig. Let us assume that each animal has two genes for hair color and that one of these two genes is "selected" in a random fashion to be passed on to each offspring. Each animal thus receives one gene for hair color from each parent. Suppose that when both genes are for black coats, the guinea pig has a black coat. Similarly, when both genes are for white coats, the animal has a white coat. When one gene is for black coats and one for white coats, a spotted coat results. All of these statements might represent a model of how we believe color of coat is hereditarily distributed among guinea pigs. A test of the model can be made by examining the proportion of the colors of coats among successive generations of animals. Pearson's chi-square statistic will provide a means of testing to see whether the proportions observed depart significantly from those expected.

Let us start with a number of pairs of guinea pigs where the male of each pair is black and the female white. According to the model, each male can be represented as *"bb"* and each female as *"ww."* Since each will contribute a gene to each offspring, all first generation offspring will receive a gene for black coats from their fathers and a gene for white coats from their mothers. All first generation offspring will therefore be spotted. When these first generation offspring are paired, what can be expected for second generation offspring? Each parent has one gene for black coats and one gene for white coats. One of the two genes is selected at random for passing on to offspring. The chance that both genes received by a second generation offspring will be black is $(^1/_2)^2$, or $^1/_4$. Similarly, the chance that both will be white is also $^1/_4$. The remainder, or $^1/_2$, will receive one gene for black coats and one gene for white coats.

If the model we have described is correct, we expect that color of coat among second generation offspring will occur in the proportions: $^1/_4$ black, $^1/_2$ spotted, and $^1/_4$ white. Suppose that the result of the first pairing is indeed a first generation of spotted guinea pigs. When these guinea pigs are paired, there are a total of 84 offspring, of which 16 are black, 48 are spotted, and 20 are white. Do these frequencies cast doubt upon our model?

To calculate X_p^2 we need to change the expected proportions into expected frequencies for each class. We do so by multiplying the expected proportion by the total number of offspring. Once more, putting expected values in parentheses in a box that includes the observed frequency, we have the following.

Black	16 (21)
Spotted	48 (42)
White	20 (21)

We calculate X_p^2 just as in the previous example:

$$X_p^2 = \frac{(16 - 21)^2}{21} + \frac{(48 - 42)^2}{42} + \frac{(20 - 21)^2}{21}$$

$$X_p^2 = \frac{25}{21} + \frac{36}{42} + \frac{1}{21}$$

$$X_p^2 = 2.09 \text{ with 2 } df$$

Referring to a table of χ^2, we find that this value is exceeded more than 25% of the time by chance and cannot be considered an unusual result. We conclude that the results are not inconsistent with the model we have proposed. Pearson's chi-square statistic has been used in a situation in which expected values are based on theory.

As an example of the use of X_p^2 in a situation in which expected values are based on *hypothetical values of interest* to the investigator, consider the following case. A hospital administrator decides that it would be feasible to go ahead with the plans for a "burn unit" if a certain level of private funding could be anticipated. A list of past donors is available. An appeal will be made for contributions of 100, 250, or 1000 dollars. His financial planner estimates that of those on the list, 50% will give nothing, 30% will give $100, 10% will give $250, and 10% will give $1000. If the responses are in these percentages, enough funds will be available. A sample of past donores is chosen and asked what they would be willing to contribute. The observed proportions are then compared with the "hoped for" (expected) proportions. A sample of 200 past donors indicates that they would be willing to contribute as follows: 125 are not interested in the project, 50 are willing to contribute $100, 15 say that they will contribute $250, and 10 indicate that they will contribute $1000. Do these frequencies show that the financial planner has been too optimistic in estimating the response to an appeal? Expected values are found by multiplying the proportions estimated by the financial planner by the sample size of 200. We calculate X_p^2 as follows.

Nothing: 125 (100)

$100: 50 (60)

$250: 15 (20)

$1000: 10 (20)

$$X_p^2 = \frac{25^2}{100} + \frac{(-10)^2}{60} + \frac{(-5)^2}{20} + \frac{(-10)^2}{20}$$

$$X_p^2 = 14.17 \text{ with 3 } df$$

When we refer to the table of χ^2, we find that for 3 df, values this large occur less than 5 times in 1000. We conclude that the estimates of the financial manager are unrealistic and decide that if the burn unit is to be built, other sources of funds will have to be found.

Consider for a moment the differences in the three examples given for the goodness of fit test. The first compares the results of a new method of treatment with empirical results using a traditional treatment. The second compares traits of offspring with results expected based on a theory of heredity. The third compares sample responses to a request for contributions with the results that an administrator hopes can be achieved. What conclusion do you draw about the goodness of fit test? If your reaction is that it appears to be applicable any time expected frequencies can be identified before the data gathering, you are not far from being correct. There are, however, two conditions that must be met for the use of the χ^2 distribution in evaluating X_p^2. These will be discussed as part of the next lesson.

SUMMARY

In this lesson you learned a method of answering questions such as the following: "Do the several outcomes of a modified method of treatment occur in proportions different from those for a traditional method of treatment?'' The procedure is a general one, which is often referred to as a goodness of fit test application of chi-square. It can be thought of as an extension of the one-sample test of a proportion to include situations in which the characteristic of interest results in multiple classes rather than just presence or absence of a trait. Observed frequencies within each class are compared with expected frequencies. The expected frequencies can arise from past empirical data, from some existing theory, or from some hypothesized proportions of interest.

The statistic computed is called Pearson's chi-square statistic, for which we have used the symbol $X_p{}^2$. The formula was given as Formula 9.1-1.

$$X_p{}^2 = \sum \frac{(f_o - f_e)^2}{f_e}$$

where f_o is the observed frequency,
f_e is the expected frequency, and
summation is over all classifications or cells.

Under certain conditions, the statistic has a distribution that approximates a known theoretical distribution called the chi-square (χ^2) distribution.

In the next lesson, the conditions for using the χ^2 distribution to evaluate $X_p{}^2$ will be explained more fully. For now, we need only know how to use the table. It was explained that to do so we once more will need to make use of the concept of *df*. To find *df* for $X_p{}^2$ for the goodness of fit test, we will ordinarily subtract 1 from the number of classes or cells. (An exception will be described in the next lesson.) Sample values of $X_p{}^2$ that exceed the tabled values of χ^2 constitute evidence against the null hypothesis.

In the next lesson you will learn more about the goodness of fit application of χ^2, including a method for testing to see whether a population has a normal distribution.

LESSON 1 EXERCISE

For several years a state that requires motorcycle riders to wear helmets has kept records of injuries involving motorcycle accidents. Injuries to riders have been classified according to severity. Injuries have occurred in the following proportions.

Result	Proportion
Death	0.05
Paralysis	0.10
Concussion	0.35
Other	0.50

An adjacent state is considering passing a helmet law. Data are available concerning motorcycle accidents for this state as well, but classifying injuries requires locating the riders. Since this will be expensive, a random sample of 1000 accidents is chosen for investigation. Of these, there were 65 deaths, 112 cases of paralysis, 359 concussions, and 464 other injuries. Is there evidence that injuries are more severe in the state without the helmet law? Should you interpret the data for a one-tail test or a two-tail test?

Lesson 2 □ More applications of the goodness of fit test

This second lesson dealing with χ^2 begins with a description of the use of X_p^2 in the case where proportions are divided between only two classes. You will see that this is an alternate solution to what we earlier called the one-sample test of a proportion. The example used will provide an opportunity to examine four important concepts regarding the use of X_p^2: (1) the interpretation of X_p^2 for directional tests, (2) the conditions under which χ^2 provides an adequate approximation for X_p^2, (3) the relationship between χ^2 and z, and (4) a modification in the calculation of X_p^2 that is appropriate in the case of 1 *df*. The lesson will conclude with a description of the use of X_p^2 to test for normality in a population.

CHI-SQUARE FOR THE CASE OF TWO CLASSES

Earlier you learned to make inferences about proportions for cases involving only two classes (for example, heads or tails, survival or death). You learned to answer questions like the following: "If the survival rate is only 0.5 for a given disease, what is the probability that of 20 randomly chosen patients, 15 or more will survive?" To answer the question, you could use the binomial formula to find the individual probabilities of 15, 16, 17, 18, 19, and 20 surviving, and then add these probabilities. The amount of work would be fairly great. Instead, you learned that it is easier to make use of the normal curve approximation to the binomial expansion. To do so, you determine where the proportion 0.75 (15 out of 20) falls on a sampling distribution that is normally distributed, with a mean of 0.5 (π) and a standard error of 0.1118 ($\sqrt{\pi(1 - \pi)/n}$). You change 0.75 to a standard score as follows:

$$z = \frac{0.75 - 0.5}{0.1118} = 2.236$$

You look up this z score in the one-tail section of the table of the normal curve (Table E-6, Appendix E) and find that a figure of 15 or more survivors will occur about 1.25% of the time.

Now see if you can do the same problem using the goodness of fit application of χ^2. First, with a sample size of 20, what are the expected frequencies for survivors and nonsurvivors? Since the null hypothesis is that $\pi \le 0.5$, the expected frequency is the same for each class, 10 and 10. Our table looks like this:

Survivors	15 (10)
Nonsurvivors	5 (10)

The computation of X_p^2 is easy.

$$X_p^2 = \frac{(15 - 10)^2}{10} + \frac{(5 - 10)^2}{10}$$

$$X_p^2 = 2.5 + 2.5 = 5.0 \text{ with } 1 \text{ } df$$

Referring to a table of χ^2, you find that values of χ^2 this great occur about 2.5% of the time. What happened? You were told that the two procedures were alternative solutions to the same problem, yet for one the probability found was 0.0125 and for the other it was 0.025.

Directional tests using X_p^2

The fact that one probability is twice the other is a clue to the difference in the two solutions. In the solution using the normal curve we were looking at a one-tail probability. It appears that the solution using Pearson's chi-square statistic must therefore be a nondirectional test. If you think about the calculation of X_p^2, you will see that this is the case. The fact that differences between observed and expected values are squared means that all values become positive. The same value for X_p^2 would result from our example if the difference were just as extreme in the other direction (5 survivors rather than 15). In this sense, a test using X_p^2 is usually a nondirectional test. If the logic of the experiment calls for a directional test, the probabilities read from the table of χ^2 are not the appropriate probabilities.

It is possible to make a directional test for this example using X_p^2. The first step is to see if the data are in a direction that might lead one to reject the directional null hypothesis. In our example this means finding out whether more than 50% survived. If the answer is no, of course nothing further needs to be done. If the answer is yes, X_p^2 is computed in the usual way and the probability from the table of χ^2 is halved. If X_p^2 is found to be significant at the 0.025 level for the nondirectional test, it is reported as significant at the 0.0125 level for the directional test.

Conditions for using χ^2 to evaluate X_p^2

Suppose that in the example we have been discussing, the sample size had been 6 rather than 20, and suppose that we had observed 5 patients surviving. We would still wish to test the null hypothesis that π equals 0.5. You learned earlier that in this case, the use of the normal curve approximation would not be appropriate. Instead, you would have to use the binomial formula to figure the probabilities for exactly 5 survivors and exactly 6 survivors. You would then add these two probabilities to find the probability of 5 or more survivors.

It is more difficult to calculate all of the exact probabilities using the binomial expansion than it is to make use of the normal curve approximation. It is therefore important to determine when the normal curve is an adequate approximation to the exact distribution resulting from the binomial expansion. The criterion proposed was that $n\pi$ and $n(1 - \pi)$ both had to be 5 or more. With extreme values of π (near 0.0 or 1.0), sample size had to be large for the binomial distribution to be approximately normal. With π equal to 0.5, the distribution would be approximately normal with an n as small as 10.

It should be noted that in the case where $n\pi$ and $n(1 - \pi)$ are not $\geqslant 5$, it is still possible to calculate a z score for any observed proportion. The result will be a standard score, but since the distribution will not be normal, it will not be a standard *normal* score. The use of the table of the normal curve to evaluate the z score for the observed proportion will be inappropriate.

An analogous situation exists for the use of $X_p{}^2$. In the case of small samples, the expected frequencies within cells will be small. The more extreme the expected proportions, the more serious the problem. In the case of small expected frequencies, we can still calculate $X_p{}^2$. The problem is that $X_p{}^2$ will not be distributed in the same way as χ^2, and the use of the χ^2 table to evaluate $X_p{}^2$ will be inappropriate. For the one-sample case with only two classes, we can find the exact probabilities using the binomial expansion. For the one-sample case with more than two classes, there is a similar solution, called the multinomial expansion. For the two-sample, two-class case, there is a rather frequently used solution, called *Fisher's exact test*. For the multisample, multiclass case, there are a number of mathematical solutions, depending on assumptions regarding the methods of gathering the data. In this text you will learn only the solution for the one-sample, two-class case (the binomial expansion) and the solution for the two-sample, two-class case (Fisher's exact test). The former you already know; the latter will be taken up in the next lesson.

At best, calculating exact probabilities even for small samples is tedious. At worst, as sample size and number of samples and classes increase, the calculations become not only tedious but also involved and difficult to understand. Fortunately, as the calculations become more difficult because of an increase in sample size, the distribution of $X_p{}^2$ approaches the distribution of χ^2. Instead of making all the calculations necessary for finding exact probabilities, you can calculate $X_p{}^2$ and find approximate probabilities from the table of χ^2.

Size of sample. The obvious question then becomes, "How large must the expected values be before the sampling distribution of $X_p{}^2$ approximates the χ^2 distribution?" Whatever standards are set will, of necessity, be arbitrary. As a result, various statistic books give differing suggestions. For the one-sample, two-class case, the same standard can be used as was used in determining whether to use the normal curve as an approximation of the exact binomial probabilities. You were told that both $n\pi$ and $n(1 - \pi)$ needed to be 5 or greater. This would be equivalent to saying that for $X_p{}^2$ to be distributed approximately as χ^2, expected values of both cells must be 5 or more. This standard is appropriate in the one-sample, two-class case involving only 1 df. It is overly conservative for problems with larger df. With 2 or more df, if all cell frequencies are about equal, the approximation is not bad even with expected frequencies as small as 3.

If expected frequencies are not equal, the distribution of $X_p{}^2$ does not approach the distribution of χ^2 as rapidly. Statistics books generally suggest that in the case of unequal expected frequencies, *most* expected cell frequencies should be 5 or more. If this is the case, the distribution of $X_p{}^2$ will be approximately that of χ^2, even if one cell has an expected frequency of only 1. Note that in all of this discussion we have been talking about expected frequencies, not observed frequencies. There is no restriction in terms of observed frequencies.

Independence of data. In addition to the need for sufficiently large expected frequencies, there is another condition for the use of a table of χ^2 as a means of evaluating $X_p{}^2$. This condition is also implicit in the use of the normal curve as an approximation to the binomial and in the calculation of exact probabilities using the binomial formula. It was stated explicitly when the multiplicative rule was introduced.

Do you remember that when the multiplicative rule was introduced, you were told that the probability of events *A* and *B* *both* occurring could be found by multiplying *P(A)* by *P(B)* *if A and B were independent events*. Similarly, to find the probability of a specific positive outcome for each of *n* consecutive trials, we can find the probability of a positive outcome on the first trial and raise this to the *n*th power, *if the outcome of each trial is an independent event* (the probability of a positive outcome remains the same from trial to trial).

The calculation of probabilities using the binomial formula is also based on the assumption that the probability of the event (for example, survival or cure) is constant for all events examined. For a particular experiment, this means that for each member of the sample, the outcome of the experiment must be independent of the outcome achieved for other members of the sample. This requirement of independence in the calculations of probabilities using the binomial formula is also assumed both for the use of the normal curve approximation for the binomial distribution and for use of the χ^2 distribution for the distribution of $X_p{}^2$.

The requirement of independence as a condition for the use of χ^2 probabilities when evaluating $X_p{}^2$ is a bit difficult to explain. The best way to make its meaning clear may be to return to the example given at the beginning of this lesson. We used the χ^2 distribution to evaluate $X_p{}^2$ calculated for a sample in which 15 out of 20 patients survived. The expected frequencies were obtained based on the null hypothesis that the proportion surviving would be 0.5. In studies such as this, the requirement for independence of the data is usually achieved through the random selection of patients to receive the modified treatment. In other words, it would have been inappropriate to choose patients in some way that increased the probability that the patients would have the same outcome for treatment.

Violation of the requirement of independence of data in a chi-square analysis is one of the most frequent errors in statistical analyses. The use of chi-square analysis is popular. The literature contains many examples of its use to compare frequencies for selected characteristics of a test population with those of a standard population. Such analyses are appropriate so long as members of the test population are independently selected at random. Often, however, samples of convenience are chosen, involving all members of some clusters such as households, apartment buildings, or places of employment. If there is any tendency for members of the clusters to be alike with regard to the characteristic examined, the $X_p{}^2$ value that is calculated for the data will be influenced. It should not be evaluated against the probabilities listed under a table of χ^2.

Relationship between $X_p{}^2$ and z

The problem we have been discussing also serves as a good example to make a point about the use of $X_p{}^2$ with 1 *df*. When we first solved the problem, using the familiar approach of calculating a *z* score, we obtained the following result:

$$z = \frac{0.75 - 0.5}{0.1118} = 2.236$$

When we used $X_p{}^2$ to solve the problem, we obtained the following result:

$$X_p{}^2 = \frac{(15 - 10)^2}{10} + \frac{(5 - 10)^2}{10} = 5.0 \text{ with } 1 \, df$$

If you have a calculator at hand, find the square root of 5. If not, estimate what you think it to be. The answer is 2.236. This is exactly the value we calculated for *z*. In others words, for the case of 1 *df*, $\sqrt{X_p{}^2} = z$, or $z^2 = X_p{}^2$. In Chapter 10, which contains more information about the χ^2 distribution, you will see that this is logical. For now, it is important to note that when $X_p{}^2$ is distributed as χ^2 with 1 *df*, you can take the square root of $X_p{}^2$ and obtain a standard normal score.

For example, if you look at Table E-8 in Appendix E, you find the number 3.84 under the column headed ''.05'' for the row corresponding to 1 *df*. If you take the square root of 3.84 you get 1.96, the point on the normal curve that leaves 2.5% in the upper tail or 5% in the two tails. This confirms our earlier conclusion that when we use the table of χ^2 to evaluate $X_p{}^2$, the probabilities obtained are appropriate for a nondirectional test.

Chi-square tables are not nearly as extensive as tables of the normal curve, since there must be a row corresponding to each value of *df*. This means that if you were to calculate a value of 2.89 for $X_p{}^2$ with 1 *df*, you would not be able to get a probability from the table of χ^2. You would only be able to conclude that it has a nondirectional probability somewhere between 0.10 and 0.05. However, taking the square root, we get 1.7, which we can look up on the table of the normal curve. We find the two-tail probability of 0.0891 or the one-tail probability of 0.0446.

Correction for continuity

We have solved the problem introduced at the start of the lesson in two different ways and have obtained identical answers. Both solutions are approximations to the exact solution, which can be obtained by use of the binomial formula. To see how well these approximations work, let us use the binomial formula and find the probability of 15 or more survivors among 20 patients when π is really 0.5. Table E-5 in Appendix E, goes up only to $n = 15$, so we must do the calculations. For the probability of exactly 15 patients' surviving, the formula is as follows:

$$P(15 \text{ of } 20) = {}_{20}C_{15}(0.5)^{15}(0.5)^5$$

The computations give a result of 0.01478. By using similar formulas for the remaining possible results and adding, we find that the probability of 15 or more survivors is 0.02069. This is the exact one-tail probability. The directional probability we calculated using the normal distribution or the χ^2 distribution was 0.0125. As an approximation, this is not too bad; however, an improvement is possible. It is called Yates' correction for continuity, and it is appropriate to use any time there is only a single degree of freedom. The correction is a method of compensating for the fact that the normal distribution and χ^2 distributions are continuous distributions. When we use them to approximate a discrete distribution such as the binomial distribution, there will be some error involved. Yates' correction for continuity minimizes the error by reducing by 0.5 the absolute value of differences between observed and expected frequencies. This is equivalent in our example to saying that our observed frequency was 14.5 rather than 15. By doing so, we are using the point halfway between the two possible discrete values of 14 and 15.

Using Yates' correction for continuity for 1 *df*, the formula for $X_p{}^2$ is as shown in Formula 9.2-1.

$$X_p{}^2 = \sum \frac{(|f_o - f_e| - 0.5)^2}{f_e} \qquad \textbf{9.2-1}$$

where   the parallel vertical lines mean "the absolute value of the difference."

The computations for our problem of 15 survivors then become:

$$X_p{}^2 = \frac{(|15 - 10| - 0.5)^2}{10} + \frac{(|5 - 10| - 0.5)^2}{10}$$

$$X_p{}^2 = \frac{4.5^2}{10} + \frac{4.5^2}{10}$$

$$X_p{}^2 = 2.025 + 2.025$$
$$X_p{}^2 = 4.05$$

We cannot find a probability listed for this value in the table of χ^2. However, since we have only a single degree of freedom, we can take the square root of 4.05 and obtain a z score to be looked up in a table of the normal curve. Taking the square root, we get a z score of 2.012. Looking for a z of 2.01 in the table of the normal curve, we find a one-tail probability of 0.0222. This probability, arrived at by using Yates' correction for continuity, is much closer to the exact value of 0.02069. The small remaining difference between the exact probability and the probability estimated by using the χ^2 distribution cannot be eliminated with finite samples.

If the correction for continuity is appropriate when we are calculating $X_p{}^2$, it stands to reason that it is also appropriate for the one-sample test of a proportion. How is the correction carried out? For our example, instead of using an observed proportion of $^{15}/_{20}$, or 0.75, we would use an observed proportion of $^{14.5}/_{20}$, or 0.725. We would then calculate z as follows:

$$z = \frac{0.725 - 0.5}{0.1118} = 2.012$$

Our z value corrected for continuity is the same as $\sqrt{X_p{}^2}$ corrected for continuity. It should be noted that the adjustment in observed frequency was one that made the observed proportion closer to the expected proportion.

At the start of the lesson, you were told that the example provided would be used to make four points about the use of $X_p{}^2$. Can you now remember these four points? They will be reviewed in the summary following the next section, on the use of $X_p{}^2$ to test for the normality of a population of scores.

TEST FOR A NORMAL DISTRIBUTION

There is a use of the goodness of fit test that relates to one of the assumptions that has been made for many of the techniques you have learned. The assumption is that the population from which a sample has been chosen has a normal distribution. This assumption was necessary for all of the comparisons involving the t distribution. It is also a basic assumption of several other techniques still to be described. By now, it may have occurred to you to ask, "How can we test whether this assumption is valid?" The logical thing to do would be to look at the shape of the distribution within a sample from the population. If the distribution of the sample were extremely skewed or bimodal, this would constitute evidence against the assumption. However, once again we face the question, "How much does the distribution of the sample have to depart from normality before we will conclude that the population is not normal?" We will also have to decide how we are to measure "departure from normality."

A solution to the problem is possible through the use of $X_p{}^2$. If the population is normal, we expect the sample to be approximately normal. If the sample is normal, we have learned to expect that about 68% of all scores will be within one standard deviation of the mean, about 95% will be within two standard deviations of the mean, and almost all of the scores will be within three standard deviations of the mean. Does this statement suggest a method for testing the assumption of normality in the population? Suppose a sample of 100 is chosen and all scores are changed to standard scores. What would the expected frequencies be for a distribution divided into standard deviations?

We know that 2.28% of the normal curve lies to the left of $z = -2$ or to the right of $z = +2$. Thus we would expect to find 2.28% of our sample, or 2.28, to be more than two standard deviations to the left or right of the means. Likewise, because 13.59% of the normal curve lies between $z = -1$ and $z = -2$ (or $z = 1$ and $z = 2$), we would expect 13.59%, or 13.59, to be within one to two standard deviations to the left or right of the means.

To save space, let us make the usual vertical table a horizontal one. It would appear as follows, where the expected frequencies in parentheses have been found from a table of the normal curve.

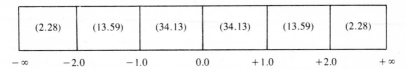

We could now find the observed frequencies for each of these classes of standard scores from the sample and compute $X_p{}^2$ in the usual fashion. The larger the value of $X_p{}^2$, the greater the evidence against our assumption that the population is normal.

There are three important points to be made from this example. The first is that the choice of points for dividing the distribution is entirely arbitrary. The particular points chosen for this example have resulted in wide variation in expected frequencies. The usual practice is to choose points so as to make expected frequencies more nearly equal. There is no need for doing so, however. The points can be chosen on any basis that is meaningful to the study, so long as expected cell frequencies are sufficiently large for $X_p{}^2$ to be distributed in the same way as χ^2.

The second point to be made is that this situation differs from the earlier example in which expected values were to come from a theoretical model. In this instance, another complexity is introduced, in that the observed frequencies depend not only on the theoretical model, but also on sample statistics. If the population mean and standard deviation were known, our problem would be similar to the example using Mendel's theory of heredity. There would be no problem in calculating standard scores and counting to get observed frequencies within each class. However, in the present example the observed frequencies depend not only on the theoretical probabilities of the normal curve, but also on two parameters (μ and σ) that must be estimated from the sample. We shall have to calculate $\bar{X}$ and s and use these as the basis for our z scores. How can this additional factor be taken into account when we are evaluating $X_p{}^2$? The solution is to subtract 1 df for each parameter that has to be estimated from a sample statistic. The table we have used will therefore not result in 5 df (number of cells minus 1). Instead, we will subtract 2 from this number and evaluate $X_p{}^2$ with only 3 df. The procedure of subtracting 1 df for each parameter estimated is a general one that is incorporated into sophisticated explanations of the concept of df. At this level, you need only be aware that this adjustment is called for whenever the data for the table are based on unknown parameter values.

The third point to be made from this example involves a general concept often overlooked in statistics books. In this example, we are testing the assumption of normality in the population. Usually this is done in a situation in which if normality exists, an analysis that depends on this assumption will be carried out. In other words, what we would like to do is to obtain evidence that the assumption is true. Unfortunately, the whole rationale of hypothesis testing is designed to provide an opportunity to demonstrate that the null hypothesis is not true. Accordingly, we generally choose an alpha level of

0.05 or 0.01; we reject the null hypothesis only when the evidence weighs heavily against it.

Consider the difference in the case of a test for normality. The null hypothesis is really a statement of what we hope is true. We would like to state that the population is normal. We do not wish to make this statement unless it is true. What alpha level should be used in evaluating $X_p{}^2$? Frequently, an alpha level of 0.05 or 0.01 is used, just as in the situation in which we hope to obtain evidence against the null hypothesis. Do you see the faulty reasoning in this procedure? Suppose that the result is one that would happen only 10% of the time when the population is indeed normal. You therefore fail to reject the null hypothesis and go ahead with the analysis on the assumption that the population is normal. How can a result that would happen by chance only 10% of the time be evidence in favor of the null hypothesis? What you hope for is a result that would not be at all unusual, say one that can be expected to happen by chance 50%, 70%, or 90% of the time. Such a result would still not prove the null hypothesis to be true. Unfortunately, statistical methods do not provide a way to do so. On the other hand, such a result at least provides some reassurance that the population does not radically depart from normality.

To summarize the third point from this example, when the null hypothesis is a statement of what you would like to prove, the process of hypothesis testing is really inadequate. Instead, report the probability of a result as extreme as the one observed. This probability is called the *P* value for the statistic. The nearer it is to 1.00, the better. By all means, do not use hypothesis testing with traditional levels for alpha.

SUMMARY

This lesson began with a demonstration that $X_p{}^2$ can be used as an alternate solution to the one-sample test of a proportion. We first used the normal curve approximation to the binomial and then the approach using $X_p{}^2$. From this example four points were made. First, it was found that the probability using z was half the probability found using $X_p{}^2$. This demonstrated that when $X_p{}^2$ is evaluated using the table of χ^2, the probabilities obtained are for a two-tail test. Second, we learned that like the normal curve approximation to the binomial, the use of χ^2 to evaluate $X_p{}^2$ is appropriate only when samples are sufficiently large and when data are independent. Third, we found that for the case of 1 *df*, the square root of the value of $X_p{}^2$ was exactly equal to the z value obtained. This allows the use of the table of the normal curve to evaluate values of $X_p{}^2$ not listed in the table of χ^2. Fourth, we learned that for the case of 1 *df*, a modification of the formula for calculating $X_p{}^2$ (the correction for continuity) would make the probability obtained from the table of χ^2 closer to the exact probability obtained by using the formula for the binomial expansion.

In the last section of the lesson, you learned how to use the goodness of fit application of $X_p{}^2$ to test for normality in a population of scores. The value of $X_p{}^2$ was calculated in the usual manner, with expected proportions coming from the table of the normal curve. You were told to subtract 1 *df* for each parameter value estimated from the sample, in this case μ and σ.

In the next lesson you will learn about an application of chi-square that can be thought of as an extension of the two-sample test of proportions. The procedure will include situations involving both multiple populations and multiple classes for the characteristic of interest.

LESSON 2 EXERCISES

1. In an experiment designed to test for mental telepathy, cards are used, each of which contains a picture of a single object. Any one of five different objects may appear on an individual card. Two persons are separated by a curtain. The "sender" is shown a card chosen at random and is asked to concentrate on the object of the card. The "receiver" is then asked to guess which of the five objects is on the card. In 100 trials, there are 28 correct responses. What is the probability of these many correct responses if the responses are simply random guesses? Use the correction for continuity and solve the problem two ways.
2. Use as division points the following values of z: -1, -0.5, 0, $+0.5$, and $+1$. Test that the following random sample comes from a population that is normally distributed.

   93, 117, 120, 122, 107, 122, 115, 110, 117, 146, 114, 114, 106, 94, 109, 117, 142, 122, 110, 133

Lesson 3 □ Tests of association

In the first two lessons of this chapter you learned how to use the statistic X_p^2 to compare proportions for a test population with proportions for a standard population. This procedure was called a goodness of fit application of χ^2. The statistic X_p^2 was shown to be a measure of the extent of discrepancies between the observed and expected frequencies. Expected frequencies were based on the null hypothesis that there were no differences in the proportions for the various classes of the test population and those for the corresponding classes of the standard population. The proportions for the standard population can be a set of proportions from some real population, a set of theoretical proportions, or simply a set of proportions that have meaning for the investigator.

If there are only two classes (absence or presence of the characteristic), the test was shown to be equivalent to a one-sample test of a proportion. The goodness of fit test using χ^2 is therefore a more general solution to the problem of comparing proportions from a test population with proportions from a standard population.

There is another application of X_p^2 that is in a sense an extension of a *two-sample* test of proportions as described in Lesson 2 of Chapter 8. Remember that a two-sample test of proportions is appropriate when there are two populations to be compared and when the characteristic of interest has only two classes. The procedure is not applicable when there are more than two populations or when there are more than two classes of the characteristic of interest.

THREE OR MORE SETS OF UNKNOWN POPULATION PROPORTIONS

There are many situations in which the research problem involves three or more study populations. One may wish to compare the proportions of patients making complete recoveries under three courses of therapy, the proportions of assault victims among those admitted to the emergency rooms of four different hospitals, the proportions of rheumatic fever patients exhibiting inflammation of the joints among five different age groups, or some other situation involving three or more study populations.

In studies such as these just described it would be possible to make all possible comparisons two at a time by using two-sample tests of proportions. Besides the work involved in such a procedure, there is another problem with making all of the possible pairwise comparisons. What is it? If you understood the first lesson in this chapter, you should recall that by doing so you would be increasing the probability of making a Type I error. The problem is even more pronounced than in the one-sample case. The number of possible comparisons increases rapidly as the number of populations increases.

Instead of making all possible comparisons two at a time, what is needed is a method of simultaneously comparing all of the sample results. Once again the appropriate statistic for making this comparison is X_p^2.

To illustrate the use of X_p^2 for comparing two or more population proportions, let us take a rather simple example. Suppose you are interested in comparing admissions to the emergency rooms of four hospitals. A question arises whether the proportions of patients admitted as assault victims differ for the four hospitals. A random sample of cases is chosen from all admissions at each hospital over a 3-year period. Although ordinarily we would choose equal-sized samples in a case such as this, let us choose different-sized samples for purposes of illustration. The sample sizes are 50, 75, 100, and 175. The numbers of assault victims are 2, 6, 16 and 56. The question we must ask is whether these observed frequencies provide evidence that the population proportions differ.

The formula for computing X_p^2 for this four-sample case is identical to that provided in the first lesson as Formula 9.1-1.

$$X_p^2 = \sum \frac{(f_o - f_e)^2}{f_e}$$

There are four hospitals. Patients from these four hospitals are placed in one of two classes: assault victims and others. As a result, instead of having a one-dimensional table (as we had in the first two lessons), we have a two-dimensional table. If a row is used for each hospital and a column for each classification, we will have a table with four rows, two columns (a four-by-two table) and, thus, eight cells. We need an observed and expected frequency for each cell. You have been given the number of patients observed at each hospital and the number who were assault victims. Subtracting will provide the observed frequency of others. But what about expected frequencies?

In the first two lessons of this chapter, involving goodness of fit, the expected frequencies were based on some external source. In this lesson, the expected frequencies are derived from the data of the experiment. Keep in mind that X_p^2 is a statistic that represents the differences between observed values and values expected under the assumption that the null hypothesis is correct. We start by asking, "If the null hypothesis is correct and the proportions of assault victims are the same for all four hospitals, what is the best estimate of this common proportion?"

Does this question sound familiar? It is the same question we asked for the test of the null hypothesis that *two* populations have identical proportions. We will use a similar solution and pool the data from the four samples to find the best estimate of the proportion of assault victims.

Out of a total of 400 cases, 80 were admitted as assault victims. Our best estimate of π is therefore 0.20. We can apply this proportion to each sample to get the expected number of assault victims in the sample from each hospital. For hospital A, with a sample of size 50, the expected number of assault victims is 50 times 0.2, or 10. For the other hospitals we also multiply the sample size by 0.2 and find the expected values: 15, 20, and 35.

We can get the expected number of other victims by multiplying sample size by the proportion of "others" in the total group (0.8). Once again we follow the practice of placing expected values in parentheses in the lower right-hand corner of each cell of a table representing the data. The resulting table appears as Table 9-2.

Totals for both rows and columns have been included in the margins of Table

Table 9-2. Observed and expected frequencies of assault victims at four hospitals

	Assault victims	Others	TOTAL
Hospital A	2 (10)	48 (40)	50
Hospital B	6 (15)	69 (60)	75
Hospital C	16 (20)	84 (80)	100
Hospital D	56 (35)	119 (140)	175
TOTAL	80	320	400

9-2. These values are helpful in carrying out the process just described for finding expected frequencies. Notice that the expected frequencies for a two-dimensional table can be found by finding proportions for marginal totals of *either* columns or rows. In the procedure described earlier, we found proportions based on marginal totals for columns. The proportion of assault victims in the total sample was found by dividing the column total (80) by the total sample size (400). The resulting proportion, 0.20, was then multiplied by each row total to get expected values for the first column for each row. A similar procedure resulted in expected frequencies for the second column of each row. Exactly the same expected frequencies would result from finding proportions for each row total and then multiplying these proportions by column totals. Try it and you will see that this is so. The practice will be beneficial, since this is the procedure that is generally used to calculate expected frequencies for a two-dimensional table.

The procedure for finding expected frequencies that has been described can be summarized by a simple formula, which can be used when X_p^2 is to be calculated for a two-dimensional table:

$$f_e = \frac{\text{(row total)(column total)}}{\text{(grand total)}}$$

Once we have the observed and expected values for each cell of the table, as shown in Table 9-2, we can calculate X_p^2 just as in the two previous lessons.

$$X_p^2 = \sum \frac{(f_o - f_e)^2}{f_e}$$

$$X_p^2 = \frac{(2 - 10)^2}{10} + \frac{(48 - 40)^2}{40} + \frac{(6 - 15)^2}{15} + \frac{(69 - 60)^2}{60} + \frac{(16 - 20)^2}{20} + \frac{(84 - 80)^2}{80} +$$

$$\frac{(56 - 35)^2}{35} + \frac{(119 - 140)^2}{140}$$

$$X_p^2 = 6.4 + 1.6 + 5.4 + 1.35 + 0.8 + 0.2 + 12.6 + 3.15$$
$$X_p^2 = 31.5$$

To evaluate this statistic, we will once again refer to the table of χ^2 (Table E-8) in Appendix E. The only question is, "How many *df* shall we use?" In the last lesson you learned to think of *df* as the number of values that are free to vary in calculating the test statistic. For our example, there were eight differences between observed and expected values. These eight differences became the basis for our statistic, X_p^2. If we regard all of the marginal totals as fixed values, how many of the observed differences are free to vary for any value of X_p^2? Consider once more a table with marginal totals that are the same as those for our example of emergency room admissions.

		TOTAL
		50
		75
		100
		175
TOTAL 80	320	400

For any row, once one cell frequency is known, the other is completely determined. Similarly, as soon as any three cell entries are known for any column, the fourth is completely determined. Can you translate this into a general formula for the number of cell entries free to vary? It is the number of cells in a row minus 1, multiplied by the number of cells in a column minus 1, or $(r - 1)(c - 1)$. For our example, this would be $(4 - 1)(2 - 1)$, or 3 *df*. Keep in mind that what we are computing is the number of cell entries free to vary when the marginal totals are fixed.

Having determined that our value of 31.5 for X_p^2 is to be evaluated with 3 *df*, we refer to the table of χ^2. The most extreme value listed is 16.27, which has a probability of 0.001. Since our value is even greater than 16.27, we conclude that the four hospitals do not admit the same proportion of assault victims in their emergency rooms.

METHODS OF GATHERING DATA FOR A TWO-WAY TABLE

In the example just presented, the four hospital populations were identified in advance and random samples of some designated size were chosen from each. The same data might be gathered in a very different way. For example, a single population of persons admitted to emergency rooms during the last 3 years might be identified. From this single population, a random sample of 400 patients might then be chosen. Each person could then be classified according to the two variables: (1) admitting hospital and (2) reason for admittance. Either method of gathering data might have resulted in the table of data that was given.

Advanced texts distinguish between the analysis of data for these two data-gathering situations. The first is called a *test of homogeneity of proportions* and the second a *test of association*. Because of the difference in the methods of gathering data, there is a difference in the mathematical model that is appropriate for finding the exact probability distribution for each situation. In this text we will not go into the procedure for finding the exact probability distribution for the case of two or more samples; instead we will confine our attention to samples large enough for the use of χ^2 to evaluate X_p^2. The one exception is the rather common two-by-two table, which will be discussed at the end of this lesson.

Because we will not be concerned with the methods for finding exact probabilities for two-dimensional tables, we need not distinguish between tests of homogeneity of proportions and tests of association. The calculation of X_p^2 is the same for both cases. The second method of gathering data is superior to the first as a way to learn about the association between "admitting hospital" and "reason for admittance." However, the first data-gathering method is, in a sense, also a test of association. For either method of gathering data, there are two variables. The first variable has four classes, corresponding to the four hospitals, and the second variable has been made into a dichotomy: assault victims and others. The null hypothesis is that there is no difference in the proportions of assault victims admitted to the four hospitals. If this null hypothesis is true, the probability that a randomly chosen patient was admitted as an assault victim is the same regardless of the hospital to which he was admitted. This is another way of saying that the variable "reason for being admitted" is not associated with the variable "admitting hospital."

The phrase *test of association* is not entirely satisfactory to describe the analysis of data for a two-dimensional table of observed and expected frequencies. However, it does have the advantage of widespread use. It will therefore be used throughout this text. Remember that this phrase is being used to describe the analysis of data for both methods of data gathering.

At this point, it should be clear that extending the chi-square analysis to data involving *more* than two classes for the characteristic of interest presents no particular problem. Instead of patients being classified as assault victims and others, there might be eight classes of reasons for admittance. The data would appear in a four-by-eight table, resulting in 32 differences between observed and expected frequencies and 3 times 7, or 21, *df*.

In an exercise at the end of this lesson you will be asked to analyze data for a three-by-four table in a test of association between severity of first heart attack and smoking history. It should also be clear that the method that has been described is appropriate for the case of a two-by-two table. Because the two-by-two table occurs so frequently, we will devote the remainder of this lesson to a discussion of the use of X_p^2 for this case and to a description of a method of figuring probabilities for the two-by-two table when dealing with small sample sizes.

TWO-BY-TWO CONTINGENCY TABLES

The χ^2 test of association has been described as an extension of a two-sample test of proportions. The extension is both in terms of the number of populations to be compared and in the number of classes for the characteristic being compared. In other words, two or more populations can be compared on a characteristic for which there are two or more classes. The test is a test of the null hypothesis that the proportions within each class are the same regardless of which population is being considered.

If the χ^2 test of association is an extension of the two-sample test of proportions, it should represent an alternative method of solving problems. Let us analyze an exercise from Chapter 8 using chi-square instead of our earlier method. Exercise 3 from Lesson 2 of that chapter involved two populations of high school students (system A and system B). School system A had a comprehensive sex education program, whereas system B did not. Random samples of 100 high school students from each system were classified as to whether they were sexually active. The data can be summarized in what is often called a two-by-two contingency table.

	Sexually active	*Not sexually active*	TOTAL
System A	22	78	100
System B	18	82	100
TOTAL	40	160	200

In our previous analysis we: (1) combined samples to find a common proportion, 0.20; (2) found $SE_{p_A - p_B}$ to be 0.056; (3) found the observed difference between the two-sample proportions, $0.22 - 0.18$, or 0.04; and (4) changed this observed difference to a z score of 0.714. When we looked this value up in the one-tail section of a table of the normal curve, we found that differences this great will happen about 23% of the time when the population proportions are the same. We concluded that there was insufficient evidence to reject the null hypothesis.

Now see if you can do the same problem by computing X_p^2. The first step will be to find expected frequencies. Once again, you will use the marginal totals as the basis for these expected frequencies. See if you can complete the problem before reading further.

Did you get expected frequencies of 20 and 80 for both systems *A* and *B?* Good. Your table, then, looked like this.

	Active	Not active
System A	22 (20)	78 (80)
System B	18 (20)	82 (80)

You then probably computed $X_p{}^2$ as follows:

$$X_p{}^2 = \frac{2^2}{20} + \frac{(-2)^2}{80} + \frac{(-2)^2}{20} + \frac{2^2}{80} = 0.50$$

You then looked up 0.5 with 1 *df* and found that it was not significant. You could not find the exact probability of this value from the χ^2 table. There is a way to find the probability, however. Remember that with 1 *df*, $\sqrt{X_p{}^2} = z$. When we take the square root of 0.5, we get a *z* score of 0.7071. This is the same *z* score we obtained previously, except for the rounding off we did when computing $SE_{p_A - p_B}$.

The demonstration works. We get the same result for this two-by-two contingency table using either our earlier method or the χ^2 method. Since the χ^2 method is both easier to remember and easier to calculate, we can look on the two-sample test of proportions as merely a teaching example.

It is hoped that some readers will want to interrupt at this point and say, "Wait a minute, you have forgotten something." If you are among those who worked the problem and remembered to correct for continuity, congratulations. We do have only a single degree of freedom and, indeed, we should have reduced the absolute value of all differences by 0.5. Doing this, we obtain a value for $X_p{}^2$ of 0.28125. Taking the square root, we get a *z* score of 0.53. Looking this up in a table of the normal curve, we find a one-tail probability of 0.2981. This is the value we should obtain as the answer to the exercise. The idea of correcting for continuity was not introduced earlier when we compared proportions. The oversight was intentional, to avoid complicating the situation while you were still learning the basics of hypothesis testing. No real harm resulted by this omission, since you now will use the χ^2 method in place of the earlier method anyway.

An alternate formula. We have shown that the basic formula for $X_p{}^2$ is used in a wide variety of situations. It is:

$$X_p{}^2 = \sum \frac{(f_o - f_e)^2}{f_e}$$

For the case of 1 *df*, we modify the formula as a correction for continuity.

In the case of a two-by-two contingency table, there is another formula that gives exactly the same result while eliminating the need to compute expected values. I personally feel that most students should stick with the basic formula. However, since you are likely to see the alternate formula for computing X_p^2 for the two-by-two table, I am presenting it. Each of the cells of the table is labeled as follows:

	Active	Not active
System A	(a) 22	(b) 78
System B	(c) 18	(d) 82

The alternate formula, with the correction for continuity, is given as Formula 9.3-1.

$$X_p^2 = \frac{(|ad - bc| - n/2)^2\, n}{(a + b)(a + c)(b + d)(c + d)}$$ **9.3-1**

The formula looks a bit imposing, but all you need to do is substitute the cell frequency corresponding to each cell letter in the appropriate place in the formula. The n in the formula is the total number observed.

$$X_p^2 = \frac{(|(22)(82) - (78)(18)| - 200/2)^2\, 200}{(22 + 78)(22 + 18)(78 + 82)(18 + 82)}$$

$$X_p^2 = \frac{(|1804 - 1404| - 100)^2\, 200}{(100)(40)(160)(100)}$$

$$X_p^2 = \frac{(300)^2\, 200}{(100)(40)(160)(100)}$$

$$X_p^2 = 0.2815$$

This value is the same one we obtained previously. It is debatable whether the calculations are any easier. This formula does have the appeal of not having to think about expected frequencies. Since the results are exactly the same, the student can choose whichever method seems easier.

SMALL EXPECTED FREQUENCIES

You have now learned that the statistic X_p^2 can be used to compare proportions in a wide variety of situations. The two basic requirements are independence of data entries and sufficiently large expected frequencies. What can be done when the first criterion is met but expected frequencies are not large enough for X_p^2 to be distributed as χ^2? You already know the answer for the case of *one* sample when the characteristic being studied has only *two* classes. For example, you know how to use the binomial expansion to find the probability of as many as six of eight patients' surviving when the probability of any one individual patient's surviving is 0.5. Exact probabilities can also be computed for the case of *one* sample with *more than two* classes. The method involves a formula for the multinomial expansion. Although the procedure is not particularly difficult to understand, it does not have wide enough application to merit inclusion in this text. If the reader runs into a situation calling for the multinomial expansion, it can be found in most advanced-level statistics texts.

In the case of *two* or more populations, computing exact probabilities is also possible. The mathematical model differs depending on how the data have been gathered; data-gathering methods include the prospective correlational study, the retrospective case control study, and the true experimental study. In general, the mathematical computations are complex and will be tedious for all but the very simple case. There is, however, one situation that has received rather widespread use. It involves the two-by-two contingency table with small expected frequencies. It is a test of association named *Fisher's exact test*. The mathematical model on which it is based is called the hypergeometric model, in which it is assumed that all marginal totals are fixed.

Fisher's exact test

A number of statistics books contain special tables that avoid many of the computations of Fisher's exact test. A detailed explanation of both the rationale and the computations is included here in the belief that, as a result, the student will find this test more understandable.

Let us begin with an explanation based on an artificial situation. Suppose a large barrel contains objects that are either balls or cubes and that are colored either red or white. Now suppose that four objects are selected and that two are balls and two are cubes, two are red and two are white. A two-by-two contingency table would have the following marginal totals.

	Red	*White*	
Ball			2
Cube			2
	2	2	4

Can you identify the cell frequencies possible under the condition that marginal totals remain unchanged? They are as follows:

Result 1

	Red	*White*	
Ball	2	0	2
Cube	0	2	2
	2	2	4

Result 2

	Red	*White*	
Ball	1	1	2
Cube	1	1	2
	2	2	4

Result 3

	Red	*White*	
Ball	0	2	2
Cube	2	0	2
	2	2	4

Results 1 and 3 would represent some indication of an association between color and shape, whereas result 2 provides no indication of association. To find out how strong the indication of association is for either result 1 or result 3, we need to know the probability of each of the three results occurring by chance when there is in fact no association. To obtain these probabilities, we go back to our basic definition of probability (n_a/n). We begin by asking how many different ways four objects can be assigned to this two-by-two table while keeping marginal totals fixed. This number will be the denominator for our definition of probability. Then we ask how many of all of these possible ways will appear as result 1, how many as result 2, and how many as result 3. These become the numerators for figuring the probability of each outcome.

For a small data set like this, it is possible to obtain the desired numbers by a process of enumeration. Let us do the problem this way first, and then examine a formula that will provide an easier way of getting the same result. We assign letters (*a, b, c,* and *d*) to the four objects so that they can be distinguished. First, for result 1, we put each possible combination of two objects in the upper left-hand box on a rotating basis, with the remaining combination in the lower right-hand box. The six possible arrangements are as follows:

ab			ac			ad			bc			bd			cd	
	cd			bd			cb			ad			ac			ab

Of course, there are six similar ways of getting result 3.

Enumerating the number of ways to obtain result 2 is a bit more difficult. If object *a* is placed in the upper left-hand corner, how many ways can the other three objects be placed, one in each corner? There are six.

a	b		a	b		a	c		a	c		a	d		a	d
c	d		d	c		b	d		d	b		b	c		c	b

Placing each of the other three objects in the upper left-hand corner would in each case result in six additional arrangements, for a total of 24 possible arrangements for result 2.

Altogether, there are 36 possible arrangements of the four objects, given the specified marginal totals. By our basic definition of probability, if the objects were assigned at random, the probabilites of results 1 and 3 are each $6/36$, or $1/6$, and the probability of result 2 is $24/36$, or $2/3$. With probabilities as high as $1/6$, neither result 1 nor result 3 would be particularly strong evidence of an association between color and shape.

With this small data set, even the two possible results that are most favorable as evidence of an association are unconvincing. Suppose, however, that there were a larger total number of objects. In this instance, the most extreme results might represent strong evidence of an association. Again, we could ask what the probability of each result is when there is, in fact, no association between color and shape. This time the process of enumeration would be rather involved. (Try it for six objects.) Fortunately, there is a formula that makes the work quite easy. To express the formula, we will have to use symbols to represent the various marginal totals and cell frequencies. Let us use f to represent cell frequencies, m to represent marginal totals, and T to represent the total number of objects. Using subscripts as identifiers, we have the following.

f_1	f_2	m_3
f_3	f_4	m_4
m_1	m_2	T

The probability of any given result can be computed from the following formula involving combinations.

$$P = \frac{[_{m_1}C_{f_1}][_{m_2}C_{f_2}]}{_TC_{m_3}}$$

9.3-2

Let us try this formula for result 1, which we figured out to have a probability of $1/6$. The use of Table E-2, in Appendix E, makes the formula easy to apply.

$$P = \frac{[_2C_2][_2C_0]}{_4C_2} = \frac{(1)(1)}{6} = 1/6$$

For result 2, we compute the value of the formula as follows:

$$P = \frac{[_2C_1][_2C_1]}{_4C_2} = \frac{(2)(2)}{6} = 4/6 \text{ or } 2/3$$

These are the same values obtained by the process of enumeration.

For the student who has access to a calculator for finding the value of factorials but not to a table of combinations such as Table E-2, there is another formula that is exactly equivalent to Formla 9.3-2. It is:

$$P = \frac{m_1! \, m_2! \, m_3! \, m_4!}{f_1! \, f_2! \, f_3! \, f_4! \, n!}$$

where n is the sum of all frequencies.

A health-related example. Now let us see how Fisher's exact test can be carried out for a set of health-related data. Suppose a retrospective study is done to determine whether there is a relationship between jogging and high blood pressure among a group of former college athletes. Ten former college athletes are located among employees of an insurance company. Five are regular joggers and five are not; four have high blood pressure and six do not. The actual result observed is as follows:

	High blood pressure	Normal blood pressure	
Joggers	1	4	5
Nonjoggers	3	2	5
	4	6	10

In this result, is there evidence against the null hypothesis that there is no association between jogging and high blood pressure among former athletes? Our first observation is that if an association does exist we would expect it to be an inverse relationship. The presence of the characteristic "jogging" would be expected to indicate the absence of the characteristic "high blood pressure." The data seem to bear this out. However, we wish to know whether the data would be a rare event or a common event, if the null hypothesis of no association is true. If the result would be a rare event (that is, have a low probability), we will reject the null hypothesis and decide that there is an association (inverse) between jogging and high blood pressure. We compute the probability of our result's happening by chance as follows:

$$P = \frac{[_4C_1][_6C_4]}{_{10}C_5} = \frac{(4)(15)}{252} = \frac{60}{252}$$

An even more extreme result keeping the same marginal totals would be:

	High blood pressure	Normal blood pressure	
Joggers	0	5	5
Nonjoggers	4	1	5
	4	6	10

For this result the probability is:

$$P = \frac{[_4C_0][_6C_5]}{_{10}C_5} = \frac{(1)(6)}{252} = \frac{6}{252}$$

As we did when figuring the exact probabilities using the binomial formula, we will add the probability for our observed result and all results that are even more extreme. This is because we want to compute the probability of a result *as extreme* as the result observed. We get a combined probability of $^{66}/_{252}$, or 0.262. Since this is not a particularly unlikely event, we conclude that there is insufficient evidence to reject the null hypothesis of *no* association. Note that if our observed result had been the most extreme possible, we would have calculated a probability of $^6/_{252}$, or 0.0238 and rejected the null hypothesis with the use of an alpha level of 0.05.

Recall that Fisher's exact test is used when expected frequencies are too small to appropriately use χ^2 to evaluate $X_p{}^2$. Just to satisfy our curiosity, let us see how far from the exact probability (0.262) we would have been if we had used χ^2. Our analysis would be as follows:

1
   (2)
	4
	   (3)
		5

3
   (2)
	2
	   (3)
		5

4	6	10

$$X_p{}^2 = \sum \frac{(|f_o - f_e| - 0.5)^2}{f_e} = \frac{(0.5)^2}{2} + \frac{(0.5)^2}{3} + \frac{(0.5)^2}{2} + \frac{(0.5)^2}{3}$$

$$X_p{}^2 = 0.4167$$

$$\sqrt{X_p{}^2} = 0.645$$

Looking up a z value of 0.64 in the one-tail section of the table of the normal curve, we find a probability of 0.261. Notice that this value agrees closely with Fisher's exact value even though χ^2 has been used to evaluate $X_p{}^2$ with 1 df with expected values for all four cells below 5. This close approximation occurs because expected values are nearly balanced and the total n is as large as 10.

SUMMARY

This lesson described an application of chi-square called a *test of association*. It is an extension of the two-sample test of proportions to include two or more populations with two or more classes for the characteristic of interest. The statistic used was again $X_p{}^2$, computed using exactly the same formula as for the goodness of fit application. The difference was that this time expected frequencies were computed based on the observed data. The data were exhibited in a table containing rows and columns. It was explained that df can be computed as $(r - 1)(c - 1)$. The two-by-two contingency table was shown to be exactly equivalent to the two-sample test of proportions. An alternative formula for computing $X_p{}^2$ for the two-by-two case was also presented. Fisher's exact test was described as a solution for those instances where a two-by-two table has expected frequencies too small for the use of $X_p{}^2$. A formula was given for calculating probabilities for this test.

LESSON 3 EXERCISES

1. A question is raised concerning the possibility of a relationship between smoking history and the severity of a first heart attack. In a retrospective study, 100 patients are chosen at random from a coronary registry. Severity of attack is categorized as mild, moderate, or severe. Smoking history is categorized on the basis of usage during the year preceding the attack. Four classes are used: (1) <pack per week, (2) <pack per day, (3) <two packs per day, and (4) two or more packs per day. Compute $X_p{}^2$ as a test of the hypothesis that there is no association between the two variables for the patients in the registry.

	<Pack per week	*<Pack per day*	*<Two packs per day*	*Two or more packs per day*
Mild	14	12	8	6
Moderate	4	1	10	25
Severe	2	2	7	9

What do you conclude? Should you make a directional or a nondirectional test?

2. A survey is taken to determine whether the public favors increasing payroll deductions to provide funds for extending Medicare under Social Security. Responses are separated for persons under 50 and persons 50 or older. Make a test of the null hypothesis that there is no association between age and response. Solve the problem both as a two-sample test of proportions and as a chi-square test of association. Use either method of computing $X_p{}^2$, but remember to correct for continuity when calculating both z and $X_p{}^2$.

	Yes	*No*
50 and older	10	5
Under 50	5	10

3. Suppose the data for problem 2 had been based on a table of only 11 responses. Use Fisher's exact test to determine the probablity of a result as extreme as the following:

	Yes	*No*
50 and older	7	1
Under 50	1	2

CHAPTER 10

Significance tests for variances using chi-square and F distributions

Lesson 1

In the previous chapter you learned of a number of applications for the chi-square (χ^2) distribution involving computation of the statistic $X_p{}^2$. It was explained that the distribution of χ^2 is a theoretical distribution and that the statistic $X_p{}^2$ under some circumstances has a similar distribution. You learned to read a table of χ^2 to find selected probabilities for cases involving selected *df*. You were given little information about the distribution of χ^2. This lesson will begin with a section describing the χ^2 distribution. It will then continue with an explanation of the use of χ^2 to make one-sample tests of variances. In the last part of the lesson another theoretical distribution, called the *F* distribution, will be introduced. The *F* distribution will be used in this chapter as a basis for making two-sample tests of variances. You have already learned that with 1 *df*, χ^2 is directly related to the normal curve. You will see that the *F* distribution is related to both the *t* distribution and the χ^2 distribution.

CHARACTERISTICS OF THE CHI-SQUARE DISTRIBUTION

Now that you have seen how useful the chi-square distribution can be, it is appropriate to examine some of the characteristics of the distribution. Earlier you learned that for 1 *df*, χ^2 equals the square of the *z* score that you would compute for the same data. From this observation you might infer that there is some basic relationship between the normal distribution and the chi-square distribution. As a matter of fact, the relationship is quite direct. Suppose that samples are taken from a population that is normal and that has a known mean and a known standard deviation. Each score in the sample is changed to a standard score and then squared. These squared standard scores are then added to obtain a sum for each sample. The sum is therefore a statistic that varies from sample to sample. This statistic is the value χ^2. Its sampling distribution is the distribution found in Table E-8 in Appendix E. Can you describe in words what χ^2 is? It is a sample statistic found by changing the scores from a random sample *from a normal population* to standard scores,

190

squaring them, and then summing the squares. In others words, $\chi^2 = \Sigma z^2$. For samples of size 1, χ^2 equals z^2, or, as we have already learned, $\sqrt{\chi^2} = z$.

The formula for χ^2 as we have just described this statistic is given as formula 10.1-1.

$$\chi^2 = \sum \left(\frac{X - \mu}{\sigma}\right)^2 \qquad\qquad \textbf{10.1-1}$$

where   sampling is from a normal distribution.

Let us see if the characteristics of the chi-square distribution can be understood intuitively. We need to know the shape, central tendency, and variability for various sizes of samples. First, what shape will the distribution have? To answer this question, let us examine the distribution expected when the samples are of size 1. In this case each sample value of χ^2 will result from squaring a single z score from a normal population. What would the distribution of many such χ^2 values look like? We know that the distribution of many randomly selected z scores from a normal population can be expected to have a normal shape, a mean of zero, and a standard deviation of 1.0. But what about the distribution of the squares of these same z scores?

Since we are squaring, negative values of χ^2 will not occur. Is it possible to get a value of 0.0? For samples of size 1 this will happen when the sample value drawn results in a z score of 0, that is, when the score drawn equals the mean of the population. If the population is a continuous distribution this is certainly a possible result. At the other extreme, how *large* a value of χ^2 is possible when samples are of size 1? Just as the theoretical distribution of the normal curve has a limit of infinity, so the χ^2 distribution has a limit of infinity. However, since almost all standard normal scores are within the limits of -3.00 and $+3.00$, almost all values of χ^2 for samples of size 1 will be between zero and nine. The probability of specific values of z decreases the further the value is from zero. The χ^2 distribution for samples of size 1 can therefore be expected to have a peak near zero and then to drop off rapidly until almost reaching the baseline at 9.00.

Now what would happen if sample size were increased to two? Remember that χ^2 is not an average. It is a sum of squared z scores. Each sample χ^2 will now be the sum of the squares of two random z scores. Once again the lower limit will be 0.00, but what about the upper limit? It is possible that both z scores will be as extreme as three standard deviations from the population mean, resulting in a χ^2 value of 18 or more. However, the probability of two such extreme score's occurring in the same random sample would be small. Similarly, a χ^2 value of zero results only when both sample values are right at the population mean. Consequently, the usual range of χ^2 values will not extend from 0 to 18.

Can you guess what happens to the shape of the distribution of χ^2 as sample size is further increased? As n increases, the lower limit remains at 0 while the upper limit gets greater. What do you think happens to the peak of the distribution? Values of χ^2 near zero happen only when *all* samples values are near the population means. Since this becomes increasingly less likely as n increases, the peak of the distribution moves to the right as sample size gets larger. What do you guess the distribution will look like when n is infinitely large? Think about what happens to the sampling distribution of the mean as sample size increases. Even though the population is skewed, as n gets larger the sampling distribution of $\overline{X}$ approaches normality. The same is true for the distribution of χ^2.

Having determined the shape of the chi-square distribution, we then ask about the central tendency and variability. Fortunately, the answer is the same no matter what the size of the sample. The mean of the sampling distribution of χ^2 equals the sample size *(n)*, and the variance is twice the sample size *(2n)*. This statement is consistent with our earlier conclusion that the lowest possible value of χ^2 is always 0.00 and that the upper limit increases as sample size increases. Proofs that χ^2 distributions have a mean equal to *n* and a variance equal to 2*n* are included in advanced texts and will not be presented here. The values themselves are easy to remember, and they provide a quick reference for evaluating X_p^2, our statistic from the last chapter, without referring to a table of chi-square. If we substitute *df* for sample size, we use the following reasoning as a rough rule of thumb. Sample X_p^2's have a mean equal to *df*. Obviously, an observed X_p^2 near the number of *df* will not be an extreme value. For the usual application it is unnecessary to look up any X_p^2 that is less than 1.5 times the *df*.

To summarize this section, if sampling is from a normal distribution with known μ and σ, and if each score of a sample is changed to a *z* score and squared, the sum of these values for each sample is called χ^2. The sampling distribution of the χ^2 values can be described by an exact formula. The shape, central tendency, and variability of the distribution depend only on the size of the samples chosen. The distribution is extremely skewed to the right for small samples and approaches the normal curve for very large samples. The mean and variance of chi-square distributions are *n* and 2*n* respectively.

Now that you have learned the basis for the χ^2 distribution, we can examine its use to make inferences about a measure of variability. You have learned the principles of inferential statistics while confining your attention to only two statistics: proportions and means. You are now ready to learn to apply these same principles to another statistic, the variance. Just as we can make a one-sample test that the population proportion (or the mean) is some specific value, so we can test the hypothesis that the variance is some specific value.

ONE-SAMPLE TESTS OF VARIANCES

Drug laboratories and product manufacturers in general are concerned about the uniformity of their products. For example, a container of aspirins is advertised as containing 12 aspirins of 325 mg each. If a container were selected at random, the average weight of the 12 tablets might be very close to 325 mg, but there would certainly be some variability in the weights of individual tablets about this mean. This variability could be expressed in terms of either a standard deviation or a variance. If we consider the variance in the weights of all the tablets produced by a certain manufacturing process, we could think of this as the population variance (σ^2); the variance for any sample set of tablets is called a sample variance (s^2). Of course, the manufacturer wants to keep σ^2 as small as possible within reasonable financial constraints.

Suppose that for years a laboratory has produced tablets where the weights have been normally distributed, with a mean of 325 mg and a standard deviation of 10 mg. Can you write the label that should appear on the box to inform the statistically naïve public of what the box contains? Perhaps something like the following would be appropriate: "12 tablets, average weight 325 mg, range approximately 295 to 355 mg."

Now suppose that a suggestion is made for modifying the production procedure. It is claimed that the modification will make the weights of tablets more uniform. How will you go about testing this claim? This is a situation calling for a one-sample, directional hypothesis test. The *alternative* hypothesis is that $\sigma_T < \sigma_S$. The *null* hypothesis is that $\sigma_T \geq \sigma_S$. As it turns out, the solution to testing this hypothesis has been worked out for variances rather than standard deviations. Since the current process has been found to have a standard deviation of 10, we square this to get the variance of 100, and our null hypothesis can be stated as follows:

$$H_o: \sigma_T^2 \geq \sigma_S^2 \qquad \text{or} \qquad \sigma_T^2 \geq 100$$

What will we do to test this hypothesis? Think back to the diagram used to symbolize the process of hypothesis testing. We have a standard population with a known parameter value of σ_S^2 and a test population with an unknown parameter value of σ_T^2. We take a single sample from the test population and compute the statistic s^2. If s^2 is so small as to be unlikely when sampling from the standard population, we will conclude that the null hypothesis is wrong. Our only problem is in determining what to expect when sampling from the standard population. In other words, we need to know the sampling distribution of variances.

As you might guess from the introduction of this discussion, the sampling distribution is directly related to the chi-square distribution. By definition:

$$\chi^2 = \sum \left(\frac{X - \mu}{\sigma} \right)^2 = \frac{\Sigma (X - \mu)^2}{\sigma^2} \text{ with } n \ df$$

If μ is unknown and a sample mean is substituted for μ in the right-hand side of the equation, the resulting expression becomes: $\Sigma (X - \overline{X})^2/\sigma^2$. The distribution of the value of this expression calculated for many samples from a normal population with known variance is distributed as χ^2 with $n - 1 \ df$. Notice that $1 \ df$ has been lost because of the substitution of the sample statistic $\overline{X}$ for the parametric value μ. If the numerator of this last expression is multiplied by $n - 1/n - 1$, we have:

$$\chi^2 = \frac{\dfrac{(n - 1)\Sigma (X - \overline{X})^2}{n - 1}}{\sigma^2} \text{ with } n - 1 \ df$$

However, $\Sigma (X - \overline{X})^2/n - 1 = s^2$. Therefore:

$$\chi^2 = (n - 1) \frac{s^2}{\sigma^2} \text{ with } n - 1 \ df \qquad\qquad \textbf{10.1-2}$$

This last formula says that if a ratio is formed by dividing a sample variance by the population variance and this ratio is then multiplied by $n - 1$, the result will be a test statistic that has the chi-square distribution. How many *df* should be used in evaluating this test statistic? If you guessed $n - 1$, you are correct.

We now return to our problem of testing the null hypothesis: $\sigma_T^2 \geqslant 100$. First, we make the modification in the production procedure. Then we take a random sample of the tablets produced under this modified procedure. Let us choose 61 tablets at random (you will see later why we choose 61 rather than 60). We now compute the variance of weights for these 61 tablets. Since this is a sample statistic, we use $n - 1$ in the denominator when we calculate the variance. Suppose that the variance turns out to be 70. The sample result is in the right direction to constitute evidence against the null hypothesis, since it is less than 100. The question is, "How often will a sample variance of 70 or less occur when we take samples of size 61 from a population in which the variance is 100?" If the answer is "rarely," or "less than alpha," we will reject the null hypothesis and permanently adopt the suggested modification.

The test statistic we will use to decide whether our sample statistic is a rare event is the ratio of the observed variance over the expected variance times the value $(n - 1)$. This statistic will have the chi-square distribution with $n - 1$ *df*. We calculate as follows:

$$\chi^2 = \frac{70}{100} (61 - 1) \text{ or } 42.0$$

How are we to interpret this value of χ^2? If the suggestion for modifying the manufacturing process does reduce variance, our ratio of observed sample variance divided by the standard population variance should be small. In other words, we are looking for a small value of χ^2, one that would occur only rarely when $\sigma_T^2 = \sigma_S^2$. We want to know where a value of 42 falls on the χ^2 distribution with 60 *df*. Can you anticipate the answer on the basis of what you have learned about the distribution of χ^2? Remember that the χ^2 distribution for 60 *df* can be expected to have a mean of 60, a variance of 120, and a distribution that approaches a normal distribution. With a variance of 120, the standard deviation will be just under 11. Where would a value of 42 fall on this distribution with a mean of 60 and a standard deviation of 11? The *z* score would be approximately 1.64. If the distribution were indeed normal, values this far below the mean would be expected to happen just over 5% of the time.

Now look at the table of χ^2 in Table E-8 in Appendix E. Remember that the table provides probabilities for values of χ^2 *greater* than those listed. From the table we find that 95% of the time χ^2 will be greater than 43.19. In other words, a value as small as our observed ratio will occur less than 5% of the time. We will probably conclude that the modified process does indeed result in less variability in the weights of tablets. (As an aside, can you see from the table why it was convenient to choose a sample of size 61 rather than 60?)

It should be noted that in this instance the result is significant even though χ^2 is less than the *df*. You remember that you were told earlier not to bother looking up $X_p{}^2$ if it was not at least one and one half times the *df*. This rule of thumb obviously does not apply in the case of evaluating χ^2 *computed to compare variances*. Whereas Pearson's chi-square statistic $(X_p{}^2)$ is only significant when it is larger than the *df*, the statistic χ^2 may be significant either because it is sufficiently *greater* or sufficiently *smaller* than the *df*. In the case of a *directional* hypothesis, the statement of the null hypothesis will indicate which tail of the distribution is of interest. In the case of a *nondirectional* hypothesis, you will have to *double* the tabled probabilities. Suppose that in our example the question had been whether the modification in this manufacturing process results in a change in the variability, without regard to the direction of the change. We would have concluded that a result as extreme as our observed χ^2 of 42 would happen less than 10% of the time rather than 5% as reported for our one-tail test.

COMPARISON OF χ^2 AND $X_p{}^2$

Before going on to a presentation of the methods for two-sample tests of variances, let us pause to compare χ^2 and $X_p{}^2$. Both of these symbols represent sample statistics. Historically, the use of χ^2 preceded the use of $X_p{}^2$. The basic use of χ^2 was to make a one-sample test of a variance, χ^2 being computed as $(n - 1)s^2/\sigma^2$. Of course, the statistic χ^2 would be of little value if its sampling distribution were unknown. Fortunately, the sampling distribution is known and appears in most statistics books.

Once χ^2 had been introduced and tables of its distribution had become available, the statistic $X_p{}^2$ (computed from the formula $\Sigma (f_o - f_e)^2/f_e$) was invented. The purpose of this statistic was completely different: to compare proportions (or frequencies) among several populations.

Once again the statistic $X_p{}^2$ would be of little value if its sampling distribution were unknown. As it turns out, the exact sampling distribution of $X_p{}^2$ is seldom used, since computations can become very complex. Fortunately, it turned out that the sampling distribution of χ^2 can be used as a fairly good approximation for the sampling distribution of $X_p{}^2$ so long as expected frequencies are sufficiently large.

Notice the difference in interpretations for $X_p{}^2$ used to compare proportions and χ^2 used for a one-sample test of a variance. Because of its method of computation, $X_p{}^2$ is usually a nondirectional test. Only large values of $X_p{}^2$ are evidence against the null hypothesis. If a directional test is to be carried out, we must first check to see if the data are in a direction to provide evidence against the null hypothesis. If the answer is yes, the probability obtained from the table of χ^2 must then be *halved*. In contrast, the χ^2 ratio, computed to test the null hypothesis that $\sigma_T = \sigma_S$, will constitute evidence against the null hypothesis *if the ratio is either too large or too small*. When an observed value of $(n - 1) s^2/\sigma^2$ is compared with a table of χ^2, the probability obtained from the table is a one-tail probability. If a two-tail probability is desired, the probability from the table must be doubled.

Over the years, the use of the statistic $X_p{}^2$ has greatly overshadowed the use of the statistic χ^2. There are many more problems in which one is interested in comparing proportions than problems in which one wishes to make a one-sample test of a variance. The result has been confusion, in which the chi-square distribution is often identified with the statistic X_p^2 rather than χ^2. Statistics books even use the symbol χ^2 for X_p^2 and then label the table of chi-square with the same symbol. It is hoped that this summary will make the distinction clear.

TWO-SAMPLE TESTS OF VARIANCES

In the previous section you learned to use the statistic χ^2 to make a one-sample test of a variance. This test statistic was shown to be identical to the ratio of the sample variance divided by the known population variance, multiplied by $n - 1$. The problem has also been solved for the case where the goal is to compare the unknown variances of *two* populations. Suppose the question is whether there is a difference between two producers (*A* and *B*) in the uniformity of the weights of the aspirin tablets they sell. The two populations are all of the tablets produced by each of the companies. There is no way to compute the two population variances, σ_A^2 and σ_B^2. We must make use of the procedures of inferential statistics to test the null hypothesis that $\sigma_A^2 = \sigma_B^2$. We take a random sample from each population and compute s_A^2 and s_B^2. If these sample statistics are very different, we would be inclined to conclude that $\sigma_A^2 \neq \sigma_B^2$. However, how different must they be before we will reach this conclusion? Suppose, for example, that a sample of size 41 from population *A* is found to have a variance of 146 and a sample of size 31 from population *B* is found to have a variance of 81. These variances are quite different, but are they so far apart as to cause you to decide that $\sigma_A^2 \neq \sigma_B^2$? Before going on, see if you can identify the steps you would follow to make the decision.

The outcome of truly understanding a principle or procedure is to be able to generalize it to a new situation. If you really understand the earlier chapter on two-sample tests for proportions and means, you should have reasoned as follows: "The first step will be to subtract one sample statistic from the other to find the difference. Then if I can identify what the sampling distribution is like for the difference between variances of pairs of samples taken from the same population, I can see where my observed difference falls on that distribution. If it is a rare event, then I will reject the null hypothesis."

If you managed to get this far in your reasoning, congratulations are in order. Your reasoning is correct. Now all we need to know is where to find a table of the sampling distribution of differences between pairs of sample variances. Unfortunately, you will not find the required table in this or other statistics books.

Having run into a dead end with this approach, can you think of another way to compare the sample variances? Think back to the one-sample test of variances. What about comparing two-sample variances by making a ratio rather than finding the difference between them? If you decide to form the ratio s_A^2/s_B^2 what kind of resulting number would represent evidence against the null hypothesis that $\sigma_A^2 = \sigma_B^2$? Either a large number or a number near zero would represent such evidence, depending on whether s_A^2 is bigger or smaller than s_B^2. For any pair of samples, the ratio of one observed variance to another is a statistic. If the sampling distribution of this statistic were known, the ratio could then be used to make two-sample tests of variances.

Think about what the sampling distribution of this ratio would be like. You should be able to guess two of the characteristics of the sampling distribution of the ratio of sample variances: central tendency and shape. If the samples were taken from populations having equal variances, what would you expect to be the mean of the sampling distribution of the ratios? If your intuition tells you that the mean of the sampling distribution will be one, you are correct. Now what are the limits for the sampling distribution? If the variance placed in the numerator is very small in relation to that in the denominator, the value of the ratio approaches zero. On the other hand, if the number placed in the numerator is very large compared with that in the denominator, the value of the ratio approaches infinity. What does this say about the shape of the sampling distribution? It must be a skewed distribution, since the mean is at one and the limits are zero in one direction and infinity in the other.

The ratio of sample variances receives widespread use. For once there is a symbol that has been uniformly used to designate the test statistic. The symbol used is the capital letter F, for its inventor, Fisher. The test statistic is commonly referred to as the F ratio, and tables of probabilities for its sampling distribution are referred to as F tables.

Before looking at the F tables, which appear as Table E-9 in Appendix E, can you visualize what form the tables will have? We have talked about the shape and the mean of the F distribution, but what about its variability? Under what circumstance would you expect sample F ratios to not only cluster about one, but to cluster *closely* about one? In other words, when will sample F ratios have *small* variability? When will sample F ratios have large variability? As a hint, when can an individual sample variance be expected to be very close to the actual population variance? Or more generally, when is a sample statistic expected to be close to the population parameter? It is hoped that by now every reader will conclude that large samples provide more precise estimates of parameter values.

The observation that the variability of the sampling distribution of F depends on sample sizes means that once again the concept of df will be used. In this instance, there is a value of df for both the numerator and the denominator of the F ratio. As a result, the table of F will have to be a matrix representing all possible combinations of df for the numerator and the denominator. There is a sampling distribution of F for each cell of the matrix; for example, 2 df for the numerator combined with 4 df for the denominator. If you now attempt to visualize what the table of F is like, you can appreciate how voluminous such a set of tables would have to be to include many points within each possible combination of df. As a result, you will not be surprised when you find that in Table E-9 the tables of F give only selected points for each combination of df. If you now turn to Table E-9, you will see that there are four tables, one for each of the following probabilities: 0.25, 0.10, 0.05, and 0.01. Within each table, the number within each cell is the number that divides off the upper 25%, 10%, 5%, or 1% of the distribution. It is traditional to give the number of *df* for the numerator first and then the *df* for the denominator. In commonly used terminology, we would "look up an observed *F* with 2 and 4 *df*." The *df* for the numerator are usually listed horizontally at the top of the table, and those for the denominator are listed vertically at the left. We find the cell corresponding to 2 and 4 *df* for the 0.05 table and learn that when population variances are equal, an *F* ratio as large as 6.94 will occur 5% of the time. Similarly, from the 0.01 table we learn that an *F* ratio as large as 18.0 will occur 1% of the time.

Notice that the tables do not give the corresponding values less than one. The need for probabilities of values below one can be avoided by making the larger variance the numerator, which always results in an $F \geq 1.00$. Care must then be used in interpreting the result. For *directional* tests, the first step is to see whether sample variances are in the proper direction. If not, no further work is necessary. There cannot be evidence against the directional null hypothesis. If the sample variances are in the right direction, the larger variance is made the numerator of the *F* ratio and the result is compared with the table to determine whether it is significant at the chosen alpha level. In the case of a *nondirectional test,* the larger sample variance is still made the numerator. However, in this case the probability obtained from the table *must be doubled* to allow for the two-tail test.

Let us now return to the example of the two companies that produce aspirin tablets. The question is whether there is a difference in the uniformity of weights of the tablets produced by companies A and B. The question as stated does not imply a directional test, so our null hypothesis is $\sigma_A^2 = \sigma_B^2$. We take a random sample from each population and compute the sample variances s_A^2 and s_B^2. Earlier you were told that the data were as follows: $n_A = 41$, $n_B = 31$, $s_A^2 = 146$, and $s_B^2 = 81$. Before reading on, see if you can use the *F* distribution to test the null hypothesis with an alpha level of 0.05.

Making the larger sample variance the numerator, we have Formula 10.1-3.

$$F = \frac{s_A^2}{s_B^2}$$

10.1-3

For our example: $F = \dfrac{146}{81} = 1.80$ with 40 and 30 *df*

Looking at the table of *F* with an alpha level of 0.05, we find that the cell corresponding to 40 and 30 *df* contains the number 1.79. Can you state in your own words the meaning of this number? The interpretation is as follows. Suppose that populations A and B have the same variance and a sample of size 41 is taken from population A and a sample of size 31 is taken from population B. If a ratio of the sample variance is formed by placing s_A^2 over s_B^2, then the ratio will exceed 1.79 5% of the time.

You were asked to test the null hypothesis at the 0.05 level. What do you conclude? Be sure you have considered everything you have been told about the *F* table before you decide. The answer is that the result is not significant at the 0.05 level. How does this conclusion match with the interpretation of the *F* table just stated? Remember that it was stated that a value *greater* than 1.79 will occur 5% of the time when the ratio is formed *by* *placing $s_A{}^2$ in the numerator*. Since we had decided to make a nondirectional test, we must also consider the other tail of the distribution, that is, extreme results when $s_A{}^2$ is less than $s_B{}^2$. The table of F does not include numbers for ratios in the direction toward zero. The solution, when a nondirectional test is to be made, is to place whichever number is larger in the numerator. When looking up the resulting F value in the table of F, we need only to remember to double the probabilities indicated for the one-tail test. In our problem we would conclude that a value as extreme as 1.79 *in either direction* will happen about 10% of the time. We therefore fail to reject the null hypothesis. The finding that one sample variance was 1.8 times greater than the other is not sufficiently rare to cause us to conclude that one company's tablets are more uniform in weight than the other's.

Notice that if we had had reason to believe in advance that $\sigma_A{}^2 > \sigma_B{}^2$ we would have made a directional test of the null hypothesis that $\sigma_A{}^2 \leq \sigma_B{}^2$. We would first check to see whether the sample variances were in the right direction. If not, there would not be any need to check further. The data could not be evidence against the null hypothesis. If the difference were in the right direction, we would once more form a ratio, making the larger variance the numerator. The one-tail probabilities listed in the table of F would then be appropriate. For our example, we would have rejected the null hypothesis at the 0.05 level.

RELATIONSHIP BETWEEN THE F DISTRIBUTION AND OTHER DISTRIBUTIONS

Before concluding this chapter on tests of variances, let us pause to examine the relationship of the F distribution to other distributions we have learned to use. You recall that like F, the statistic χ^2 can also be interpreted as a ratio: the χ^2 ratio involves a sample variance in the numerator and an actual population variance in the denominator. This ratio is multiplied by df. Under what circumstances would the F ratio be equivalent to the χ^2 ratio? It would be equivalent only if the sample variance in the denominator of the F ratio were equal to the population variance. When would this happen? It happens when the sample variance in the denominator is based on a sample of infinite size. As a result, there should be a direct relationship between the last row of the F table ($df = \infty$) and the columns of the χ^2 table. Remembering that χ^2 is a ratio multiplied by $n - 1$, we see that the expected relationship holds. If any entry in the 0.05 column of the table of χ^2 is divided by the df ($n - 1$), you will see that the result corresponds to the 0.05 value for the corresponding entry in the last row of the table of F. For example, for χ^2 with 10 df, the value under the 0.05 column is 18.31. Dividing by 10, we get 1.83. When we look at the F table for 10 and ∞ df, the value for the 0.05 level is 1.83.

The F distribution is also related to the t distribution. The relationship is similar to that between χ^2 and the normal distribution. Remember that with 1 df, $\sqrt{\chi^2} = z$. In the next chapter you will see that when there is 1 df for the numerator, $\sqrt{F} = t$. It appears that all of the distributions you have learned to use are interrelated. This is indeed the case, with the t distribution, the χ^2 distribution, and the F distribution all being derived from the normal distribution.

SUMMARY

In this lesson you learned that the distribution of χ^2 is also a sampling distribution of a statistic. The statistic is the sum of squared standard normal scores. It has a distribution that is extremely skewed to the right for small samples and approaches normality as sample size approaches infinity. Its mean and variance are n and $2n$ respectively. You then learned that the original use of χ^2 was to make a one-sample test of a variance. The test statistic χ^2 was shown to be equal to $(n - 1)$ times the ratio of the sample variance to the population variance. For any sample, this test statistic can be computed and used to test the null hypothesis, $\sigma_T^2 = \sigma_S^2$.

Subsequently, you learned that for the two-sample test of variances, you form the ratio of the sample variances. This ratio is called the F ratio. Its sampling distribution is a skewed distribution with a mean of one. The variability of the distribution of F depends on the size of samples chosen. Accordingly, a complete table of F would be a matrix with cells corresponding to various combinations of df for the numerator and denominator. A complete set of probabilities would be listed for each cell. Instead, an abridged set of tables is provided, with F values for the various combinations of df for only four levels of probability. The tables give one-tail probabilities, which can be doubled for the case of nondirectional tests.

At the end of the chapter, it was stated that the F table is related to the χ^2 distribution when the F is based on a sample of infinite size for the denominator. The next chapter will demonstrate another similarity between χ^2 and F. The statistic χ^2 was developed to make a one-sample test of a variance. Later, this use was overshadowed by the use of χ^2 as an approximation for the distribution of X_p^2, which is used to compare proportions. Similarly, although F is a statistic that can be used to compare two population variances, you will discover that this use is overshadowed by its use as a test statistic to make a simultaneous comparison of several means in a process called analysis of variance.

LESSON 1 EXERCISES

1. Reliability of a measurement or evaluation process can be thought of in terms of the extent to which the same result occurs where the same "thing" is measured on several occasions. In the case when several judges are asked to evaluate each of several objects, the goal is to have agreement among the judges. An index of this degree of agreement would be the variance of the ratings for any object. Suppose that as part of a state dental clinical examination, each candidate is asked to do a certain type of restorative procedure. When the candidate is finished, ten different examiners independently rate the result using a scale of 1 to 9. During the years that this procedure has been followed, the ratings for any one candidate have been found to be approximately normal, with a variance of 0.8. One year, because of a snowstorm, five of the regular examiners are not able to be present. However, replacements are available from among members of the local dental society. There is some concern that these inexperienced persons will reduce the reliability of the ratings. When the first candidate was rated, the results were: 7, 7, 5, 6, 8, 8, 7, 7, 7, 8. Is there evidence that there is less agreement than there had been previously when all judges were experienced? What is the null hypothesis, and what would be an appropriate alpha level? Carry out the appropriate inferential procedure.

2. An experimenter wishes to test whether two populations have the same mean. He plans to use the t test and remembers that the formulas differ depending on whether variances are assumed to be equal. The experimenter decides to test the null hypothesis that $\sigma_A^2 = \sigma_B^2$. He takes a sample of 41 from each population and gets: $s_A^2 = 100$ and $s_B^2 = 200$. Do you think he should use the formula that assumes that variances are equal? What alpha level is appropriate?

CHAPTER 11

Comparison of several means

Lesson 1 □ One-way analysis of variance

In the previous lesson, you learned about the F distribution and its use in comparing the variances of two populations. In this chapter, you will see how the F distribution can be used to compare the *means* of *several* populations. The process used is called analysis of variance; it is abbreviated ANOVA. As the rationale of ANOVA is presented, you will see why the designation ANOVA is appropriate even though the goal is to compare means or to make an "analysis of means."

This first lesson will explain how to compare the means of several populations that differ in a single factor. The process is called a *one-way analysis of variance*. As an example, your interest might be in comparing first-, second-, third-, and fourth-year medical students on some variable, such as a measure of altruism. The four groups or populations of interest are distinguished from each other only by their current class standing. If all four groups were divided on the basis of a second factor, such as sex, there would be two factors, and a two-way ANOVA would be required. You will discover that both the rationale and the basic computational procedures for the one-way ANOVA are very much the same as for more complex designs.

THE LOGIC OF ANOVA

Many students learning how to carry out an ANOVA become familiar with formulas for the computations without ever understanding why the process works. This is because an examination of the computational formulas does not readily lead to an understanding of the logic. To make the process of ANOVA meaningful, we will start with an explanation that makes the rationale clear, even though the computations are laborious. Then in a later section we will show a simplified computational routine that gives the same result. To make the process concrete, we will use an example with the same set of numbers for both explanations.

Suppose that the goal of a research project is to discover whether there are differences in the mean altruism scores for the four classes of students in a college of medicine. Suppose also that a reliable, valid measure of altruism can be obtained through a process that includes both the completion of an attitude scale and an interview by a trained psychologist. The result is a score that has a distribution which is approximately normal and ranges from 0 to 9. If the measurement process were less involved, an altrusim score would be obtained for each of the students in all four classes and the means would then be compared. Because the process to be used to measure altruism is rather costly and time-consuming, however, only a sample of students from each class is selected. To use these sample results to make an inference about the entire set of classes, the samples are chosen using a table of random numbers. For our purposes, to keep the computations at a manageable level, ten students are chosen from each class.

The results might appear as in Table 11-1. The top part of the table contains the ten scores for the four random samples. The bottom part of the table contains the summary statistics for these scores. All of the statistics should be familiar to you. Because the explanation that follows requires that you understand what these statistics are, the last column contains question marks. You should calculate the value that is to replace each question mark. See if you can do so before proceeding.

The first incomplete row, with the symbol n, contains the number of scores in each sample. Accordingly, the first question mark can be replaced by the number 10. The second such row contains the sums of the scores within each sample. For the sample from the fourth year, the sum is 28. The third incomplete row contains the sum of the squares of each of the scores within each sample. If you square each of the scores from the fourth year and add the results, you find the result to be 110.

Replacing the remaining question marks by the appropriate values requires you to recall concepts learned much earlier. The value to replace the question mark in the fourth

Table 11-1. Altruism scores for students chosen at random from four medical school classes

i	First year	Second year	Third year	Fourth year
1	3	3	2	4
2	5	0	1	4
3	4	4	2	1
4	2	3	3	5
5	4	4	4	0
6	3	1	4	2
7	4	4	3	0
8	7	1	6	4
9	2	1	0	4
10	6	6	4	4
n	10	10	10	?
ΣX	40	27	29	?
ΣX^2	184	105	111	?
Σx^2	24	32.1	26.9	?
$\overline{X}$	4.0	2.7	2.9	?
s^2	2.67	3.57	2.99	?
$SE_{\bar{x}}^2$	0.267	0.357	0.299	?

row of summary statistics was given a name when it was first introduced. Do you remember that we referred to the sum of the squares of deviations as the "brick," because it is involved in so many statistical computations? We gave both a definitional and a computational formula for this concept. The definitional formula is $\Sigma x^2 = \Sigma(X - \overline{X})^2$. The computational formula is $\Sigma x^2 = \Sigma X^2 - (\Sigma X)^2/n$. The value for sample scores for the fourth year is 31.6. The symbol $\overline{X}$ for unfinished row 5 refers to the sample mean, which for the last column is 2.8.

The concepts called for in the last two incomplete rows should also be familiar, even though you may not immediately recall the formulas for their computation. The symbol s^2 refers to the sample variance. It is the square of the standard deviation. The formula is $s^2 = \Sigma x^2/n - 1$. (Do you remember why $n - 1$ is used?) The result for the fourth-year class is 3.51.

The symbol in the last row stands for the variance of the means. It is the square of the standard error of the mean. You learned earlier that the standard error of the mean is found by dividing the population standard deviation (σ) by the square root of the sample size. When the population σ is unknown, the sample standard deviation *(s)* is used to estimate it. For the value that takes the place of the last question mark, we divide the sample variance (3.51) by the sample size and get 0.351.

Now that you have a good understanding of each of the statistics in Table 11-1, we return to our original interest: possible differences among the four class means. If there were only two classes involved, the null hypothesis would state that the two population means were equal. We would consider any large difference between the two observed sample means as evidence against the null hypothesis. To determine whether the observed difference is greater than would be expected by chance, we would form a ratio by dividing the observed difference by the standard error of the difference between means. To determine statistical significance, the ratio would be compared with either the values of the normal curve or the *t* distribution, depending on whether σ were known or unknown.

It was mentioned earlier that the general inferential procedures you have learned will be applicable to more complex cases. Can you "discover" the process of ANOVA for this problem, in which there are four groups rather than two? What will the null hypothesis be? How would you decide whether sample means differ from each other more than one would expect by chance? What table will be used to determine statistical significance?

Let us examine these questions one at a time. First, as you no doubt guessed, the null hypothesis will be that all four classes have the same population mean. Similarly, you should know intuitively that the greater the differences among sample means, the stronger the evidence against the null hypothesis. The problem is how we will measure the extent of differences among the four means. We could take all possible pairs of means and do numerous two-sample *t* tests. However, you remember from Chapter 9, on chi-square, that making multiple tests increases the probability of making a Type I error. What we need is a single number that summarizes the differences among the four sample means, that is, a number that tells on the average how far they are apart. If we used the average distance of each mean from the others, the process would probably be called "analysis of average absolute deviations." Instead, we use the variance of the means as the number that indicates the distances between sample means. The process is called "analysis of variance." In other words, the process involves computing the variance of the four sample means. The larger the variance of the means, the stronger the evidence against the null hypothesis.

The next question to be answered is, "How large a variance of sample means can be expected just by chance when the null hypothesis is true?" You learned earlier that the standard error of the mean $(SE_{\bar{x}})$ represents the standard deviation of the means of many samples of the same size all taken at random from a single population. The square of $SE_{\bar{x}}$ would represent the variance of these many sample means. The answer to our question is then obvious: find $SE_{\bar{x}}$ and square it to get the expected variance of the means. The only problem is how do we find $SE_{\bar{x}}$? The formula is $SE_{\bar{x}} = \sigma/\sqrt{n}$. Squaring, we get the formula for the expected variance of means: $SE_{\bar{x}}^2 = \sigma^2/n$. The difficulty is that we usually do not know σ^2. Instead, we use the approximate formula, $SE_{\bar{x}}^2 \approx s^2/n$, where s^2 stands for sample variance. In our problem, there are four sample variances and we therefore will get four separate estimates of $SE_{\bar{x}}^2$. In cases where all sample sizes are the same, there is no reason to consider any of the estimates of $SE_{\bar{x}}^2$ better than any other. Since we need to arrive at a single estimate of the expected variance of the sample means, we find the average of the four estimates.

Completing these calculations, we then have an observed variance of the sample means and an expected variance of sample means (expected under the assumption that the null hypothesis is correct). If the observed variance is greater than the expected variance, we have evidence against the null hypothesis. Our final question is what table to use to determine whether the observed variance is "significantly" greater than the expected variance. Since we want to compare two variances, we will form an F ratio, placing the observed variance of sample means over the expected variance of sample means. To look this F value up in the tables, we will need to know the df for the numerator and the denominator. The df for the numerator is the number of groups less one, in our case $4 - 1$, or 3. For the denominator the df is found by finding the df for each sample estimate of $SE_{\bar{x}}^2$ and then adding the results from all samples. For each sample, the df equals the sample size less one (in our case df equals 9). Adding df for the four sample estimates of $SE_{\bar{x}}^2$ gives 36 df for the denominator.

Now that we have tracked through the entire process, we are ready to go back and do a one-way ANOVA for the data of Table 11-1. Looking at the data, we see that the sample means for first-year students appears to be higher than for the other three classes. This is consistent with a common belief that students enter medical school motivated by a strong desire to be of service to others, but that this altruistic motivation declines while they are in training. However, are the differences among these sample means greater than one would expect if samples of this size were taken from populations having a common mean? To find out, we will make an F ratio of the observed variance of means over the expected variance of means.

To find the observed variance of means, we find the variance of the four means: 4.0, 2.7, 2.9, and 2.8. The computational formula for variance is:

$$s^2 = \frac{\Sigma X^2 - \frac{(\Sigma X)^2}{n}}{n - 1}$$

In our case, there are four sample means; so n is 4. The sum of the four means is 12.4, and the sum of the squares of the four means is 39.54. Placing these values in the formula, we find the observed variance of the means to be 0.3667.

You will remember that the last row of Table 11-1 gives the four sample estimates of the variance of the means: 0.267, 0.357, 0.299, and 0.351. The average of these four estimates gives us our best single estimate of the expected variance of the means, or 0.3185.

Our resulting F ratio is therefore:

$$F = \frac{\text{obs.var.}}{\text{exp.var.}} = \frac{0.3667}{0.3185} = 1.15 \text{ with 3 and 36 } df$$

Referring to the table of the F distribution, we find that there is no entry corresponding to 3 and 36 df. To be conservative, we go down to the next lower number on the table. For 3 and 30 df, we find that an F value ≥ 2.92 occurs 5% of the time. The F value we computed is well below this value, and in fact it is quite near 1.0. Our conclusion is that we do not have sufficient evidence to reject the null hypothesis that all four classes have the same mean score for altruism.

Before going to the computational methods for a one-way ANOVA, let us summarize the logic of the method as it has been presented. The goal is to compare a number of population means. Samples are chosen randomly from each population. Evidence against the null hypothesis results when the means are far apart. The variance is used to measure this variability of sample means. We have called this the observed variance of the means. This observed variance of sample means is compared with the variance expected under the assumption that the null hypothesis is true. Each sample provides an estimate of the expected variance of means. These estimates are averaged to find a best single estimate. The observed and expected variances are compared by forming an F ratio with the observed variance as the numerator. If k is used to represent the number of samples all of size n, the F distribution computed has $k - 1$ and $(n - 1)k$ df.

COMPUTATIONAL PROCEDURES

The procedure just described is appealing, in that it is based on the concept of a standard error of the mean, a concept already used for the one- and two-sample case. Furthermore, in the one-way ANOVA with samples of equal size, the computations are not too difficult. However, when sample sizes vary, or when two or three factors are involved, the method described becomes inadequate. A computational method that may not be as intuitively appealing, but that makes it easier to identify the effects of each factor, is needed.

Because the rationale that underlies the computational method can be difficult to grasp, the computations will be presented first. Once you have understood the computations, perhaps an explanation of the rationale behind the computational method will make sense.

In a *multifactor* experiment there are quite a number of computations to be made. Fortunately, there is an algorithm that applies no matter how complex the design. Once you learn the algorithm, you will be able to do the basic computations for any ANOVA. The central concept underlying the computational method is to divide the total variability of scores into ''between groups'' variability and ''within groups'' variability and to use the ratio of these two as a test of significance.

For initial computations, rather than using the variance as a measure of variability, the ''sum of squared deviations'' (Σx^2) is used. In analysis of variance, the symbol generally used is SS; SS_T stands for the total sum of squared deviations, SS_{bcells} stands for the sum of squared deviations between groups, and SS_{wcells} stands for the sum of squared deviations within groups. The reason for using the sum of squared deviations rather than the variance as a measure of variability is that SS_{bcells} plus SS_{wcells} equals SS_T. This additivity is useful, since SS_T and SS_{bcells} are relatively easy to compute. Then SS_{wcells}, which requires more calculations, can be obtained by subtraction.

$$SS_{wcells} = SS_T - SS_{bcells} \qquad \textbf{11.1-1}$$

To make concrete the computations of SS_T, SS_{bcells}, and SS_{wcells}, let us return to the example of the four samples of altruism scores in Table 11-1. To find SS_T we pool all 40 scores and find Σx^2 as though the scores were a single group. We will find it easy to use the computational formula for the sum of squared deviations:

$$\Sigma x^2 = \Sigma X^2 - \frac{(\Sigma X)^2}{n}$$

When showing the application of this formula in ANOVA. most books use double summation signs to indicate that summation is for all scores for all groups. Since many students have trouble with the complex formulas that result, we will use the capital letter T to represent total and the single subscript j to represent groups. The letter T without a subscript represents the total of all scores; similarly the letter n without a subscript represents the total number of scores. Accordingly:

$$SS_T = \Sigma X^2 - \frac{T^2}{n} \qquad \textbf{11.1-2}$$

where T equals the sum of all scores and
       n is the number of all scores.

This formula tells us that to find SS_T, we need to know the sum of all the scores and the sum of squares of all the scores. The last term of this equation, T^2/n, is used repeatedly in subsequent computations. To refer to it easily, it is often given a name, the *correction factor*. For convenience, we will use the capital letter C to represent this value.

The formula for finding SS_{bcells} can be expressed using the symbols we have defined.

$$SS_{bcells} = \left(\sum \frac{(T_j)^2}{n_j} \right) - C \qquad \textbf{11.1-3}$$

It is probably harder to remember the formula than it is to remember what it says to do. First, find the total for each group; square that total and divide by the number in the group. Add all of these values and subtract the correction factor.

As indicated in Formula 11.1-1, once SS_T and SS_{bcells} are known, SS_{wcells} can be found by subtraction.

Now see if you can find SS_T, SS_{bcells}, and SS_{wcells} for the data in Table 11-1. Using the statistics in the last seven rows of the table makes the work easy. Before proceeding, see if you can get these values. For SS_T, we need to know the total number of scores, the sum of all scores, and the sum of squares of all scores. By adding across the first three rows, we find that $n = 40$, $T = 124$, and $\Sigma X^2 = 510$. We then compute SS_T as follows:

$$SS_T = 510 - \frac{(124)^2}{40}$$

$$SS_T = 510 - 384.4$$
$$SS_T = 125.6$$

We note that $C = 384.4$. Then, to find SS_{bcells} and SS_{wcells}, we calculate as follows:

$$SS_{bcells} = \frac{(40)^2}{10} + \frac{(27)^2}{10} + \frac{(29)^2}{10} + \frac{(28)^2}{10} - 384.4$$

$$SS_{bcells} = 160 + 72.9 + 84.1 + 78.4 - 384.4$$
$$SS_{bcells} = 395.4 - 384.4$$
$$SS_{bcells} = 11.0$$

By subtraction:

$$SS_{wcells} = 125.6 - 11.0 = 114.6$$

You remember that the fourth row from the bottom of Table 11-1 contains the sum of squared deviations within each group. Just to satisfy yourself that subtracting to get SS_{wcells} is appropriate, add up the numbers in this row. You should get the same number, 114.6.

We have now completed the major computations for the ANOVA table, which appears frequently in scientific publications. The remainder of the table converts the SS to variances by dividing by df, and then provides the value of the F ratio for the test of significance. Let us see what the table looks like and how it is arrived at for our data. The table appears as Table 11-2.

Table 11-2. ANOVA for altruism scores of students chosen at random from
four medical school classes

Source	SS	df	ms	F
Between	11.0	3	3.67	1.15
Within	114.6	36	3.18	
TOTAL	125.6	39		

Notice that for the column for *df* we use the same values as for the earlier analysis. Four groups less one gives 3 *df* between; ten scores less one for the four groups gives 36 *df* within. Notice also that the total *df* is the total *n* less one.

The column headed *ms* refers to mean square, or average sum of squared deviations. It is obtained by dividing *SS* by *df*. Finally, *F* is obtained by dividing ms_{bcells} by ms_{wcells}. As you can see, the value of *F* is exactly the same as that which we obtained earlier, and it is again interpreted with 3 and 36 *df*.

Now that we have gone through all of the computations, let us summarize the steps.

1. Pool all the scores, and find *n*, *T*, and ΣX^2.
2. Find *C* by squaring *T* and dividing by *n*.
3. Find SS_T by subtracting *C* from ΣX^2.
4. Find the total for each of the *j* groups.
5. Square each group total, and divide by the number in the group.
6. Find SS_{bcells} by adding the results from step 5 and subtracting *C*.
7. Find SS_{wcells} by subtracting SS_{bcells} from SS_T.
8. Find *df* for between, within, and total.
9. Divide *SS* by *df* to get *ms*.
10. Divide ms_{bcells} by ms_{wcells} to get *F*.

You will see in a later lesson that these steps are a general outline for the calculations of ANOVA. In a multifactor design, steps 4, 5, and 6 become an algorithm applied once for *each* of the ways of grouping scores.

The computations of steps 1 through 10 should now be clear. However, you probably have little notion as to why the process works. Unfortunately, many students understand the rationale as presented in the earlier section but have great difficulty in seeing the logic of the computational method. Let us try to make the process understandable.

Just as the variance is a measure of variability, so also is the sum of squared deviations *(SS_T)*. The only difference is that whereas SS_T is just a sum of squared deviations from the mean, the variance is an average, since the sum has been divided by $n - 1$. Nevertheless, the greater the differences among individual scores, the greater the value of SS_T.

The computational formulas for ANOVA are based on a partitioning of the sum of squares of the scores. The question asked is, "What explanations are available for the observed differences among the total set of scores?" If the total group of scores can be divided into subgroups, and it is found that these subgroups have different means, the question is partially answered. In other words, part of SS_T arises because of group differences. The remaining portion of SS_T is unexplained. It represents the variability of scores within the subgroups, or SS_{wcells}.

The concepts SS_T, SS_{bcells}, and SS_{wcells} can best be illustrated by going back to the scores of Table 11-1 and considering what happens under each of two circumstances. First, let us see what happens if there is no variability within groups; that is, if *every score is at the group mean*. In such a case, all ten scores for first-year students would be 4.0, all ten scores for second-year students would be 2.7, and so forth. Finding SS_T for this set of scores can be done rather easily:

$$n = 40$$
$$\Sigma X = 10(4.0) + 10(2.7) + 10(2.9) + 10(2.8) = 124.0$$
$$\Sigma X^2 = 10(4.0^2) + 10(2.7^2) + 10(2.9^2) + 10(2.8^2) = 395.4$$
$$SS_T = 395.4 - \frac{(124)^2}{40}$$
$$SS_T = 395.4 - 384.4$$
$$SS_T = 11.0$$

Notice that SS_T now equals 11.0, the same value we previously obtained for SS_{bcells}. This makes sense because SS_T must equal SS_{bcells} plus SS_{wcells}; but SS_{wcells} equals zero because there is no variability within groups. You can verify that using the algorithm described earlier to find SS_{bcells} will also result in the value 11.0.

Now consider a second circumstance, in which the scores have the original differences within groups, but *all groups have the same mean*. The data of Table 11-1 could be modified by adding constants to the scores within each group so as to make all group means equal. Assume that all group means are to be made equal to 4.0. We need do nothing with the scores of first-year students, since the mean is already 4.0, but we add 1.3, 1.1, and 1.2 respectively to the scores of second-, third-, and fourth-year students. Adding these constants will not affect the variability of scores, only the means. The modified scores will now appear as in Table 11-3.

We again find SS_T, using the numbers at the bottom of the table.

$$SS_T = \Sigma X^2 - \frac{(\Sigma X)^2}{n}$$
$$SS_T = 754.6 - \frac{(160)^2}{40}$$
$$SS_T = 754.6 - 640$$
$$SS_T = 114.6$$

Notice that this time SS_T equals 114.6, the same value we previously figured for SS_{wcells}. Again, this makes sense because in this instance there are no differences between group means, making SS_{bcells} equal to zero.

If you now think about what happened to the relationships between SS_T, SS_{bcells}, and

Table 11-3. Modified altruism scores for students chosen at random from four medical school classes

i	First year		Second year		Third year		Fourth year		
1	3		4.3		3.1		5.2		
2	5		1.3		2.1		5.2		
3	4		5.3		3.1		2.2		
4	2		4.3		4.1		6.2		
5	4		5.3		5.1		1.2		
6	5		2.3		5.1		3.2		
7	4		5.3		4.1		1.2		
8	7		2.3		7.1		5.2		
9	2		2.3		1.1		5.2		
10	6		7.3		5.1		5.2		
									TOTALS
n	10	+	10	+	10	+	10	=	40
ΣX	40	+	40	+	40	+	40	=	160
ΣX^2	184	+	192.1	+	186.9	+	191.6	=	754.6

SS_{wcells}, in each of these two circumstances, you will gain insight into the computational methods of ANOVA. When all scores within each group were equal to the group mean, there was no variability within. We then found that SS_T equaled SS_{bcells}. The concept of SS_{bcells} can therefore be thought of as the variability arising solely from *differences in group means*. On the other hand, when all group means were made equal without affecting the variability within groups, SS_T equaled SS_{wcells} and, of course, SS_{bcells} equaled zero. In other words, the concept of SS_{wcells} can be thought of as the *variability within groups, independent of any differences* that may exist *among group means*.

From these two illustrations, it should be apparent that indeed $SS_T = SS_{bcells} + SS_{wcells}$. It should also be clear that SS_{bcells} is a measure of the portion of the total variability that results from difference among group means and that SS_{wcells} is the remaining portion resulting from differences within groups. Clearly, the larger the value for SS_{bcells}, the greater the evidence of real differences among populations means. Because sample means can be expected to vary by chance depending on (1) the sizes of samples and (2) the variability of individual scores within the population, it makes sense to compute the F ratio by first dividing by df and then making a ratio of ms_{bcells} over ms_{wcells}.

SUMMARY

In the previous lesson you learned to compare population variances by making a ratio of sample variances. The ratio was called F and was compared with a table of the F distribution. The current lesson explained how the F ratio can be used to compare several population means in a process called ANOVA.

The process of ANOVA was described using two different approaches for the same set of data. In what might be called the conceptual approach, the F ratio is formed by placing the observed variance among sample means as the numerator of the ratio and the expected variance of the means as the denominator. The observed variance was the variance computed for the observed sample means. The expected variance was the average of the square of $SE_{\bar{x}}$ computed for each sample.

The computational process for ANOVA is based on a partitioning of the sum of squared deviations for the total set of scores (SS_T) into a sum of squared deviations related to differences among groups (SS_{bcells}) and a sum of squared deviations within groups (SS_{wcells}). Since $SS_T = SS_{bcells} + SS_{wcells}$, it is possible to compute SS_T and SS_{bcells} and then obtain SS_{wcells} by subtraction. The computation of SS_T is straightforward. It is the sum of squared deviations for the total set of scores. The computation of SS_{bcells} follows an algorithm that works both for the one-variable case presented in this chapter and for more complex situations to be presented subsequently. Once SS_T, SS_{bcells}, and SS_{wcells} have been computed, the next step is to figure df for each of these. Then each sum of squares is divided by df to arrive at a mean square (ms). Finally, ms_{bcells} is divided by ms_{wcells} to obtain an F ratio. This F ratio is compared with a table of the F distribution. If the F value is larger than the tabled value, there is evidence against the null hypothesis that all group means are equal. In the next lesson, you will learn of procedures for determining which means differ from the others. You will also learn about the assumptions for the use of ANOVA. In a third lesson, you will learn how to use ANOVA for situations involving more than one variable. Finally, in a fourth lesson, you will learn how to handle some complications that arise in certain research designs.

LESSON 1 EXERCISE

The survival of automobile accident victims is greatly affected by the length of time that elapses before medical care is available. Ideally, emergency care would be uniformly available quickly throughout a metropolitan area. However, there may be differences among precincts, depending on traffic patterns, ambulance services, and location of hospitals. To determine whether differences do exist in a certain city, all accidents were investigated for a 1-week period for the four precincts of the metropolitan area. The variable studied was the number of minutes elapsing from the time of an accident until the arrival of an emergency vehicle. The number of accidents requiring emergency medical care for the four precincts were respectively 5, 6, 8, and 9. The number of minutes elapsed before the emergency vehicles arrived were as follows:

Precinct A	Precinct B	Precinct C	Precinct D
6	9	16	11
10	10	13	17
7	7	15	12
5	12	7	9
11	13	19	14
	8	12	16
		12	21
		14	14
			15

Use the computational method for ANOVA to test the null hypothesis that all four precincts have the same average response time. Make up an ANOVA table as part of your answer.

Lesson 2 □ Individual comparisons, assumptions, and models

For the exercise at the end of the previous lesson, you should find that the F value computed is significant at the 0.01 level. The null hypothesis can be rejected and the conclusion reached that the average length of time before an emergency vehicle arrives at the accident scene is not the same in the four precincts. However, having merely determined that all four means are not equal is usually not the end point in the analysis. We want to know which means differ from which others.

In an experiment involving a number of groups, it is possible to make comparisons between any two groups or between any combinations of groups. For example, with four precincts, A, B, C, and D, we could compare A with B, A with C, A with D, B with C, B with D, and C with D. However, we could also compare the mean of A with the average of the means of B and C, or we could compare the average of the means of A and B with the average of the means of C and D, and so forth. As the number of groups increases, the number of possible comparisons increases rapidly. These comparisons are not all independent of each other. For example, if the mean of A is greater than the mean of B, and the mean of B is greater than the mean of C, then the mean of A must be greater than the mean of C.

Any method for comparing individual means or combinations of means must make allowance for the number of comparisons being made and the independence of the comparisons. A distinction also needs to be made between those comparisons planned in advance of the experiment and comparisons made simply to explore the data. Unfortunately, there is no single, universally accepted procedure that accommodates both of these considerations. Instead, a number of techniques exist, both for the one-factor and multifactor case and for planned and post hoc comparisons. A discussion of the merits of each of these methods is beyond the scope of this book. Instead, a single method, proposed by Scheffé, will be described. It has wide applicability, and even if it is applied in a situation in which another technique would be more appropriate, it errs on the side of being conservative. Unfortunately, the procedure is a bit complicated and will require a rather lengthy explanation.

THE SCHEFFÉ METHOD

As indicated earlier, if there are a large number of groups in an experiment and all possible comparisons are made using two-sample t tests, it becomes almost certain that some will be found to be significant. The Scheffé procedure is a method of making allowance for the multiple comparisons being made. The method is designed for post hoc examinations of the data after an overall F test has shown that somewhere among the means there are real differences.

The basic rationale of the Scheffé method is to set up a confidence interval for any comparison that is desired. For example, if the mean of group A is to be compared with the mean of group B, the technique involves establishing a confidence interval for the difference between population means A and B. If the confidence interval includes zero, the comparison is not considered to be significant. If the confidence interval does not include zero, the null hypothesis is rejected and the conclusion is reached that mean A does not equal mean B.

Earlier, you learned to set a confidence interval for the difference between the means of two groups. The procedure was the same as for any confidence interval:

Observed statistic ± (appropriate tabled value)(standard error of the observed statistic)

The observed statistic in this case is the difference between the two means. The appropriate tabled value is obtained either from the normal curve or the t distribution, depending on whether σ is known or unknown. The standard error is the standard error of the difference between means.

In the Scheffé procedure, the basic concept of the confidence interval remains unchanged. The only differences are:

1. The observed statistic may be a difference based on a comparison of a *combination* of means.
2. The appropriate tabled value will be obtained from the F distribution and will be adjusted for the multiple comparisons.
3. The standard error of the observed statistic will be the ms_{wcells} from the analysis of variance, appropriately weighted for the comparison desired.

The easiest way to make the meaning of these statements clear is to go to a concrete example. Let us use the results of the exercise at the end of the previous lesson. The problem was to determine whether there were differences in the average lengths of time between the occurrence of an accident and the arrival of an emergency vehicle, among four precincts of a city. A solution to the problem yields the following means for the four precincts: 7.8, 9.83, 13.5, and 14.33. The ANOVA table appears as follows:

Source	SS	df	ms	F
Between	183.37	3	61.12	6.12
Within	239.63	24	9.98	
TOTAL	423	27		

With an F of 6.12 with 3 and 24 *df*, the result is found to be significant at the 0.01 level. We conclude that the means are not the same for all four precincts.

Once we have concluded that differences exist, we then ask, "Where are the differences?" Is the mean for precinct A different from that of B? Is the mean for precinct C different from that of precinct D, and so forth. In addition, suppose that precincts A and B are on the east side of the city and are served by one set of ambulance services and precincts C and D are on the west side of the city and are served by another ambulance service. We may not have thought of it before the experiment, but now we may see that there appears to be a difference between the east side precincts and the west side precincts. We wish to test to see whether the difference is significant. Our question becomes, "Is there a difference between the average of the means of precincts A and B and the average of the means of precincts C and D?'' Of course, many other comparisons are possible as well.

It is convenient to be able to represent any desired comparison by a system of coefficients, or weights, to be applied to the respective means to make the comparison. For example, to compare the means of precincts A and B, you will subtract one from the other to obtain an observed difference. This comparison can be represented by the coefficients $+1$ and -1. In other words, if the mean of A is multiplied by $+1$ and the mean of B by -1, and the results are then added, the final result will be the difference between means A and B. To compare the average of the combination of A and B with the average of the combination of C and D, we would take $^1/_2$ the mean of A plus $^1/_2$ the mean of B, then subtract $^1/_2$ the mean of C and $^1/_2$ the mean of D. Our coefficients are therefore $+^1/_2$, $+^1/_2$, $-^1/_2$, and $-^1/_2$. Can you figure out what the coefficients would be to compare the mean of precinct A with the average of the means of the other three precincts? Since there are three other precincts, taking an average involves dividing by 3 or, alternatively, multiplying by $^1/_3$. Our coefficients are therefore $+1$, $-^1/_3$, $-^1/_3$, and $-^1/_3$. Notice that in each case the sum of the positive coefficients is $+1.00$ and the sum of the negative coefficients is -1.00. These coefficients will be used in the calculation of the standard error when we set up our confidence interval for any desired comparison.

Now let us examine the data from our example. Precinct A has a lower sample mean than Precinct B. We decide to see whether the difference is significant. To obtain a positive difference, we subtract the mean of A from the mean of B. Our coefficients are -1 and $+1$. By multiplying the means for A and B by -1 and $+1$ respectively and then adding, we get an observed difference equal to 2.03. Our confidence interval for the difference between the mean of A and the mean of B is therefore:

$$2.03 \pm \text{ (appropriate tabled value)(standard error of the statistic)}$$

To complete the analysis, we need to compute the values to go into the two sets of parentheses. For the first set of parentheses, we obtain a value from the F distribution corresponding to the chosen level of significance. The df for choosing the appropriate F are those listed in the ANOVA table. Suppose that we decide to make our tests at the 0.01 level. Looking at a table of F with 3 and 24 df, we find that the value of F that is exceeded only 1% of the time is 4.72. Now we must modify this value based on the number of comparisons possible. It can be shown that for k groups, there are $k - 1$ independent comparisons possible. Accordingly, we multiply the F value required for significance by $k - 1$. In our case k is 4, so we multiply by 3 and get 14.16. Recall that in our earlier discussion of confidence intervals, the tabled value was z or t. However, what we have is a modified value of F. Fortunately, there is a direct relationship between the F value appropriate for a comparison of two means and the coresponding t value. If you examine the column of the F table for 1 df in the numerator, you will see that the numbers are the squares of the numbers listed in a table of t. In others words, for the case of 1 df in the numerator, the square root of F is the same as t. Our final step in calculating the number to go in the first set of parentheses is to take the square root of the adjusted F value. Taking the square root of 14.16, we get 3.76 as our appropriate tabled value. To summarize, the calculation of an appropriate tabled value is made as follows: (1) find the F value for the chosen level of significance, using the *df* for between and within from the ANOVA table, (2) adjust this F value by multiplying by the number of groups less 1, and (3) take the square root of the result.

The next step is to calculate the number for the second set of parentheses, the standard error for the observed difference. In an earlier lesson, the formula for the standard error of a difference between means was given as follows:

$$SE_{\bar{x}_A - \bar{x}_B} = \sqrt{\frac{\sigma_A^2}{n_A} + \frac{\sigma_B^2}{n_B}}$$

If we assume that $\sigma_A = \sigma_B$, the formula can be written:

$$SE_{\bar{x}_A - \bar{x}_B} = \sqrt{\sigma^2 \left(\frac{1}{n_A} + \frac{1}{n_B}\right)}$$

You also learned earlier that in the two-sample case where it is assumed that $\sigma_A = \sigma_B$, the best estimate of the common σ is found as follows:

$$s = \sqrt{\frac{\Sigma x_A^2 + \Sigma x_B^2}{n_A + n_B - 2}}$$

If we eliminate the square root sign, we have our best estimate of σ^2.

$$s^2 = \frac{\Sigma x_A^2 + \Sigma x_B^2}{n_A + n_B - 2}$$

Substituting this value for σ^2 in the previous formula, we have an imposing-looking formula for the standard error of the difference between means.

$$SE_{\bar{x}_A - \bar{x}_B} = \sqrt{\left(\frac{\Sigma x_A^2 + \Sigma x_B^2}{n_A - n_B - 2}\right)\left(\frac{1}{n_A} + \frac{1}{n_B}\right)}$$

Of course, we also need to be aware that the computation of Σx_A^2 and Σx_B^2 involves additional steps. We use the computational formula:

$$\Sigma x^2 = \Sigma X^2 - \frac{(\Sigma X)^2}{n}$$

At this point, you have no doubt concluded that the calculation of the number to go in the second set of parentheses involves an overwhelming amount of work. You will be pleasantly surprised to learn that you already did the majority of the work when you carried out the computations for ANOVA. The value calculated for ms_{wcells} is exactly equal to the result of carrying out all of the calculations indicated by the expression:

$$\frac{\Sigma x_A^2 + \Sigma x_B^2}{n_A - n_B - 2}$$

You arrived at ms_{wcells} while avoiding much of the work through use of the computational method of ANOVA.

Substituting ms_{wcells} for σ^2 in the formula for the standard error of the difference between means, we have:

$$SE_{\bar{x}_A - \bar{x}_B} = \sqrt{ms_{wcells} \left(\frac{1}{n_A} + \frac{1}{n_B} \right)}$$

This formula is appropriate for the case of comparisons between any two means. However, if we want to make it appropriate for comparisons between any combination of means, we will have to allow for the coefficients corresponding to the desired comparison. Instead of having only two fractions in the parentheses in the formula, we will have a fraction for each group involved in the desired comparison. Also, instead of having 1 as the numerator for each fraction, we will have the square of the coefficient that is to be applied to each sample mean. If we use the symbol c to represent the coefficients and the letter j to designate any particular group, the general formula for the number to go in the second set of parentheses becomes:

$$SE = \sqrt{ms_{wcells} \left(\sum \frac{c_j^2}{n_j} \right)} \qquad \qquad \text{11.2-1}$$

Now let us return to our example. We want to compare means A and B. The coefficients are -1 and $+1$, and the sample sizes are 5 and 6. From the ANOVA table we find ms_{wcells} to be 9.98. The standard error for our desired comparison is therefore:

$$SE_{\bar{x}_A - \bar{x}_B} = \sqrt{9.98 \left[\frac{(-1)^2}{5} + \frac{(+1)^2}{6} \right]} = \sqrt{9.98 \left(\frac{1}{5} + \frac{1}{6} \right)} = 1.91$$

Finally, we are ready to calculate the confidence interval for the difference between means A and B. The 99% confidence interval is:

Observed difference $\pm$ (appropriate tabled value)(standard error) or
$2.03 \pm (3.76)(1.91)$ or 2.03 ± 7.18 or -5.15 to 9.21

Since the confidence interval extends from -5.15 to $+9.21$, it includes the value zero, and we conclude that there is not sufficient evidence to reject the null hypothesis at the 0.01 level.

The long explanation that has been given for the Scheffé procedure has obscured its basic simplicity. To demonstrate how efficient the procedure is, let us take two other examples, one for another comparison of two means and the other for a comparison of the combination of pairs of means. First, let us look at the two most extreme means, those for precincts A and D. Again, to obtain a positive difference, we subtract mean A from mean D. Our coefficients are once again -1 and $+1$. The observed difference is $14.33 - 7.8$, or 6.53. The number to go into the parentheses for the appropriate tabled value remains unchanged at 3.76. The number for the second set of parentheses is changed only by changing the n in the second fraction.

$$SE_{\bar{x}_A - \bar{x}_B} = \sqrt{9.98 \left(\frac{1}{5} + \frac{1}{9} \right)} = 1.76$$

The 99% confidence interval for the difference between mean A and mean D is:

$$6.53 \pm (3.76)(1.76) \qquad \text{or} \qquad 6.53 \pm 6.62 \qquad \text{or} \qquad -0.09 \text{ to } 13.15$$

Once more the confidence interval includes zero and we fail to reject the null hypothesis at the 0.01 level.

Now let us try a comparison involving combinations of means. We decide to compare the means of precincts A and B with the means of precincts C and D. Can you carry out the Scheffé procedure for this comparison? What are the appropriate coefficients if you want to end up with a positive difference? What is the observed difference? Has the number for the first set of parentheses changed? What has changed in the calculation of the number for the second set of parentheses?

We want to compare the average of the first two means with the average of the second two. Since we are going to add the first two means together and divide by 2, the appropriate coefficients are $1/2$ and $1/2$. The same is true for the second two means. To make our result positive, we make the first two coefficients negative. Our coefficients are therefore $-1/2$, $-1/2$, $+1/2$, and $+1/2$. To get our observed difference, we multiply the means by the coefficients and then add.

Observed difference $= (-1/2)(7.8) + (-1/2)(9.83) + (+1/2)(13.5) + (+1/2)(14.33)$
Observed difference $= 5.1$

Once again there is no change in the number for the first set of parentheses, the appropriate tabled value. It remains 3.76. For the second set of parentheses, the *ms* of course remains the same, but we now have four fractions and the coefficients are no longer 1. Since each numerator is squared, the signs all become positive and each numerator becomes $(1/2)^2$, or $1/4$, or 0.25.

$$SE = \sqrt{ms\left(\sum \frac{c_j^2}{n_j}\right)}$$

$$SE = \sqrt{9.98\left(\frac{0.25}{5} + \frac{0.25}{6} + \frac{0.25}{8} + \frac{0.25}{9}\right)}$$

$$SE = 1.23$$

Our confidence interval is:

$$5.1 \pm (3.76)(1.23) \qquad \text{or} \qquad 5.1 \pm 4.62 \qquad \text{or} \qquad 0.48 \text{ to } 9.72$$

Since the confidence interval does not include zero, we reject the null hypothesis and conclude that there is a difference in the combination of means A and B compared with the combination of means C and D.

PLANNED COMPARISONS

It was indicated earlier that the Scheffé procedure is intended for exploration of the data by examining post hoc hypotheses about possible differences between means or combinations of means. It becomes a conservative procedure when used to make comparisons planned in advance of an experiment. In other words, the use of the Scheffé procedure when the comparisons to be made were planned in advance results in an increased risk of a Type II error. When we examine planned comparisons, a minor modification in the Scheffé procedure provides a more powerful test. The modification is only in the choice of a number for the appropriate tabled value. You remember that in the Scheffé procedure the number is taken from the F table and then modified for the number of independent comparisons possible for the data set under consideration. In the case of planned comparisons, it is appropriate to use a value from the t distribution based on the within df from the ANOVA table. If the planned comparison is nondirectional, a value of t corresponding to the two-tail probability is used. If the experimenter planned in advance to test a directional hypothesis, then a t corresponding to a one-tail probability is appropriate.

Again, let us return to our example. Suppose that before the experiment began, the experimenter had some basis for hypothesizing that the mean for precinct A would be less than that of an average of the other three precincts. He wants to test this hypothesis at an alpha level of 0.01. This is a directional hypothesis planned in advance. To keep the difference positive, the coefficients become -1, $+1/3$, $+1/3$, and $+1/3$, or -1, $+0.33$, $+0.33$, and $+0.33$. The observed difference is calculated as follows:

Observed difference $= (-1)(7.8) + (+1/3)(9.83) + (+1/3)(13.5) + (+1/3)(14.33)$
Observed difference $= 4.75$

The appropriate tabled value is taken from the t distribution with 24 df. The one-tail column for an alpha level of 0.01 contains the value 2.492. This is the value that goes in the first set of parentheses.

The standard error is as follows:

$$ SE = \sqrt{9.98 \left[\frac{(-1)^2}{5} + \frac{(0.33)^2}{6} + \frac{(0.33)^2}{8} + \frac{(0.33)^2}{9} \right]} $$

$$ SE = 1.56 $$

The 99% confidence interval becomes:

$4.75 \pm (2.492)(1.56)$ or 4.75 ± 3.88 or   0.87 to 8.63

Since the confidence interval does not include zero, we reject the directional hypothesis and conclude that the average length of time before the arrival of an emergency vehicle is indeed less for precinct A than for an average of the other three precincts.

Notice that if this comparison had not been planned in advance, the only change would be in the tabled value. As in the previous calculations, it would have been the modified F value of 3.76. The 99% confidence interval would have been:

$4.75 \pm (3.76)(1.56)$ or 4.75 ± 5.86 or -1.11 to 10.61

The confidence interval would contain zero, and the null hypothesis would not be rejected. The procedure using the *t* value is more powerful in this case for two reasons. First, the *t* value has not been adjusted to make allowance for multiple tests, and second, a directional test has been used. Parenthetically, it should be noted that if the same directional hypothesis had been planned in advance, and the mean of *A* had turned out to be greater than the average of means *B, C,* and *D,* no further calculations would have been called for. The data could not possibly represent evidence against the null hypothesis.

ASSUMPTIONS OF ANOVA

In the discussion of the procedure of Scheffé, reference was made to an estimate of the σ that is common to the groups within an experiment. This reference implies one of the assumptions that underlies the process of ANOVA and leads to a discussion of other assumptions basic to the process. The assumption that there is a common variance within groups (homogeneity of variance) is a requirement not only for the individual comparison of means but for the overall test as well. If you think back to the introduction to ANOVA and the conceptual description of the process, you will see that this statement makes sense. You remember that the *F* ratio was formed by placing the observed variance of means over the estimated variance of means. However, the estimated variance of means resulted from averaging several different estimates of the standard error of the mean, one for each sample. Since each of these estimates is in turn based on a sample standard deviation, averaging only makes sense if indeed the standard deviations observed in the samples are estimates of some common σ in the several populations.

A second requirement of ANOVA may be inferred from the previous descriptions of assumptions for the two-sample *t* test. Sigma is needed for the calculation of the standard error of the difference between means. However, generally σ is unknown and must be estimated from sample standard deviations. You learned that this estimate is reasonably good if the populations are normally distributed. If the populations are not normally distributed, large samples are required before *s* becomes a good estimate of σ. Accordingly, it is generally stated that a *t* test for difference between means is only valid when the populations are normally distributed. The same assumption is basic to ANOVA.

Finally, when carrying out an analysis of variance, one assumes that any errors of sampling or measurement are randomly distributed among the scores within the experiment. An explanation of the full meaning of this statement is beyond the scope of this text. It is sufficient here to warn you to use a random process whenever possible in planning an experiment. If subjects are to be chosen from a number of populations, choose randomly. If a group of subjects is to be assigned to a number of different treatments, assign them randomly. If measurements are to be made over a period of time, make the measurements in a random order among subjects. If measurements are to be made by a number of different observers, again assign the observers randomly. Whenever choices are to be made, whether in the selection or assignment of subjects or in the order or method of measurement, randomization ensures against systematic errors. The importance of ensuring against systematic errors cannot be overstated. A systematic error that results in a bias affecting group means obviously is undesirable in any experiment. In ANOVA, a systematic error that does not affect the means but that results in an increase or decrease in the variability about one mean while leaving other groups unaffected is also serious. Such a systematic bias would invalidate the pooling of data to find the ms_{wcells} that is to be used as a common measure of error variance.

In summary, the process of ANOVA is based on the assumption that the measurements to be analyzed represent random selections from populations that are normally distributed and that have similar variances. Having stated these conditions for the use of ANOVA, most statistics books then admit that the conditions are rarely met in real experiments. Populations are rarely truly normally distributed and seldom have exactly the same variances. With the advent of computers, it has been possible to determine the effects of violating the assumptions of normality and homogeneity of the variance. Fortunately, these modeling studies have shown the process to be fairly insensitive to violations in the assumptions of the mathematical model. If the sample sizes are fairly large and samples are of equal sizes, the probabilities of the F table will be nearly correct even with fairly sizable departures from normality or with rather different population σ's. Of course, tests are available for checking these assumptions should this appear desirable. If there is doubt about the normality of the data, the statistic $X_p{}^2$ can be calculated as described in Lesson 2 of Chapter 9, and a test of normality can be made by referring to the table of χ^2. If there is doubt about homogeneity of variance, an F ratio of the two most extreme variances can be formed as described in the lesson explaining tests for equality of variances. If, following one of these tests, there is a strong indication that one of the assumptions is being seriously violated, you have two choices: use a score transformation procedure or use a nonparametric test. Since a chapter will be devoted to a discussion of nonparametric tests, only the method of score transformations will be discussed at this point.

SCORE TRANSFORMATION

Often a set of data that has a skewed distribution is transformed (as opposed to coded). Earlier, you learned that the mean and standard deviation of a set of scores can be modified by adding, subtracting, multiplying, or dividing. This coding by one or a combination of these arithmetic operations does not change the basic shape of the distribution, since each score is acted on by the same constant. Instead of using a constant to code scores, researchers often try to eliminate the skewness in a set of scores by taking the logarithm of the scores or by taking some power of each score. The ANOVA is then carried out on the transformed scores, which are more nearly normal. I am not particularly enthusiastic about this procedure, unless the transformed scores themselves have some intrinsic meaning. The statement that ANOVA has shown that the means of the logarithms of the lengths of time until recovery differ for four different treatments may or may not be of interest to potential recipients of the four treatments. In most instances the same data can be analyzed by a nonparametric technique, which does not require the assumption of normality.

MODELS FOR ANOVA

It is common in statistics books to identify different models for ANOVA. For example, in the first model, it is assumed that if there are four treatments, these four treatments represent all of the treatments that are of interest. The grouping based on these treatments is called a *fixed-effects factor*. The experiment results in an inference to the means of the four populations of persons who might receive one of the four treatments. In a second model, the treatments or experimental groups do not represent all possible instances of interest. Instead, they represent a random selection from among some larger set. For example, an experimenter might be interested in the average caloric intake among school children in a large school system. Rather than taking a random sample from each of the 13 grades, he or she will take random samples from only four grades, which are themselves randomly chosen. In this case, the grouping based on treatments is called a *random-effects factor*. The inference to be made in the experiment is not only to the four grades from which the samples have been taken, but to the remaining grades as well.

The concepts of fixed-effect factors and random-effects factors are being introduced here even though, in the single-variable case, distinctions between the two are relatively unimportant. Computations are the same in either case, as are the F tests. In the next lesson, in which the two-variable case is discussed, the distinction between fixed-effects factors and random-effects factors does become important. A combination of the two factors in which one variable is fixed and the other random also becomes possible. The result is called a *mixed model*. The important thing for now is merely to become familiar with the terminology and the difference between the concepts. Remember that the fixed-effects factor implies that the groups included in the experiment represent all of the groups of potential interest, whereas for the random-effects factor, the groups represent a random selection from among the groups of potential interest.

RELATIONSHIP OF ANOVA TO THE INDEPENDENT-SAMPLE t TEST

As was noted in the explanation of the Scheffé procedure, there is a direct relationship between the F value for comparing two means and the corresponding t value. In an earlier lesson you learned to use a t test for the difference between means under two different assumptions: equal variances and unequal variances. In this lesson you learned that the process of ANOVA is a procedure for comparing means under the assumption that population variances are equal. Accordingly, if there are only two groups and it is assumed that population variances are equal, the t test and an ANOVA should give equivalent results. Let us look at an example. In the lesson on two-sample experiments you were given an example in which two different study manuals were used in preparation for taking the Medical College Admission Test (MCAT) exam. The data were as follows:

	n	ΣX	ΣX^2	Σx^2	$\bar{X}$	s	SE
A	15	7950	4,423,500	210,000	530	122.5	31.6
B	12	6120	3,271,200	150,000	510	116.8	33.7

A two-tail test based on the assumption of equal variances resulted in a t value of $^{20}/_{46.5}$, or 0.43, with 25 df.

Now let us do the same problem using ANOVA. First, we need to calculate SS_T. To do so, we need the total of all the scores, which we will call T, and the total of the squares of all the scores. We can get these by adding the numbers for the two samples.

$$\Sigma X^2 = 4,423,500 + 3,271,200 = 7,694,700$$
$$T = 7950 + 6120 = 14,070$$
$$n = 15 + 12 = 27$$

$$SS_T = 7,694,700 - \frac{(14,070)^2}{27}$$

$$SS_T = 7,694,700 - 7,332,033.33$$
$$SS_T = 362,666.67$$

We note that the correction factor, C, equals 7,332,033.33.

Our next step is to obtain SS_{bcells}. Can you remember the procedure? For each group, we will square the total and divide by the number in the group. We will add all of these results and subtract the correction factor.

$$SS_{bcells} = \frac{7950^2}{15} + \frac{6120^2}{12} - 7,332,033.33$$

$$SS_{bcells} = 7,334,700 - 7,332,033.33$$
$$SS_{bcells} = 2666.67$$

Now to obtain SS_{wcells} we need only subtract SS_{bcells} from SS_T.

$$SS_{wcells} = 362,666.67 - 2666.67 = 360,000$$

Notice that the SS_{wcells} we were able to get by subtraction is the same value we would get by calculating Σx^2 for each sample and then adding. In others words, looking at the table, we find the values 210,000 and 150,000, which add up to 360,000.

We are now ready to construct the ANOVA table.

Source	SS	df	ms	F
Between	2,666.67	1	2666.67	0.1852
Within	360,000	25	14400	
TOTAL	362,666.67	26		

Now we can compare the results of the t test and the ANOVA. Of course, a t of 0.43 with 25 df is not significant. Similarly, an F value of 0.1852 with 1 and 25 df is also not significant. However, you learned earlier that the F value with 1 df for the numerator is the square of the t value. Taking the square root of 0.1852, we get 0.4303, which is the same value obtained for t.

This example demonstrates that when it is assumed that variances are equal, the two-sample test of differences between means can be solved either by a t test or by ANOVA. Generally, the computations are easier for ANOVA; however, ANOVA results in an

F value for which only very limited probabilities are available. The best course of action is therefore to do the problem as an ANOVA and then to take the square root of F to obtain a t value, which can be compared with the expanded tables of the t distribution.

SUMMARY

In this lesson, you learned how to make comparisons among individual means in an experiment involving several groups. In the case where the comparisons are to involve all possible comparisons or comparisons chosen on a post hoc basis, the first step is to carry out an ANOVA. If the result is not significant, the analysis is terminated. If the result is significant, a procedure proposed by Scheffé is used. The Scheffé procedure involves finding a confidence interval for the difference between any combination of means to be compared. If the confidence interval includes zero, the result is not significant. The confidence interval is based on the rationale learned earlier. The appropriate tabled value turns out to be the square root of the F value modified for the number of possible independent comparisons. The standard error turns out to be the square root of ms_{wcells}, appropriately weighted for any combination of means that is desired. In the case where a comparison is planned in advance of the experiment, the process is much the same. There are only two differences. The overall F test is omitted, and the appropriate tabled value is obtained from the table of t. This t value is not modified for the multiple tests that could be carried out.

While learning how to make individual comparisons, you learned that the standard error term used for the many confidence intervals is always based on ms_{wcells}. This implied one of the assumptions of ANOVA, homogeneity of variance among the populations. Two other assumptions are: normal distributions in the populations and random errors among measures. It was pointed out that minor violations of the assumptions of normality and homogeneity of variance have been shown to have little effect, especially for fairly large samples of equal size.

The concepts of fixed-effects factors and random-effects factors were then introduced. These concepts will be discussed more extensively in the next lesson, involving two or more variables.

Finally, the chapter concluded by demonstrating that in the case where variances are assumed to be equal, the t test and ANOVA are alternative approaches that give identical results. The exercise that follows provides an opportunity to determine how well you understood the material in this lesson. See if you can arrive at the ANOVA table without reference to the explanations in the previous lesson. If so, you will find the computations in the next chapter relatively easy to learn.

LESSON 2 EXERCISE

Students taking a state board examination at a particular testing site earned their degrees at five different schools. Schools *A, C,* and *D* were public, whereas schools *B* and *E* were private. Performances are listed.

School A	School B	School C	School D	School E
97	75	93	155	90
75	52	106	89	77
100	84	96	124	88
109	93	132	100	108
103	88	101	96	80
	59	66		66
	66	118		43
	86			97
				49
				109

Is there evidence of differences in the mean levels of performance by the students from the five schools? If you find that there are differences somewhere in the data, use the Scheffé procedure to look for them. Suppose that all you had decided in advance was to compare the mean of school *D* with the others, since school *D* had a radically different curriculum and educational philosophy. On a post hoc basis you also decide to compare public schools with private schools. Use an alpha level of 0.05 for the comparison.

Lesson 3 □ Two-way ANOVA (basic concepts)

In the two previous lessons on ANOVA, examples were restricted to problems in which a single variable was being investigated. The single variable (factor) could have any number of categories (levels). The procedures of ANOVA would certainly be useful if they were only applicable in these single-factor experiments. However, the usefulness of ANOVA is multiplied many times by its extension to multifactor designs. The use of ANOVA to analyze data representing two or more factors with two or more levels provides the possibility of two important advantages: a test of interaction effects and an increased power for the test of the main effects. To illustrate these two advantages, let us look at a hypothetical research problem.

Suppose that two alternate methods of therapy (*A* and *B*) are to be compared. However, it is believed that the effectiveness of the two therapies depends on the age of the patient. Accordingly, the patients for whom therapy is planned are divided into three age groups. Ideally, an equal number of patients from each age group are randomly assigned to each therapy. For our purpose, let us say that ten persons are assigned to each of the six combinations of age groups and therapies. The two factors in the experiment are: type of therapy and age group. The first factor has two levels and the second three. Suppose that some measure of the result of therapy is available for each of the 60 subjects. Suppose further that the measure is expected to have approximately a normal distribution, with equal variances among the six populations represented by the six sample groups.

The design of the experiment just described suggests several means that can be compared. The mean of all subjects receiving therapy *A* could be computed and compared with the mean for all subjects receiving therapy *B*. Similarly, the mean for all subjects in the youngest age group could be compared with the means of the middle and oldest age groups. Finally, any one of the six individual groups' means could be compared with any other, or with a combination of any others.

The process of ANOVA provides a method for making any or all of these comparisons. The comparison of means for the three age groups and the comparison of means of the two therapies are referred to as tests of *main effects*. If a test of a main effect is found to be significant, it then becomes appropriate to examine the data further to find the source of the differences. For example, if it is found that there is a significant main effect for age, one could then ask whether all three age groups have different means or whether the significant main effect is due to only one of the groups being different from the other two.

Even when neither main effect is found to be significant, it is possible that an effect called an *interaction effect* exists. This brings us to the first of the two important advantages of multifactor ANOVA. The experiment just described could be conducted without a breakdown by age to answer the question, "Is the mean for therapy *A* the same as the mean for therapy *B*?" Similarly, it could be conducted without a breakdown by therapy to answer the question, "Are the means different for the three age groups?" Suppose, however, that therapy A gives the best results for the young, that for the middle age group there is no difference in the results of the two therapies, and that therapy B gives best results for the old. If the advantage for A among the young is offset by the advantage for B among the old, the result of comparing therapies without a breakdown by age will be a nonsignificant F for the single-factor, or one-way, ANOVA. In the situation just described, the relative effectiveness of therapies A and B does not remain constant throughout all levels of the second factor, age. This is what is referred to as an *interaction effect*. A multifactor ANOVA (in this case a two-way ANOVA) provides a method of testing for the significance of an interaction effect. We will have more to say later about such tests and their interpretation. For now, it is important merely to be aware that the test for an interaction is possible in a multifactor ANOVA.

A second major advantage of the multifactor ANOVA is the possibility of an increase in the power of the test to be made. Suppose that in the above experiment, it is already known that the effects of therapy change with age. A main effect for age is expected and is of little interest. Suppose also that there is no reason to expect an interaction effect. In other words, whichever therapy is found to be more effective, this difference can be anticipated to hold true for all three age groups, both in terms of the direction of the difference and the amount of the difference. In this situation the two-way ANOVA still has an advantage over combining all three age groups in a one-way ANOVA. As stated at the beginning of this paragraph, the two-way ANOVA will provide a more powerful test. Recall that the F ratio in the one-way ANOVA uses the ms_{wcells} as the denominator. The more variability within groups, the larger the ms_{wcells} and the less powerful the test. If, indeed, age is related to therapy results, variability within groups will be less in the two-factor experiment in which groups are similar in age than in the one-factor experiment in which the groups encompass all ages. As a result, the ms_{wcells} will also be less for the two-way ANOVA than for the one-way. Given an equal number of experimental subjects, the two-way ANOVA will then be more likely to identify any difference that does exist between the two methods of therapy.

Having been told that multifactor experiments: (1) provide information about interaction effects that cannot be identified in one-factor experiments and (2) provide a more powerful test of the main effects, you should now be eager to add the procedures of multifactor ANOVA to your statistical tools. Unfortunately, there is a drawback. The procedures for multifactor ANOVA can be very complicated. Voluminous textbooks have been written dealing with the subject. For *some* multifactor designs, an understanding of the rationale of the analysis and its interpretation has to be left to the professional statistician. Similarly, for *some* designs, the computations are so involved as to require the use of computers for solutions.

Having given this bleak outlook regarding your chances of becoming an expert on multifactor ANOVA, I hasten to add that the remainder of this lesson and the one that follows are both understandable and useful. Because of the complexities that arise in some applications of multifactor ANOVA, many textbooks written at this level skip the subject and refer the reader to advanced texts. This is unfortunate, because with advance planning many experiments can be designed in such a way as to result in a design for which computations and interpretations are readily understood. Most of the complications of multifactor ANOVA arise when samples are not equal in size. If an experiment can be planned so that it fits a two-factor, fixed-effects model, with equal numbers in each group, you will find the analysis and interpretation of the results rather straightforward. Even in the case where a random-effects model is appropriate, the complications are not too great.

For those experiments that do not result in equal numbers or that involve a complication referred to as "nesting," you will usually need to seek help. For those with a strong mathematical background, advanced textbooks are available. For most readers of this text, help from a statistical consultant will be required. In either case, knowledge of the vocabulary that relates to multifactor ANOVA will be essential. The remainder of this lesson, therefore, consists of two parts. First, a number of terms that are used to describe various types of multifactor experiments will be introduced. Once these terms are understood, it will then be possible to describe those types of designs for which the analysis is straightforward. The last part of the lesson will then describe the computations and interpretation of results for these designs.

VOCABULARY OF MULTIFACTOR ANOVA

As indicated earlier, when an experiment is designed to study simultaneously the effects of two or more variables, these variables are referred to as *factors*. A factor may be any variable that results in the subjects of the experiment being divided into categories, either as the result of direct assignment to an experimental condition or as the result of measurement of some characteristic or attribute of the subject. In either case, the categories are referred to as *levels* of the factor. It is common to describe multifactor experimental designs by giving the number of levels in each factor, for example, 2×3, 3×4. (The $\times$ is read "by.") An experiment that involves sex, three age groups, and two methods of therapy would be referred to as a $2 \times 3 \times 2$ design.

As indicated at the end of the last lesson, if the levels of a factor include all of the categories of interest, the factor is referred to as a fixed-effects factor. If the levels represent a random sampling of the categories of interest, the factor is referred to as a random-effects factor. If all of the factors in an experiment are considered fixed, the design is referred to as a *fixed-effects model* or, in some texts, as "Model I." If all the factors are considered to be random, the design is referred to as a *random-effects model* or as "Model II." Some texts also refer to the *random-effects model* as a "components of variance" model. Finally, if some factors are fixed and others are random, the design is referred to as a *mixed model* or as "Model III."

From the standpoint of analysis, the ideal multifactor design is a completely *crossed,* completely *balanced* design. This is a design in which every level of each factor is present for each level of every other factor (completely crossed) and in which the number of subjects is exactly the same within each group that results from these combinations

⌐ly balanced). As an example, if a study involves four age groups and three ⌐⌐⌐⌐⌐⌐⌐⌐⌐⌐ crossed, completely balanced design, each method of ⌐⌐⌐⌐⌐⌐ categories. ⌐⌐ctor, the first factor ⌐⌐ a *nested* factor. Suppose that two forms of therapy are proposed and an ⌐riment is to be conducted to compare their effectiveness. Six physicians participate in ⌐ne experiment, three of whom practice therapy *A* and three of whom practice therapy *B*. Each physician treats a number of patients, and the results of therapy constitute the data for the experiment. In this experiment, the two factors are: method of therapy (with two levels), and individual physicians (with six levels). If each physician used each method of therapy with an identical number of patients, we would have the ideal completely crossed, completely balanced design. However, if each physician uses only the one method of therapy that he personally believes is best, we have a nested design in which the factor, individual physicians, is nested within the factor, method of therapy. It then becomes impossible to separate the effect of a particular method of therapy from the effect of treatment by the particular group of physicians associated with that method of therapy. A factor may be completely nested, as in this example, or *partially nested*. If some, but not all, of the physicians used both methods of therapy, the factor would be partially nested.

In the introduction to this lesson, it was explained that multifactor designs are sometimes used even though interest is really confined to a single variable, such as the effect of various dosages of a drug. The experiment is conducted as a multifactor study because one or more variables can be identified that are related to the variable of real interest. Including these variables as factors in the study increases the power of the test of significance for the variable of interest. For example, a study designed to compare the effects of four levels of drug dosage may include as a variable the weights of the laboratory mice used in the study. If the mice are divided into ten weight groups before the experiment, and animals from each weight group are then randomly assigned to each drug dosage, the design is referred to as a *randomized block design*. Each weight group is a block from which random assignments are made. The advantage is that response to dosage will be more uniform within a single weight group than for a group of mice of widely differing weights. The result will be a smaller ms_{wcells} and a more powerful *F* test.

Just as a momentary digression, suppose that there had been only two dosage levels in the above example and that a single animal from each block had been assigned to each dosage level. Can you think of a method of analysis that was described earlier that would be appropriate? You would have ten scores for one dosage level and ten scores for the other. With only two groups, you could do a *t* test for difference between means. But what kind of *t* test would be appropriate? Are the ten scores in one group independent of the ten scores in the other? The answer, of course, is no. The scores can be paired according to weight blocks, and the paired *t* test for matched samples would be appropriate. The randomized block design in ANOVA can therefore be seen to be an extension of the paired *t* test to include the situation in which there are more than two groups.

In a randomized block design, the blocks can be formed in any way desired. They may be based on personal attributes, geographical groupings, etc., or on any combination of any of these. The guiding principle is to form blocks that will result in the scores within blocks being as nearly the same as is possible.

Within any multifactor design, whether crossed or nested, it is possible to have a single score or several scores for each combination of levels. If there is more than one score for each combination of levels, the design is said to involve multiple measurements. This usage is sometimes the source of misunderstanding when it is confused with repeated measurements made on the same subject. Repeated measurements on the *same* subject may occur either within a treatment combination or as the result of exposing the same subject to each treatment combination. In the latter instance, "subjects" becomes a factor in the design, with the number of levels for subjects corresponding to the number of subjects. In such designs, the order of experiencing the various experimental conditions becomes an important consideration. The performance of an individual subject is not likely to be independent from one measurement to the next. A discussion of designs for such repeated measurement experiments is found in advanced texts.

A final term you are likely to encounter when speaking with a consultant or referring to advanced texts is *orthogonal*. Unfortunately, there is no simple explanation that completely describes what the statistician means when he uses this term. The nearest synonym would be *independent*. When the statistician speaks of orthogonal contrasts or comparisons between means, he is referring to contrasts that are independent of each other. In other words, if two tests of differences between means are orthogonal, the finding of significance for one test would not in any way influence the likelihood of finding significance for the second test. Similarly, when the statistician speaks about an orthogonal multifactor design, he is speaking about a design in which the effects of each factor can be independently evaluated. To accomplish this, an equal number of experimental subjects are randomly chosen from a common population for assignment to each of the possible treatment combinations in a completely crossed, completely balanced design.

Now that the vocabulary of multifactor ANOVA has been introduced, it is possible to identify those types of designs that result in analyses that even the beginning student can carry out and interpret. In essence, they include those designs in which factors are completely crossed and completely balanced. In other words, so long as there is no nesting of factors and so long as there are an equal number of scores per treatment combination, the analysis is relatively straightforward. Whether the model is a fixed-effects, random-effects, or mixed model, the initial computations are uncomplicated and follow directly from the procedures learned for the one-way ANOVA. In the case of the fixed-effects model, the F ratios computed are also exactly what one would expect based on the procedures of the one-way ANOVA. When we get to the random-effects model or the mixed model, the choice of the denominator of the F ratio gets a bit complicated. This problem will be discussed in the final lesson in this chapter on ANOVA.

COMPUTATIONS OF A TWO-WAY ANOVA
A simple example

To show the computations and the interpretation of the results in a two-way ANO
it is best to use a very simple example with small samples and small, even numbers. Since many of the applications of multifactor ANOVA grew out of agricultural research, it is perhaps appropriate to choose this area for an example. The procedures for this greatly oversimplified example will be identical to those for real data so long as the design is completely crossed and completely balanced.

Suppose that the goal of a research project is to attempt to discover whether the yield of pea plants is affected by the application of: (1) fertilizer and (2) weed control chemicals. A small plot of ground with uniform soil and uniform sun and moisture exposure is selected for the experiment. The plot is then divided into four parts, with two pea seeds planted in each part. One part of the plot gets neither fertilizer nor weed control, one gets fertilizer only, one gets weed control only, and the remaining one gets both fertilizer and weed control. Keeping in mind that our data are completely artificial both in the amount of yield and the means of measurement, suppose that the two plants in the untreated part of the plot yield respectively 1 cup and 2 cups of peas. Our plot, with the yields for the untreated part, now looks like this.

	No weed control	*Weed control*
No fertilizer	1 cup 2 cups	
Fertilizer		

Now suppose that the effect of fertilizing is to increase the yield of a plant by 1 cup, and that the same is true for the effect of weed control. If the same two seeds had been planted in one of the other parts of the plot, what is your conclusion as to the number of cups each would have yielded? Adding 1 cup to each plant for the effect of weed control, we have yields of 2 cups and 3 cups for the upper right part of the plot. Similarly, adding 1 cup to each for the effect of fertilizer, we have 2 cups and 3 cups for the lower left part of the plot. But what about the lower right-hand box, which represents both fertilizer and weed control? Now we must decide whether the effects of fertilizer and weed control result in a simple additive effect. If so, we would add 2 cups to the yield of each plant, 1 for fertilizer and 1 for weed control, and get yields of 3 and 4 cups. It is conceivable, however, that the two in combination would have a synergistic effect and increase the yield by more than 2 cups. It is also possible that the two in combination would result in an increase of less than 2 cups, or even in a decrease in the yield. Any result other than a simple additive effect is referred to as an interaction effect and is measured as part of the analysis. Notice that this interpretation of an interaction effect corresponds to our earlier explanation, in which it was said that an interaction effect exists if the effect of one factor is not constant over all levels of another factor.

For our first analysis, let us assume that there is a simple additive effect for the two treatments. Our plot now looks as follows. The sum of the scores within each box has been included in parentheses. Totals have also been included for each row, each column, and the entire plot.

	No weed control	Weed control	TOTAL
No fertilizer	1,2 (3)	2,3 (5)	8
Fertilizer	2,3 (5)	3,4 (7)	12
TOTAL	8	12	20

We are now ready to do the computations of ANOVA for this data. Earlier it was mentioned that if you learned the computational routine for the one-way ANOVA, you would find it easy to extend the process to include the multifactor ANOVA. Remember the computational procedure? The first goal is to obtain the total sum of squares, the second is to find the sum of squares between groups, and the third is to find the sum of squares within groups.

The total sum of squares is based on a familiar concept that we earlier called the "brick." In others words, we treat the entire set of scores as one large group and find the sum of squared deviations from the grand mean. We do not need to actually find the grand mean, however. Instead, we use the computational formula:

$$SS_T = \Sigma X^2 - T^2/n$$

To find the ΣX^2 for the data in the table, we need to square the yields of each of the eight plants and add. Doing so, we obtain the value 56. From the lower right-hand magin, we see that T is 20. The term T^2/n, which we will call the correction factor, is therefore $400/8$, or 50. Subtracting 50 from 56, we find that SS_T equals 6.

Now can you remember how to find the sum of squares between groups? You start by squaring the total of each group and dividing by the number in the group. You then sum all of these results and subtract the correction factor. For our example, there are four groups, each with two scores.

$$SS_{bcells} = 3^2/2 + 5^2/2 + 5^2/2 + 7^2/2 - 50 = 54 - 50 = 4$$

Now to find the sum of squares within we subtract.

$$SS_{wcells} = SS_T - SS_{bcells}$$
$$SS_{wcells} = 6 - 4 = 2$$

We have now carried out all of the computations just as though this were a one-way ANOVA with four unrelated groups. However, in a multifactor design, the sum of squares between groups can be further broken down. We can group the scores into those with fertilizer and those without and find SS_{rows}. We can also group the scores into those with weed control and those without and find $SS_{columns}$. The same algorithm learned earlier is used to find the sum of squares for *any* desired grouping. Find the total for each group, square, divide by n, sum the results, and subtract the correction factor. Let us find the sum of squares for rows, or for the effect of fertilizer. This time there are four scores in each group.

$$SS_{rows} = 8^2/4 + 12^2/4 - 50$$
$$SS_{rows} = 16 + 36 - 50$$
$$SS_{rows} = 2$$

For this particular example, the effect of weed control was the same as for fertilizer, and the numbers are identical. Therefore, $SS_{columns}$ also equals 2.

The SS_{bcells} that we first computed is a sum of squares based on each of the possible combinations of levels for the two factors of the experiment. The SS_{bcells} in a two-way ANOVA represents the total "between-groups" sum of squares. This total between-groups sum of squares can be divided into three components. We have already calculated two: SS_{rows} and $SS_{columns}$. The third is $SS_{interaction}$ and is found by subtraction. In other words, $SS_{interaction} = SS_{bcells} - SS_{rows} - SS_{columns}$. For our example, $SS_{rows} + SS_{columns} = SS_{bcells}$. The interaction effect must therefore be zero.

We now complete our ANOVA table for a fixed-effects model. Everything except the F column would be the same for the random-effects model. This difference will be explained in the next lesson. Our ANOVA table appears as Table 11-4.

Notice that df for between cells is one less than the number of groups. The same is true for rows and columns, each of which had two groups. The df for interaction can be obtained either by subtracting df for rows and columns from df for cells or by multiplying

Table 11-4. Two-way ANOVA comparing the effects of fertilizer and weed control

Source	SS	df	ms	F
Rows	2	1	2	4
Columns	2	1	2	4
Interaction	0	1	0	0
Between cells	4	3		
Within cells	2	4	0.5	
TOTAL	6	7		

df for rows by *df* for columns. The *df* for within groups is obtained as in the one-way ANOVA by subtracting one from the number in each group and adding. Once again, the total *df* equals the total number of scores less one.

We started out this example by saying that the purpose of the experiment was to determine whether an effect exists for fertilizer and whether an effect exists for weed control. These are referred to as *main effects*. Because this is a multifactor experiment, we are also able to test for the significance of an interaction effect. We need to compute an *F* ratio for each of these effects. As in the one-factor case, we divide the *ms* for the effect to be tested by the *ms* within cells. We obtain *F* values of 4, 4, and 0, and 1 and 4 *df* in each case. Referring to a table of *F*, we find that none is significant at the 0.05 level. This finding does not mean that main effects do not exist. (Remember that we deliberately built in an effect when constructing the data.) It means that the effect was too small to detect with our very small samples. On the other hand, the finding of no significant interaction was expected, since none was built into the data.

In our contrived example, $SS_{interaction}$ was exactly zero. We would be very surprised if $SS_{interaction}$ equaled exactly zero for the case of real data. The $SS_{interaction}$ is a statistic, as are SS_{bcells}, SS_{rows}, and $SS_{columns}$. All will vary from population values because of random sampling. Accordingly, even if there is zero interaction effect in a population, it would be rare for SS_{rows} and $SS_{columns}$ to exactly equal SS_{bcells} for a particular random sample.

Both to review the process and to provide additional insight into the relationship between SS_{bcells}, SS_{rows}, $SS_{columns}$, and $SS_{interaction}$, let us see what happens when we modify the data to represent the presence of an interaction. Let us suppose that rather than being synergistic, the actions of fertilizer and weed control used in combination neutralize each other. In other words, fertilizer by itself increases yield by 1 cup; weed control by itself increases yield by 1 cup; however, when they are used in combination, they neutralize each other, and there is no effect at all. Starting with the same data as before in the upper left part of the plot, we would anticipate the following results:

	No weed control	Weed control	TOTAL
No fertilizer	1,2 (3)	2,3 (5)	8
Fertilizer	2,3 (5)	1,2 (3)	8
TOTAL	8	8	16

See if you can carry out the ANOVA for this data. Before even beginning, can you anticipate the result? What do you expect for the effect for rows and columns? How will the SS_{wcells} compare with the previous example? To answer this last question, think about the meaning of SS_{wcells}. The value of SS_{wcells} is a measure of variability within cells. Since we have not changed the distance between scores within cells, we expect SS_{wcells} to remain the same as before. If you carry out the analysis correctly, you obtain the following results:

$$SS_T = 4 \qquad SS_{rows} = 0$$
$$SS_{bcells} = 2 \qquad SS_{columns} = 0$$
$$SS_{wcells} = 2 \qquad SS_{interaction} = 2$$

You should be able to complete an ANOVA table for these results, getting an F of zero for each main effect and an F of four for interaction. As before, the F value of four for interaction is not significant even though an interaction effect was built into the data. With the small samples used, the test is not powerful enough to identify the effect. A Type II error will result.

Again, it should be noted that the results are based on a contrived example in which we constructed the data in such a way that row and column effects would be zero. With real data, we would expect some differences among row means and among columns means as the result of random sampling.

The example that has been used to show the computations of a multifactor ANOVA involves two factors with two levels for each factor. Exactly the same computations are involved in a two-way ANOVA with more than two levels for each variable. Let us now consider an example of a 2 × 4 design using numbers that are more like those found in actual studies.

A more realistic example

Suppose that four publishers produce materials designed for a self-study health course for eighth-grade students. A research project is designed to compare the four sets of materials. There is a separate breakdown of results for girls and for boys. A group of 40 girls and 40 boys is available for testing the materials. You randomly assign ten girls and ten boys to study each set of curriculum materials. Following a period of study, each student is given a test involving concepts related to health.

Table 11-5 contains the means for the various ways in which scores can be grouped.

Table 11-5. Mean scores on a test of health concepts

	Curriculum				
	A	**B**	**C**	**D**	**All curricula**
Girls	76.3	73.4	78.4	79.3	76.8
Boys	68.9	66.0	71.6	87.7	73.55
Both	72.6	69.7	75.0	83.5	75.2

One of the main effects that can be tested in this experiment can be stated as follows: "Do any of the curriculum materials result in superior performance on the health test, or are all curricula equal?" A comparison of means suggests that D may be best, followed by C, then A, and finally B. The question is whether these observed differences are statistically significant. The second main-effects question is: "Does one sex score higher on the test than the other sex?" Here an examination of the means reveals a bit of a problem. If we look only at the combined means in the last column of the table, girls seem to perform better than boys. However, if we compare the means of girls and boys for each curriculum, we find that girls have higher means using curricula A, B, and C, but that boys have a higher mean using curricula D. This last statement implies that an interaction effect may be present and suggests another way of interpreting the term *interaction*. We wanted an answer to the main-effects question: "Do girls perform better than boys?" However, to answer the question we must respond: "It depends on what curricula you are talking about." Anytime we must respond to one of the main-effects questions with the words "It depends on which level of the other factor you are considering," there is some evidence for an interaction effect. As seen in the example given earlier, we can test for both main effects and interaction effects by the process of ANOVA.

Table 11-6. Scores on a test of health concepts following study of four separate sets of curriculum materials

	Curriculum				
	A	**B**	**C**	**D**	**Total**
Girls	75	75	85	61	
	79	94	71	69	
	95	79	85	98	
	67	51	67	79	
	74	86	77	86	
	75	64	87	75	
	80	75	76	71	
	73	68	82	90	
	64	76	69	85	
	81	66	85	79	
Subtotal	763 +	734 +	784 +	793 =	3074
Boys	54	64	68	77	
	69	67	79	87	
	59	70	68	92	
	75	79	89	95	
	87	86	69	100	
	75	68	79	77	
	62	54	60	85	
	71	68	62	88	
	76	45	64	87	
	61	59	78	89	
Subtotal	689 +	660 +	716 +	877 =	2942
TOTAL	1452 +	1394 +	1500 +	1670 =	6016

To carry out the ANOVA, we will need to compute SS_T. To do so, we need all of the scores. The score of each student and the sums for the relevant groupings of scores appear in Table 11-6.

Once we square each score and find the sum of these squares (ΣX^2), the rest of the calculations are uncomplicated. The ΣX^2 can be found to equal 462,524. You should now have no difficulty following the remaining calculations. They are identical to those in the example given earlier.

Correction factor $= (6016)^2/80 = 452,403.2$

$$SS_T = \Sigma X^2 - C = 462,524 - 452,403.2 = 10,120.8$$

$$SS_{bcells} = \frac{(763)^2}{10} + \frac{(689)^2}{10} + \frac{(734)^2}{10} + \frac{(660)^2}{10} + \frac{(784)^2}{10} + \frac{(716)^2}{10} + \frac{(793)^2}{10} + \frac{(877)^2}{10} - C$$

$$SS_{bcells} = 455,653.6 - 452,403.2 = 3250.4$$

$$SS_{wcells} = SS_T - SS_{bcells}$$

$$SS_{wcells} = 10,120.8 - 3250.4 = 6870.4$$

$$SS_{columns} = \frac{(1452)^2}{20} + \frac{(1394)^2}{20} + \frac{(1500)^2}{20} + \frac{(1670)^2}{20} - C$$

$$SS_{columns} = 454,522 - 452,403.2 = 2118.8$$

$$SS_{rows} = \frac{(3074)^2}{40} + \frac{(2942)^2}{40} - C$$

$$SS_{rows} = 452,621 - 452,403.2 = 217.8$$

$$SS_{interaction} = SS_{bcells} - SS_{rows} - SS_{columns}$$

$$SS_{interaction} = 3250.4 - 2118.8 - 217.8 = 913.8$$

Construction of the table of ANOVA (Table 11-7) is also straightforward.

Note that we have considered the experiment to involve fixed-effects factors and therefore have used the ms_{wcells} as the denominator of each F ratio.

What, then, do we now conclude as the result of our experiment? First, looking at the significant interaction effect, we state that there is evidence that girls do better than boys when curricula A, B, and C are used but that boys do better than girls when curriculum D is used. A closer examination of the sets of curriculum materials might provide an explanation for this interaction effect. We might find, for instance, that examples and

Table 11-7. Table of ANOVA comparing health concepts test scores

Source	SS	df	ms	F
Rows	217.8	1	217.8	2.28
Columns	2118.8	3	706.26	7.4*
Interaction	913.8	3	304.6	3.18†
Between cells	3250.4	7		
Within cells	6870.4	72	95.42	
TOTAL	10,120.8	79		

*P < 0.01.
†P < 0.05.

illustrations used in the materials have a bias toward being of more interest to one sex than the other. If the bias is one way for curricula A, B, and C and the other way for curriculum D, this would be a possible explanation for the significant interaction.

Before we leave the discussion of the interaction effects, it should be emphasized that an interaction can exist even when there is not a reversal of direction of means as in this example. We could have had data in which girls scored higher than boys for all four curricula and still have had a significant interaction effect. This would occur if the difference in favor of girls were much more pronounced or much less pronounced for a particular curriculum.

Now let us turn our attention to the other significant effect. Can you anticipate the direction the discussion will take? We found a significant F value for columns. This finding causes us to reject the null hypothesis that all four curricula result in equal means. What comes next? Are we satisfied to end our analysis with this statement? If not, where do we go from here? You should be able to anticipate both the next question and the procedure to be used in answering it.

We, of course, are not satisfied merely to know that there are real differences somewhere among the four means corresponding to the four curricula. Just as in the one-way ANOVA, we wish to know where these differences lie. Also, just as in the one-way ANOVA, we will use the procedure of Scheffé to search for the differences. By referring to the summary of the Scheffé procedure, you should be able to make any comparison in which you are interested. The only question that might arise relates to the df to be used in obtaining the F value needed for the procedure. If you were to guess the answer, you would probably be correct. Since it is the main effect for columns that has been found to be significant, we will use the df associated with the F ratio for columns. In our case, the df will be 3 and 72. Since our table does not give the value for 72 df for the denominator, we choose to be conservative and use the value for 3 and 60 df. The value for the 0.01 level of probability is 4.13. Before reading on, see if you can use the Scheffé procedure to compare the mean for D with the average of the other three means. This will serve as a good review of the procedure.

The coefficients appropriate for the desired comparison are $-1/3$, $-1/3$, $-1/3$, and $+1$. The observed difference between the average of means A, B, and C, and mean D is found to be 11.07. The confidence interval for the comparison is therefore:

$$11.07 \pm (\text{appropriate tabled value})(\text{standard error})$$

To get the value for the first set of parentheses, we multiply the F value of 4.13 by the number of groups, less 1; we then take the square root. By the number of groups, we mean the numbers of groups involved in the significant main effect, not the number of groups in the experiment. Accordingly, we multiply 4.13 by 3 and then take the square root. We obtain a value of 3.52 for the first set of parentheses.

For the second set of parentheses, we take the square root of the ms_{wcells}, properly weighted for the number and sizes of the groups involved in the comparison. Here, we make use of the coefficients stated earlier.

$$SE = \sqrt{95.42 \left[\frac{(-0.33)^2}{20} + \frac{(-0.33)^2}{20} + \frac{(-0.33)^2}{20} + \frac{(+1)^2}{20} \right]}$$

$$SE = 2.52$$

Our confidence interval is therefore:

$$11.07 \pm (3.52)(2.52) \quad \text{or} \quad 11.07 \pm 8.87 \quad \text{or} \quad 2.2 \text{ to } 19.94$$

Since the confidence interval does not include zero, we conclude that the mean for curriculum *D* is indeed different from the average of the other three means. We can use the Scheffé procedure to make any other comparison of curricula means based on combined groups of girls and boys. You will be asked to do so in one of the exercises that follow.

SUMMARY

This lesson began with a description of the multifactor experiment and its two advantages. The first advantage is that the multifactor experiment allows for the identification of an interaction effect. It was explained that an interaction effect occurs when there is not a simple additive relationship between study variables. In other words, an interaction effect is present when the effect of the combination of two variables is either greater or less than one would expect to get as a result of combining the variables. The second advantage of the multifactor experiment is an increase in the power of the test based on a reduction of variability of measurements within groups.

The lesson then continued with an explanation of some of the terminology associated with ANOVA. Earlier you learned about fixed-effects factors and random-effects factors for the one-way ANOVA. When the experiment includes more than one factor, there is also the possibility of a mixed model, in which one variable is fixed and the other is random. In describing multifactor experiments, it is common to designate the number of levels for each variable. For example, an experiment with three factors each having four levels would be a 4 × 4 × 4 design. If all possible combinations of levels of all variables occur in the experiment, the design is said to be completely crossed. If, in addition, all combinations have the same number of observations or scores, the design is completely balanced. It was explained that the computational procedures that are included in this lesson are appropriate for the completely balanced, completely crossed design. As opposed to a completely crossed design, some experiments result in nesting of factors, in which not all levels of one factor occur at all levels of another factor. When a combination of levels includes more than one score, the design is said to include multiple measurements. If the subjects of an experiment are grouped on some variable before an experiment, and equal numbers are then assigned at random from each group or block to each combination of factor levels, the design is called a randomized block design. The last term to be mentioned was the term *orthogonal,* which can be roughly translated as "independent."

The computational procedures of a two-way ANOVA were then described for a contrived example in which it was shown that the sum of squares for between cells is made up of three components: the sum of squares for rows, the sum of squares for columns, and the sum of squares for interaction. The computations for the two-way ANOVA were shown to be exactly the same as for the one-way ANOVA, except that the algorithm for computing SS_{bcells} is repeated for cells, rows, and columns. The computations were then shown for a more real situation—a 4×2 design in which the interaction and column effects were found to be significant. The procedure of Scheffé was then used to make a comparison of column means.

The next lesson will discuss: (1) the denominator of the F ratio for the random-effects model, (2) the two-way ANOVA for the experiment in which there is one entry per cell, (3) the ANOVA for more than two factors, and (4) an approximate procedure for a two-way ANOVA with unequal sample sizes.

LESSON 3 EXERCISES

1. Refer to the health curricula example in the lesson and test to see whether the average of means A and B is different from the average of means C and D. Consider this to be a post hoc test, and use an overall alpha level of 0.05.
2. A virus that infects household cats and diminishes their vision is treated by two different therapies, A and B. Therapy A involves hospitalizing the animal and is quite costly. Therapy B involves office visits and is much less expensive. There is some indication that the effectiveness of treatment may be related to how quickly therapy follows exposure. Data from an experiment to compare the two methods of therapy are contained in the following table. There is a breakdown for the length of time between exposure and treatment. The effectiveness of treatment is measured by the percentage of vision retained 2 months after infection. The percentages are assumed to be normally distributed.

	Within 48 hours		After 48 hours	
	86	74	75	93
	68	97	74	74
Treatment A	44	58	75	49
	60	54	80	83
	94	78	34	70
	71	96	57	50
	35	80	24	53
Treatment B	81	91	65	15
	60	68	59	61
	70	53	32	56

What do you conclude from these data?

Lesson 4 □ Multifactor ANOVA (complications)

This final lesson on ANOVA is concerned with some of the complications that arise when the design of an experiment does not conform to a conventional completely crossed, completely balanced, fixed-effects model. We begin with a discussion of the *F* ratio to be used in testing the main effects in a two-way ANOVA for the random-effects model. Following this will be a discussion of three additional topics: designs with one entry per cell, designs with more than two factors, and, finally, designs with unequal cell frequencies.

RANDOM-EFFECTS MODELS

You will recall that in a random-effects model, the levels of the factors present in an experiment do not include all levels of potential interest but a random sampling of those levels. As an illustration, suppose that an educator wishes to investigate the effects of varying the length of time between the taking of classroom tests and the return of the graded tests to the students. A large metropolitan public school system that has a uniform curriculum in personal health provides an opportunity for examining the question. Since the school system contains many junior high schools, and since the length of time before return of classroom tests could be set at many different time intervals, the experiment becomes a random-effects design. Suppose that five schools and four time intervals are selected at random. Within each school selected for participation in the experiment, each of 120 eighth-grade students is randomly assigned to one of four classes, each of which uses one of the different time intervals for return of classroom tests. Following the experiment, each student takes a comprehensive standardized test of personal health concepts.

The two main effects in this experiment are schools and time intervals. The experiment is of a random-effects design, since results are to be generalized beyond the specific schools and the time intervals chosen for the experiment. Can you visualize what the ANOVA table will look like for this 5 × 4 factorial design? The initial steps in the analysis will be identical to those of the fixed-effects model described in the last lesson. The values for SS_T, SS_{bcells}, and SS_{wcells} will be obtained first. Then SS_{bcells} will be further divided into SS_{rows}, $SS_{columns}$, and $SS_{interaction}$.

Can you figure out the *df* appropriate for each of the sum of squares? Altogether, there are 20 classes with 30 students in each class, for a total of 600 students. The total *df* will therefore be 599. The 20 classes correspond to 20 cells in the design, so there are 19 *df* for between cells. These 19 *df* are divided among rows, columns, and interaction. If rows correspond to the five schools and columns to the four time intervals, there will be 4 and 3 *df* respectively for rows and columns. The *df* for interaction can then be obtained either by subtracting the *df* for rows and columns from the *df* for between cells or by multiplying the *df* for rows by the *df* for columns. In either case, we have 12 *df* for the interaction sum of squares. Finally, the *df* for within cells is obtained by subtracting one

Table 11-8. Two-way ANOVA comparing health examination scores for time intervals* and schools

Source	SS	df	ms	F
Rows (schools)	2156	4	539	?
Columns (time intervals)	1574	3	524.7	?
Interaction	3486	12	290.5	?
Between cells	7216	19		
Within cells	52,648	580	90.8	
TOTAL	59,864	599		

*Time intervals refers to the length of time between the taking of tests and their return to students.

from the number in each group and multiplying by the number of groups. We multiply 29 by 20 and get 580. To obtain the *ms* for each effect, we divide the sum of squares by the *df*. An ANOVA table with some possible results appears as Table 11-8.

If the design of this experiment were appropriate for a fixed-effects model, you could readily replace the question marks in the *F* column with the appropriate values. Each of the mean squares would be divided by the ms_{wcells}. In the random-effects model, the calculation of *F* values is a bit more involved. Although it is relatively easy to learn what to do, it is difficult to understand why the procedure is appropriate. Therefore, let us start with the "what" and reserve the "why" until later.

The *F* value for the interaction is calculated first; it is identical to its calculation in the fixed-effects model. We therefore divide 290.5 by 90.8 and obtain a value of 3.2. This *F* value is then compared with a table of *F* for significance. If the *F* value for interaction is found to be significant, the *ms* for interaction will then be used as the denominator of the *F* ratios for rows and columns. With 12 and 580 *df,* our value of 3.2 is highly significant. Accordingly, we use 290.5 as the denominator of our *F* ratio for rows and columns. For rows, we divide 539 by 290.5 and obtain an *F* of 1.85 with 4 and 12 *df.* Notice that *df* for the denominator corresponds to *df* for interaction, not *df* for within cells. For columns, we divide 524.7 by 290.5 and obtain a value of 1.81 with 3 and 12 *df.* Comparing these two *F* values with the tables of the *F* distribution, we find that neither is significant.

What then do we conclude as the result of our experiment? Remember that our conclusions are to apply to all schools within the system and all time intervals from which random selections have been made. The finding of a significant interaction effect with no significant row or column effect indicates that we cannot make a general statement that there are differences among the many schools within the system in the performance to be expected on the personal health examination. Neither can we conclude that different time intervals for return of a classroom test will, in general, make a difference in performance on the subsequent personal health examination. Instead, when comparing any two schools, we will have to make reference to what might be expected for a particular time interval, rather than for all time intervals combined. Similarly, when comparing any two time intervals, we will have to make reference to what might be expected for a particular school, rather than for all schools combined.

Obviously, the researcher has a much easier time interpreting the results of his experiment if interaction effects are not found to be significant. There is a trade-off, however; the calculations for F become more involved. It has been explained that for the random-effects model, you first calculate F for the interaction effect. Then, if the F is significant, you use the $ms_{interaction}$ as the denominator of the F ratio for rows and columns. What about the case where it is decided that no interaction effect exists? In this instance, a new ms is calculated, based on a pooling of information for interaction and within. We first add $SS_{interaction}$ and SS_{wcells} to get a pooled SS for error. In a similar manner we then add df for interaction and within to get a pooled df for error. The pooled SS_{error} is then divided by the pooled df for error to get a pooled ms_{error}. This pooled ms_{error} is used as the denominator of the F ratio for rows and columns. When the resulting F values are compared with the tabled values of the F distribution, the df for the denominator is based on the pooled df for error.

All of this sounds more complicated than it actually is. If, in our experiment, the F of 3.2 obtained for interaction had not been significant, we would have added 3486 to 52,648 to get a pooled SS_{error} of 56,134. We would add 12 and 580 to get a pooled df for error of 592. We would then divide 56,134 by 592 and get 94.82 as our pooled ms_{error} to use as the denominator in forming the F ratios for rows and columns.

As stated earlier, learning what to do in calculating F values for the random-effects model is not too difficult. Understanding why the procedure is appropriate is difficult and may elude some (many?) readers. Mathematical arguments are contained in numerous advanced texts and may be of use to some readers. For most readers, however, the following explanation is probably sufficient.

The basic difference between the fixed-effects model and the random-effects model is that in the first, all combinations of levels are included in the design, whereas in the second, only a sampling of combinations are included. This difference is important when we are interpreting the observed sums of squares for rows and columns. In the fixed-effects model, any interaction effects that may exist will not contribute to the observed differences between row or column means. Row and column sums of squares will be independent of any existing interaction effects. This is because each of these values results from sums that include all levels of the remaining factor.

An example may help to make this point clear. Examine the *population* means in the diagram for a 3×3 model design shown at the top of the following page.

	A_1	A_2	A_3	
B_1	0	1	2	1
B_2	1	1	1	1
B_3	2	1	0	1
	1	1	1	1

To avoid confusing the situation by including sampling variability, let us specify that we are dealing with entire populations. Since there are different population means for the different cells, the sum of squares for between cells will not be zero. However, what will happen when this SS_{bcells} is divided into SS_{rows}, $SS_{columns}$, and $SS_{interaction}$? Since all row means are equal and all column means are equal, SS_{rows} and $SS_{columns}$ will both be zero. The SS_{bcells} results entirely from $SS_{interaction}$. In other words, the presence of an interaction effect has not affected SS_{rows} or $SS_{columns}$. Because SS_{rows} and $SS_{columns}$ are not affected by interaction effects in the fixed-effects model, the F ratios for testing row and column effects do not include the $ms_{interaction}$.

Notice now what would happen if instead of a 3 $\times$ 3 fixed-effects model design, two of the levels of A and two of the levels of B were randomly chosen for inclusion in the experiment, resulting in a 2 $\times$ 2 random-effects design. Each column and row mean is still the same over *all* levels of the other factor. However, what will happen to *observed* row and column means based on only two levels of A and two levels of B. For example, suppose that A_1 and A_2 are chosen, along with B_1 and B_2. The cell means remain the same as before, but row and column means need no longer be equal.

	A_1	A_2	
B_1	0	1	0.5
B_2	1	1	1
	0.5	1	

Since each row and column mean is no longer based on *all* levels of the remaining factor, the interaction effect present in the complete set of data has contributed to observed row and column effects. Accordingly, in the random-effects model, the *ms* for rows or for columns is a *combination* of interaction effects and row or column effects. It is therefore appropriate first to test to see whether an interaction effect is present. If so, an F ratio that has the $ms_{interaction}$ as the denominator is formed for testing row and column effects. The resulting F ratio for rows will be:

$$F = \frac{ms_{row}\ \text{(which includes interaction)}}{ms_{interaction}}$$

Of course, the value of this F ratio will depend on whether there is a real effect for rows; it will therefore be an appropriate test of the main effect for rows. A similar argument can be made for including the $ms_{interaction}$ in testing the main effect for columns.

Now what about the case where the interaction effect is not found to be significant? Even if there is no interaction effect in the *total* data set, one cannot expect the $ms_{interaction}$ to be exactly zero for a set of samples. Instead, it will vary because of the particular set of combinations of levels randomly selected and the particular samples randomly chosen within each combination. In other words, in the case where there is not a real interaction effect present in the total data set, the observed $ms_{interaction}$ provides another estimate, along with ms_{within}, of the effects of random sampling. It is therefore appropriate to combine these two estimates of sampling variability into a single ms for use as the denominator of the F ratio for testing main effects.

Learning how to do an analysis without gaining any intuitive feel for why the process works does not provide much satisfaction. Based on the instructions given at the outset of this lesson, *all* readers should know *what* to do to carry out a two-way ANOVA for the random-effects model. It is hoped that the explanation just given has provided most readers with some insight as to *why* the procedure is appropriate.

ONE ENTRY PER CELL

In many experiments, the cost of sampling is great. The need for economy may dictate a design in which there are no multiple measurements within cells. In others words, a design may be needed in which there is a single subject for each of the combinations of levels in the experiment. Let us consider an example. Suppose an experimenter wishes to study the effects of various levels of carbohydrates in the diets of rhesus monkeys. He is interested in whether carbohydrates have an effect on the frequency of cavities in the monkeys' teeth. Because he suspects that frequency of cavities is related to the age of the monkey, he includes age as a variable in the experiment. He chooses four age levels: 6 months, 12 months, 18 months, and 24 months. A number of monkeys are selected at each age level and randomly assigned to one of three experimental conditions relating to carbohydrate level in the diet. The experiment is a randomized block experiment in which there has been blocking on the variable age. With four age levels and three treatment conditions, there will be twelve combinations of age and carbohydrate levels.

Because of the expense of obtaining and caring for rhesus monkeys, only a single monkey is used for each of the 12 cells. After 1 year of the experiment, the following data are obtained with regard to the number of cavities per animal.

Carbohydrate level

		Low	Medium	High	TOTALS
	6	2	4	4	10
	12	3	2	4	9
Age at start (in months)	*18*	1	3	2	6
	24	2	3	5	10
	TOTALS	8	12	15	35

Can you carry out a two-way ANOVA for these data? You begin by squaring each number and adding to find ΣX^2. This turns out to be 117. You also need the correction factor, which is $(35)^2/12$, or 102.08. Next you find SS_T, SS_{bcells}, and SS_{wcells}. Finding SS_T is easy.

$$SS_T = 117 - 102.08 = 14.92$$

To find SS_{bcells} we will square the total within each cell, divide by the number of entries within each cell, add, then subtract the correction factor. What happens when this is done? Since there is only one entry per cell, what this entails is exactly what has been done to find SS_T. When SS_{bcells} equals SS_T, what happens to SS_{wcells}? We should not be surprised to find that SS_{wcells} equals zero. With only one entry per cell, there cannot be variability within. What then will we use as the denominator of the F ratio for testing main effects? The explanation just given for F ratios in the random-effects model provides a clue to the solution. We will find SS_{rows} and $SS_{columns}$ in the usual manner. We then subtract these from the SS_T and call the remainder $SS_{residual}$. The ms for residuals will then be the denominator for our F ratios for testing main effects.

What about the test for an interaction effect in this one entry per cell case? Notice that what we have decided to call $SS_{residual}$ is obtained by subtracting SS_{rows} and $SS_{columns}$ from SS_T. However, SS_T equals SS_{bcells}. In fact then, what we have called $SS_{residuals}$ for the one entry per cell case is calculated in exactly the same way as $SS_{interaction}$ in the multiple-measurements case. Since there is no way to calculate SS_{wcells}, there is no way to test the significance of the $SS_{interaction}$ (or $SS_{residual}$). The economy provided by employing only one entry per cell has therefore cost us the possibility of testing for an interaction effect.

An ANOVA table for the example just given appears as Table 11-9.

Reference to a table of the F distribution results in the conclusion that this experiment fails to provide evidence of an effect for either age or carbohydrate level. Since the means for carbohydrate level are in the order one would expect, and since the F value is

Table 11-9. ANOVA for number of cavities in Rhesus monkeys in relation to age and carbohydrates in diet

Source	SS	df	ms	F
Rows (age)	3.59	3	1.2	1.4
Columns (carbohydrates)	6.17	2	3.08	3.58
Residual	5.16	6	0.86	
TOTAL	14.92	11		

near the F required for significance at the 0.05 level, we may well be making a Type II error. Depending on the importance of the finding and the cost of further research, we would probably wish to extend the experiment to include additional animals.

MORE THAN TWO FACTORS

Once you recognize the advantages of two factor-designs, you might conclude that three- or four-factor designs would be even better. Indeed, there are many situations in which variability within cells could be reduced by including in the study design such variables as age, economic status, educational level, or sex. As you have learned, the reduction in variability within cells results in a more powerful test of the main effects. A word of caution is appropriate, however. First, each additional variable included in the design of a completely crossed, completely balanced design increases the number of subjects required for the study. This increase is not a simple additive effect. With 10 subjects per cell in a two-level, single-factor design, 20 subjects are required. As additional factors, each with two levels, are included in the design, the total number of subjects required in order to maintain 10 subjects per cell goes from 20 to 40, to 80, to 160, and so forth. If new factors have more than two levels, the relative increase is even greater.

Besides increasing the number of subjects required, adding factors to a study design also greatly complicates the calculations and the interpretation of the ANOVA. This is especially true in regard to interaction effects. As a result, the three-factor experiment probably represents the limit in terms of practical applications for most researchers. Even the three-factor experiment can lead to complicated interpretations if what are called ''second-order'' interaction effects are found to be significant.

You will probably want to merely read over the description of the three-factor ANOVA that follows without attempting to make a detailed study of the calculations in the example. In the future, if the need to do the calculations for a three-factor ANOVA arises, the example given should provide an adequate pattern. An exercise at the end of the lesson provides an opportunity for practice.

In place of the two-dimensional diagram used to display the results of a two-factor study design, we will need to use a three-dimensional drawing for the three-factor design. To keep things as simple as possible, we choose as our example a $2 \times 2 \times 2$ design. Let us label the factors A, B, and C. The diagram to represent the groups in the study appears as a cube that is itself composed of eight smaller cubes, or cells. Each of the eight cells represents a group from the study. Instead of rows and columns, the main effects can now be equated to comparisons of means for: layers (factor A), slices (factor B), and sections (factor C).

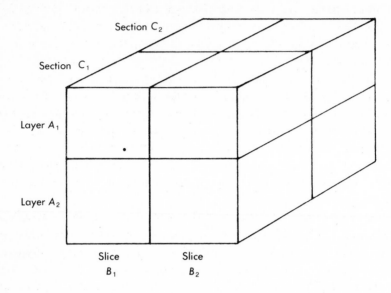

The calculations for the three-way ANOVA represented by this cube follow directly from what you have learned for the two-factor case. You begin by finding the total of all of the scores, squaring this total, and dividing by n to obtain the correction factor. You then square each individual score and add to find the ΣX^2. Subtracting the correction factor from ΣX^2 provides SS_T.

Finding SS_{bcells} is exactly the same as for the two-variable case. The total within each cell is squared and divided by the number within the cell. These results are then added, and the correction factor is subtracted to provide SS_{bcells}. To find the SS_{wcells} we need only to subtract SS_{bcells} from SS_T.

As in the two-factor case, we now need to go back and subdivide SS_{bcells} into its component parts. This time we will have sums of squares for three main effects—SS_A, SS_B, and SS_C—plus interaction effects. To find the sum of squares for main effects, we again use the familiar algorithm. For SS_A (layers), we will find the total of all scores in the top layer and the total of all scores in the bottom layer, square these totals, divide by the number in each layer, add the results, and subtract the correction factor. The process of obtaining SS_B and SS_C is identical, except that for B, the totals are for the right and left slices and that for C, the totals are for front and back sections.

It is when we get to interaction effects that complications arise. Instead of having a single interaction effect to test for, we have three first-order interaction effects and one second-order interaction effect. First-order interaction effects involve two variables at a time. We therefore have the possibility of an *AB* interaction effect, an *AC* interaction effect, and a *BC* interaction effect. When an *AB* interaction is found, we are saying that if factor *C* had not been included in the experiment, so that the scores in the two sections were combined, the analysis would have resulted in a significant interaction effect between factors *A* and *B*. The same kind of interpretation is appropriate for *AC* and *BC* interaction effects. The researcher who is familiar with the meaning and relationships among the variables will probably not have difficulty interpreting the meaning of a significant first-order interaction effect.

If a significant second-order interaction effect is found, the interpretation is more complicated. At least one of the first-order interaction effects is not the same for all levels of the third factor. For example, the *AB* interaction present in the front section of the cube may not be present in the back section, or it may take a different form. Finding a meaningful interpretation of a significant second-order interaction effect and explaining that interpretation in the discussion section of the research paper constitute a real challenge for the investigator unfortunate enough to discover a significant second-order interaction effect.

Just as the interpretation of interaction effects becomes complicated in the three-way ANOVA, so the calculations become more involved. Remember that in the two-way ANOVA, we were able to find SS_{bcells}, SS_{rows}, and $SS_{columns}$. Then we found $SS_{interaction}$ by subtracting SS_{rows} and $SS_{columns}$ from SS_{bcells}. To break down the SS_{bcells} in a three-way ANOVA, we will calculate SS_A, SS_B, and SS_C as explained earlier. We then calculate first-order interactions SS_{AB}, SS_{AC}, and SS_{BC} separately. Finally, we can then obtain SS_{ABC} by subtracting SS_A, SS_B, SS_C, SS_{AB}, SS_{AC}, and SS_{BC} from SS_{bcells}. Once this is done, we have all of the sums of squares, and the remainder of the ANOVA table corresponds to our earlier explanation under the two-factor case.

It can therefore be seen that although the work is greater, the calculations for a three-way ANOVA follow directly from the procedures learned for the two-factor case. The only new calculation to be learned is that involved in finding the sums of squares for first-order interactions. Although these calculations are not difficult to make, they are a bit difficult to explain. Let us try by using the *AB* interaction as an example. We ignore for the time being the division of the scores into front and back sections by combining the scores of section C_1 and C_2, reducing the number of cells from 8 to 4. In essence, we now have a 2×2 design, with only four cells for the combination of variables *A* and *B*. We now find the total for each of these cells, square this total, divide by the number in each cell, and add the results. From this total, we subtract the correction factor and the *previously calculated SS_A and SS_B*. The result is SS_{AB} interaction. We can follow a similar procedure to obtain SS_{AC} and SS_{BC}. As stated earlier, you need not be concerned at this point about these calculations. They will be shown in the example that follows. The example, together with the preceding explanation, should provide sufficient information to provide a pattern to follow in the event of a future need. For now, the reader should be content with a general understanding of the process.

Example of a three-way ANOVA

We would like a simple example that shows the calculations for a three-way ANOVA for a completely crossed, completely balanced fixed-effects factor design. An expansion of the data used earlier to present the two-factor ANOVA will serve the purpose. Remember the example involving the effects of fertilizer and weed control on the output of pea plants? Let us use the data set in which an interaction effect was built into the data (p. 234). To introduce a third variable, we include a second strain of seeds. The two strains of seeds will be represented as sections in our diagram. The front section will contain the data from the example as presented earlier. The data for the back section will be based on the supposition that the second strain will, in general, produce 1 or 2 more cups of peas per plant than those of the first strain. To make the example a bit more realistic than before, the data for the second strain will be made a little less ''perfect.'' With two plants per cell and with totals included in parentheses, our diagram appears as follows:

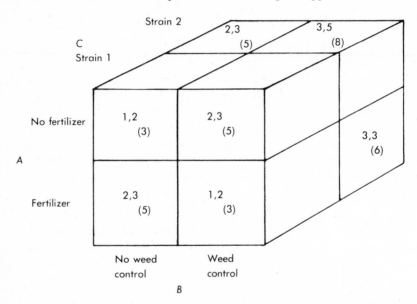

The outputs for the two plants in the cell in the back lower left-hand corner (which cannot be seen) are 3 and 4, with a total of 7. Notice that to obtain the numbers for the back section, we have added either 1 or 2 to each of the numbers for corresponding cells of the front section.

It is very difficult to show the calculations for a three-way ANOVA, because the scores must be combined in three different ways. Table 11-10 should help. The top set of 2 × 2 tables present the scores by individual layer and for the combination of layers. The middle set presents the scores by slices and by a combination of slices. The bottom set shows the scores by section and for a combination of sections. Referring to these sets of 2 × 2 tables while you trace through the following steps should make the calculations understandable.

Table 11-10. Combination of scores for a three-way ANOVA

Top layer (no fertilizer)

	No weed control	Weed control	
Strain 2	2,3 (5)	3,5 (8)	13
Strain 1	1,2 (3)	2,3 (5)	8
			21

Lower layer (fertilizer)

	No weed control	Weed control	
Strain 2	3,4 (7)	3,3 (6)	13
Strain 1	2,3 (5)	1,2 (3)	8
			21

Combined

	No weed control	Weed control	
Strain 2	2,3,3,4 (12)	3,5,3,3 (14)	26
Strain 1	1,2,2,3 (8)	2,3,1,2 (8)	16
			42

Left slice (no weed control)

	Strain 1	Strain 2	
No fertilizer	1,2 (3)	2,3 (5)	8
Fertilizer	2,3 (5)	3,4 (7)	12
			20

Right slice (weed control)

	Strain 1	Strain 2	
No fertilizer	2,3 (5)	3,5 (8)	13
Fertilizer	1,2 (3)	3,3 (6)	9
			22

Combined

	Strain 1	Strain 2	
	1,2,2,3 (8)	2,3,3,5 (13)	21
	2,3,1,2 (8)	3,4,3,3 (13)	21
			42

Continued.

Front section (strain 1)

	No weed control	Weed control	
No ferti-lizer	1,2 (3)	2,3 (5)	8
Ferti-lizer	2,3 (5)	1,2 (3)	8
			16

Back section (strain 2)

	No weed control	Weed control	
No ferti-lizer	2,3 (5)	3,5 (8)	13
Ferti-lizer	3,4 (7)	3,3 (6)	13
			26

Combined

	No weed control	Weed control	
No ferti-lizer	1,2,2,3 (8)	2,3,3,5 (13)	21
Ferti-lizer	2,3,3,4 (12)	1,2,3,3 (9)	21
			42

The steps in the calculation are as follows:

1. Find T, the total of all scores. $T = 42$.

2. Square T and divide by n to obtain the correction factor. $C = \dfrac{42^2}{16} = 110.25$.

3. Square each score and add to find ΣX^2. $\Sigma X^2 = 126$.

4. Subtract the correction factor from ΣX^2 to obtain SS_T.
   $SS_T = 126 - 110.25 = 15.75$.

5. Square the total in each cell, divide by the number in each cell, add the results, and subtract the correction factor to obtain SS_{bcells}.
$$SS_{bcells} = \frac{5^2}{2} + \frac{8^2}{2} + \frac{3^2}{2} + \frac{5^2}{2} + \frac{7^2}{2} + \frac{6^2}{2} + \frac{5^2}{2} + \frac{3^2}{2} - 110.25 = 10.75.$$

6. Subtract SS_{bcells} from SS_T to obtain SS_{wcells}. $SS_{wcells} = 15.75 - 10.75 - 5.0$.

7. Find the total for each layer, square the result, divide by the number in each layer, add the results, and subtract the correction factor to obtain SS_A.
$$SS_A = \frac{21^2}{8} + \frac{21^2}{8} - 110.25 = 0.$$

8. Repeat step 7 for slices to obtain SS_B. $SS_B = \dfrac{20^2}{8} + \dfrac{22^2}{8} - 110.25 = 0.25$.

9. Repeat step 7 for sections to obtain SS_C. $SS_C = \dfrac{16^2}{8} + \dfrac{26^2}{8} - 110.25 = 6.25$.

10. Combine the scores of the front and back sections, find the total for each of the four resulting cells, square the result, divide by the number in the cell, add the results, and subtract SS_A, SS_B, and the correction factor to obtain SS_{AB}. (See the bottom row of 2 × 2 tables.)
$$SS_{AB} = \frac{8^2}{4} + \frac{13^2}{4} + \frac{12^2}{4} + \frac{9^2}{4} - 0 - 0.25 - 110.25 = 4.0.$$

11. Carry out step 10, combining the scores of the right and left slices, and in the end subtracting SS_A, SS_C, and the correction factor to obtain SS_{AC}. (See the middle row of 2 × 2 tables.)
$$SS_{AC} = \frac{8^2}{4} + \frac{13^2}{4} + \frac{8^2}{4} + \frac{13^2}{4} - 0 - 6.25 - 110.25 = 0.$$

12. Carry out step 10, combining the scores of the upper and lower layers, and in the end subtracting SS_B, SS_C, and the correction factor to obtain SS_{BC}. (See the top row of 2 × 2 tables.)
$$SS_{BC} = \frac{12^2}{4} + \frac{14^2}{4} + \frac{8^2}{4} + \frac{8^2}{4} - 0.25 - 6.25 - 110.25 = 0.25.$$

13. Finally, subtract SS_A, SS_B, SS_C, SS_{AB}, SS_{AC}, and SS_{BC} from SS_{bcells} to obtain SS_{ABC}. $SS_{ABC} = 10.75 - 0 - 0.25 - 6.25 - 4.0 - 0 - 0.25 = 0$.

We now have calculated all of the sums of squares and are ready to construct our ANOVA table in the usual fashion. Before doing so, let us review the calculations for our data set. The sum of the 16 scores is 42. Squaring this figure and dividing by 16 gives a correction factor of 110.25. When each score is squared and added, the resulting ΣX^2 is found to be 126. Subtracting the correction factor yields an SS_T of 15.75. As indicated in step 5, we next square each cell total, divide by 2, and add the results, obtaining a value of 121. Subtracting the correction factor, we have an SS_{bcells} equal to 10.75. Subtracting SS_{bcells} from SS_T, we find that SS_{wcells} equals 5.0. To find SS_A we get layer totals of 21 and 21. Squaring these totals, dividing by 8, and adding the results, we get 100.25. Subtracting the correction factor, we have an SS_A equal to 0. To get SS_B, we get slice totals of 20 and 22. Squaring, dividing by 8, and adding, we get 110.5. Subtracting the correction factor, we find SS_B to be 0.25. To get SS_C, we find section totals of 16 and 26. Squaring, dividing by 8, and adding, we get 116.5. Subtracting the correction factor gives an SS_C of 6.25.

Now, to obtain SS_{AB}, we combine the scores of the front and back sections and get cell totals of 8, 13, 12, and 9. Squaring, dividing by 4, and adding, we get 114.5. Subtracting SS_A, SS_B, and the correction factor, we find SS_{AB} to be 4.0. To obtain SS_{AC}, we combine right and left slices and get cell totals of 8, 8, 13, and 13. Squaring, dividing by 4, and adding, we get 116.5. Subtracting SS_A, SS_C, and the correction factor, we find an SS_{AC} of 0. To obtain SS_{BC}, we combine upper and lower layers and get cell totals of 8, 8, 12, and 14. Squaring, dividing by 4, and adding, we get 117. Subtracting SS_B, SS_C, and the correction factor, we find that SS_{BC} equals 0.25. Finally, to obtain SS_{ABC}, we subtract SS_A, SS_B, SS_C, SS_{AB}, SS_{AC}, and SS_{BC} from SS_{bcells}. We find that the second-order interaction sum of squares equals 0.

We now have calculated all of the sums of squares and are ready to construct our ANOVA table (Table 11-11) in the usual fashion.

The only number in the ANOVA table that requires explanation is that for df for the second-order interaction. As indicated earlier, to find df for first-order interaction, you find the product of the df for the two factors involved in the interaction. Similarly, to find

Table 11-11. ANOVA table comparing the effects of three factors* on the productivity of pea plants

Source	SS	df	ms	F
Factor A (layers)	0	1	0	0
Factor B (slices)	0.25	1	0.25	0.4
Factor C (sections)	6.25	1	6.25	10.0†
AB interaction	4	1	4	6.4†
AC interaction	0	1	0	0
BC interaction	0.25	1	0.25	0.4
ABC interaction	0	1	0	0
Between cells	10.75	7		
Within cells	5.0	8	0.625	
TOTAL	15.75	15		

*Factor A equals fertilizer, factor B equals weed control, and factor C equals seed strain.
†P < 0.05.

df for the second-order interaction, you find the product of the *df* for the three factors involved in the interaction.

Notice the *F* value for the *AB* interaction effect. In the earlier two-factor analysis, this is equivalent to the row by column interaction, which was not found to be significant. With the amount of data being doubled by the addition of a second strain of peas, the test has become more powerful and the interaction effect, which we deliberately built into the data, is now found to be significant. The only other effect that is found to be significant is that of factor *C*, which was the strain of peas. This finding is not surprising, since, when constructing the data for the second strain, we added either 1 or 2 cups to the output of each plant in the original data.

UNEQUAL CELL FREQUENCIES

In some experiments, it is not possible to arrange for equal numbers of subjects within each cell. One example is an experiment in which subjects are "lost" during the course of the experiment. For example, an experimental animal may die, a measurement may be invalid, or a person may move. If it can be assumed that the loss of the subject was unrelated to the outcome of the experiment, the process of ANOVA can still be used to analyze the data. However, because the data are not completely balanced, the sums of squares will not be additive; that is, $SS_{bcells} \neq SS_{rows} + SS_{columns} + SS_{interaction}$. You may wish to verify this fact by attempting an ANOVA for the following data.

1,2	2,3
2,3	3,4 3,4

Using the procedures you have learned, you find the following results:

$$\Sigma X = 27$$
$$\Sigma X^2 = 81$$
$$C = 72.9$$
$$SS_T = 8.1$$
$$SS_{bcells} = 5.6$$
$$SS_{wcells} = 2.5$$
$$SS_{rows} = 3.267$$
$$SS_{columns} = 3.267$$
$$SS_{interaction} = -0.934$$

Of course, it is not possible for any sum of squares to be negative, since squaring any number always results in a positive value.

When sums of squares are not additive, an exact solution for ANOVA requires complex calculations. Usually it will be necessary to make use of a computer program to carry out the many steps. However, if the cell frequencies are nearly equal, there is a procedure that gives a close approximation to the exact solution. It is called an *unweighted means*

ANOVA. As you will see shortly, the calculations for an unweighted means ANOVA follow directly from the procedures you are already familiar with.

Before the calculations for an unweighted means ANOVA are described, the requirement for additivity of sums of squares should be stated more precisely. Along with most other authors of textbooks at this level, I have implied that sums of squares are additive only when cell frequencies are equal. Actually, this overstates the requirement. Frequencies only need to be *proportional,* within either rows or columns. In others words, the sums of squares would also be additive for a design with the following *proportional* frequencies between columns.

	a_1	a_2
b_1	$n = 2$	$n = 4$
b_2	$n = 3$	$n = 6$
b_3	$n = 4$	$n = 8$

Table 11-12. Data for an unweighted means ANOVA

	Curriculum				
	A	B	C	D	Total
Girls	75	75	85	61	
	79	94	71	69	
	95	79	85	98	
	67	51	67	79	
	74	86	77	86	
	75	64	87	75	
	80	75	76	71	
	73	68	82	90	
	64	76	69	85	
		66		79	
Subtotal	682 +	734 +	699 +	793 =	2908
Boys	54	64	68	77	
	69	67	79	87	
	59	70	68	92	
	75	79	89	95	
	87	86	69	100	
	75	68	79	77	
	62	54	60	85	
	71	68	62	88	
	76	45	64	87	
		59	78		
Subtotal	628 +	660 +	716 +	788 =	2792
TOTAL	1310 +	1394 +	1415 +	1581 =	5700

Since research designs with proportional frequencies are not very common, many textbooks ignore this case.

Unweighted means ANOVA

Earlier, an example was used to describe the calculations for a two-way ANOVA that involved four sets of self-study health curriculum materials. There were test scores for ten boys and ten girls for each curriculum. Let us use the same set of data to show the calculations for an unweighted means ANOVA. To make cell sizes unequal, we will drop the last score from groups A and C for girls and from groups A and D for boys. Our data appear as Table 11-12.

In an unweighted means ANOVA, the scores within each cell are replaced by a single number, the mean. The analysis then is very similar to a multifactor ANOVA with a single entry per cell. The only difference is that in this case it is not necessary to use the residual as a measure of expected variation. Instead, it is possible to go back to the scores within each cell to get SS_{wcells}, which is then appropriately weighted to obtain an ms_{wcells}. The calculation of the ms_{wcells} will be described shortly. First, let us complete the rest of the calculations. Table 11-13 contains the means and the sums of the means for rows and columns.

Treating the data as a factorial design with a single entry per cell, we get the following results:

$$\Sigma X = 601.1$$
$$C = 601.09^2/8 = 45,163.65$$
$$SS_{bcells} = (75.78)^2 + (69.78)^2 + (73.4)^2 \ . \ . \ . \ (87.56)^2 - C =$$
$$45,469.85 - 45,163.65 = 306.20$$

$$SS_{columns} = \frac{(145.56)^2}{2} + \frac{(139.4)^2}{2} + \frac{(149.27)^2}{2} + \frac{(166.86)^2}{2} - C =$$
$$45,371.93 - 45,163.65 = 208.28$$

$$SS_{rows} = \frac{(306.15)^2}{4} + \frac{(294.94)^2}{4} - C = 45,179.36 - 45,163.65 = 15.71$$

$$SS_{interaction} = SS_{bcells} - SS_{columns} - SS_{rows} = 306.20 - 208.28 - 15.71 = 82.21$$

Notice that we have not listed SS_T, which would be the same as SS_{bcells}. We will not need SS_T anyway, since we do not plan to obtain SS_{wcells} by subtraction.

The only other value needed to complete the analysis is a mean square within as a measure of error variance. To obtain this value, we will make use of the original scores

Table 11-13. Cell means for the unweighted means ANOVA

	Curriculum				
	A	B	C	D	Total
Girls	75.78	73.40	77.67	79.30	306.15
Boys	69.78	66.00	71.60	87.56	294.94
Both	145.56	139.40	149.27	166.86	601.09

within each cell. First, we find the sum of squared deviations separately *within each group* using the computational formula:

$$\Sigma x^2 = \Sigma X^2 - \frac{(\Sigma X)^2}{n}$$

These results are then added, and the total is divided by the *df* within. (The value of *df* for within is found by subtracting 1 from the frequency within each cell and then adding the results.) The result of these calculations provides an *unweighted measure of error variance*. It must be weighted to reflect the sizes of the various cells. To obtain the weight, we add the reciprocals of the cell frequencies and then multiply the sum by the reciprocal of the number of cells. This weight is then multiplied by the error variance to obtain the appropriate mean square within. As usual, it is difficult to describe the process of finding the mean square within. It is much easier to illustrate the process using the example we have chosen. Step 1 is to find the sum of squared deviations within each cell of Table 11-12. Using the computational formula for the first cell, we obtain the following:

$$\Sigma x^2 = \Sigma X^2 - \frac{(\Sigma X)^2}{n} = 52{,}306 - \frac{(682)^2}{9} = 625.5$$

Using the same procedure for the remaining cells and adding the results, we find the sum of deviations squared within cells is 6726.23. The *df* within add up to 68. Dividing 6726.23 by 68, we get an unweighted ms_{wcells}, or unweighted error variance, of 98.915. To find the weight, we first add the reciprocals of cell frequencies:

$$^1/_9 + {}^1/_9 + {}^1/_{10} + {}^1/_{10} + {}^1/_9 + {}^1/_{10} + {}^1/_{10} + {}^1/_9 = 0.8444$$

We then multiply by the reciprocal of the number of the cells to obtain the proper weighting factor.

$$W = {}^1/_8(0.8444) = 0.10555$$

The weighted mean square within is therefore:

$$ms_{within} = (98.915)(0.10555) = 10.44$$

We are now ready to make up the ANOVA table. It is as shown in Table 11-14.

Table 11-14. Comprehensive ANOVA table for health concepts test scores

Source	SS	df	ms	F
Rows	15.71	1	15.71	1.50
Columns	208.28	3	69.43	6.65*
Interaction	82.21	3	27.40	2.62†
Between cells	306.20	7		
Within cells		68	10.44	

*$P < 0.01$.
†$P < 0.05$.

Notice that the F values obtained for the unweighted means ANOVA do not differ much from those obtained earlier for the data when there were equal numbers for each cell (Table 11-7). The results agree so well because the cell frequencies were indeed nearly equal. The process just described is called an unweighted means analysis because each of the means counts as a single number when we figure sums of squares for the main effects and the interaction effects. A similar process can be used when frequencies are nearly *proportional* rather than nearly equal. In this case, each mean is weighted according to cell or column proportions. As stated earlier, we will not be concerned with the case of exactly proportional or nearly proportional cell frequencies.

SUMMARY

This final lesson on ANOVA contained a discussion of four situations in which the procedures of multifactor ANOVA need modification. The first situation involves the random-effects model in which the interaction mean square has to be included in the F ratio for testing main effects. The second is a situation in which there is a single entry for each treatment combination, preventing the use of the mean square within as a measure of error variance. The third is a design involving three factors, giving rise to the possibility of three first-order interaction effects and one second-order interaction effect. The last situation is one in which the number of measurements within each cell is not equal. In this instance, instructions were given for doing an unweighted means ANOVA in which the mean for each cell is treated as a single entry and the variability of scores within each cell is weighted to arrive at a weighted error mean square. In each of these cases, a detailed solution to a problem was presented so that all readers will know *what* to do when the circumstance arises. Some explanation for the logic behind each solution was also given in the hope that most readers will also understand *why* the analysis is appropriate. The exercises that follow will provide a test of your ability to solve problems similar to those in this lesson. In actual practice, most readers will use a computer program to do the calculations for complex ANOVA problems. An understanding of the material in this and the preceding lessons should allow a student to make a proper interpretation of the output from such a program. The following chapter, dealing with correlation and regression, will be similar. Concepts will be presented and explained using a relatively simple problem. These concepts can then be applied directly to the output of computer programs designed to carry out complicated multipredictor regression problems.

LESSON 4 EXERCISE

All of the questions make use of the following data set.

	B1	B2
A1	3.043	6.984
	4.525	8.417
	2.135	8.056
	3.727	8.561
A2	6.371	9.956
	5.965	5.643
	5.298	7.719
	5.568	8.932
A3	8.120	7.943
	7.535	6.971
	7.762	8.479
	7.131	10.709

1. Consider A and B random-effects factors and do a two-way ANOVA. Interpret your results.
2. Consider A and B fixed-effects factors. Use only the first score from each group and do a single entry, two-way ANOVA. Interpret your results.
3. Consider A and B fixed-effects factors. The first two scores in each group are those of males, and the second two are those of females. Do a three-way ANOVA, including sex as variable C.
4. Consider A and B fixed-effects factors. Disregarding sex as a variable, drop the last score from groups $A1$-$B1$, $A2$-$B2$, and $A3$-$B1$ and do an unweighted means ANOVA for cells of unequal size.

CHAPTER 12

Correlation and regression

Lesson 1 □ The correlation coefficient r

In an early lesson dealing with descriptive statistics, you were introduced to a formula for measuring the relationship between two variables. The resulting value was called a correlation coefficient and was given the symbol r. At that time, you were provided with a limited amount of information about the derivation of the formula, its interpretation, and its use. The present chapter contains four lessons devoted to a discussion of correlation and a closely related topic called regression analysis. In the first lesson you will be shown the logic behind the formula for the correlation coefficient r, and you will learn how to make inferences regarding this coefficient. In the second lesson you will learn about some other measures of relationship that are appropriate when r is not. In the third lesson, the procedures and rationale of regression analysis will be introduced for the relatively simple one-predictor case. Then, in the last lesson, you will learn how the procedures of regression analysis can be extended to two or more predictors. Anyone who intends to become involved in education or personnel choices will readily identify applications for these concepts.

THE DERIVATION OF r

When the correlation coefficient r was introduced, you were given two formulas, one a definitional formula and the other a computational formula. These formulas, which follow, are Formulas 3.3-2 and 3.3-5 respectively.

$$r = \frac{\Sigma z_X z_Y}{n}$$

or

$$r = \frac{\Sigma XY - \frac{(\Sigma X)(\Sigma Y)}{n}}{\sqrt{\left[\Sigma X^2 - \frac{(\Sigma X)^2}{n}\right]\left[\Sigma Y^2 - \frac{(\Sigma Y)^2}{n}\right]}}$$

For either formula, the basic data required are pairs of scores for two variables, X and Y. For the first formula, you need to calculate the means and standard deviations for each variable and then change each X and Y score to a z score. Once this has been done, the correlation coefficient is then the average of the products of the pairs of z scores. Use of the computational formula avoids the necessity of changing each score to a z score, allowing the work to be carried out in a single operation on sophisticated calculators.

The logic behind the definitional formula for r is usually not presented in statistics books, even though it is quite straightforward. The rationale for the formula will be illustrated here by the use of a set of examples. Suppose that you are given three X scores and three Y scores without being told how the scores are paired. To keep things very simple, suppose that the set of X scores includes the numbers 1, 2, and 3 and that the set of Y scores also includes the numbers 1, 2, and 3. Not knowing how the scores are paired, you are told to pair them in any way you want in an effort to achieve the highest possible value for the sum of the products of the pairs (ΣXY). If it is not immediately apparent, arranging the X and Y scores in different pairings will soon demonstrate that ΣXY is largest when the scores are paired largest with largest and smallest with smallest.

X	Y	XY
1	1	1
2	2	4
3	3	9
		14

In a similar fashion, it should be obvious that ΣXY is smallest when the largest X score is paired with smallest Y score and vice versa.

X	Y	XY
1	3	3
2	2	4
3	1	3
		10

If these same scores are paired in a haphazard fashion, ΣXY will be a value somewhere between 14 and 10. Based on this example, we can conclude that the magnitude of ΣXY tells us something about the relationship between X and Y.

If ΣXY is a measure of the relationship between X and Y, why then do we need the complicated formulas used to calculate r? The answer is that, in addition to the relationship between X and Y, three other factors influence the magnitude of ΣXY. You can probably think of at least two. First, it is obvious that the addition of more pairs of scores would increase ΣXY. We can avoid this problem by finding the *average* of the products of the pairs of scores. Our measure of relationship would then be $\Sigma XY/n$ where n is the number of pairs of scores. In addition, you may have guessed that the size of the X and Y scores also influences both ΣXY and $\Sigma XY/n$. If, on average, X and Y scores are large, $\Sigma XY/n$ will be large; if X and Y scores are small, $\Sigma XY/n$ will be small. To devise an index of relationship that is comparable from one data set to another, we need somehow to remove the influence of the average magnitude of X and Y scores. Can you think of a solution to the problem? Suppose that instead of using the actual X and Y scores, we change each to a deviation score by subtracting the mean. Since deviation scores always have a mean of zero, we will be removing the effect of the magnitude of the original X and Y scores. Our index of relationship is now $\Sigma xy/n$ and is unaffected by the number of pairs of scores or the magnitude of the original scores.

By now you may have identified the third factor to be eliminated. If not, compare the definitional formula for r with the index we have derived. The definitional formula says to find the average of the products of z *scores*, whereas our index says to find the average of the products of *deviation scores*. What is the difference? In the definitional formula, the deviation scores have been divided by the respective standard deviations to obtain z scores. Dividing any set of scores by the standard deviation of those scores results in a new set of scores with a standard deviation of one. In other words, the final step in deriving the formula for r is to eliminate differences in the variability of the X and Y scores from one data set to another by coding all scores to have a standard deviation of one.

To demonstrate that the variability of scores does indeed affect ΣXY, consider two examples.

Example A			Example B		
Mean = 20, n = 3			Mean = 20, n = 3		
X	Y	XY	X	Y	XY
19	19	361	10	10	100
20	20	400	20	20	400
21	21	441	30	30	900
		1202			1400

For both examples, the number of pairs of scores are equal and the means of X and Y are equal. However, the greater variability of scores in example B has resulted in a larger ΣXY.

To summarize, we first learned that the magnitude ΣXY provides some information as to how X is related to Y within a set of pairs of scores. However, ΣXY is also affected by the numbers of pairs of scores, the average size of the scores, and the variability of the scores. To standardize our index of correlation, we make it an average by dividing by n; we convert to deviation scores by subtracting the mean; and finally we change to standard scores by dividing by the standard deviation. Our formula is therefore: $r = \Sigma z_X z_Y/n$, or its equivalent computational form.

Now that you have learned the logic behind the formula for the correlation coefficient r, you can see why coding has no effect on the correlation coefficient. Any coding that affects the mean or the standard deviation will have no effect on r. The formula itself will code all X and Y scores to a set of scores with a mean of zero and a standard deviation of one.

INTERPRETATION OF r

At this point, some additional insight into the interpretation of correlation coefficients can be provided. When the topic of correlation coefficients was first introduced, it was explained that the values for r could range from $+1.00$ to -1.00. It was further explained that an r of 1.00 (either $+$ or $-$) indicates a perfect relationship between variables X and Y. What is meant by a perfect relationship? By this we mean that it is possible to find a single formula that can be used to translate any X score to the exact value of its paired Y score or any Y score to the exact value of its paired X score. If r is some value between 0.00 and ± 1.00 it indicates that a relationship exists but that it is not perfect. In this case, knowledge of the score for X provides *some* information as to what value to expect for Y, but the information is inexact. When r is 0.00, there is no relationship between X and Y, and knowledge of one score of a pair provides no information concerning the other score. In this latter case, no matter what X score is observed, the best estimate to be made as to the value of Y is the mean for Y.

The process of finding the formula for translating any X score into the best estimate of its paired Y score is called *regression analysis*. Of course, the same process applies to translating any Y score into the best estimate of its paired X score. If the formula being sought is restricted to one that provides for a constant change in the estimate of Y for each increment in X, the process is called *linear* regression analysis, as opposed to *curvilinear* regression analysis. An example of a formula resulting from a linear regression analysis might be $Y = 2X$. For every increase of 1 unit in X, the estimate of Y $(\hat{Y})$ increases 2 units. In curvilinear regression analysis, the increment in the estimate of Y for each increment in X follows a regular pattern, but the change need not be constant. An example of a formula resulting from curvilinear regression would be $Y = X^2$.

The discussion of regression analysis in the lessons that follow will be restricted to linear regression. This is equivalent to saying that when X and Y scores are plotted on a scatter diagram, the formula for estimating Y scores from X scores can be represented by a straight line (called a regression line) through the swarm of dots. With this restriction in mind, it is possible to present another interpretation of the meaning of the correlation coefficient r. The nearer r is to either $+1.00$ or -1.00, the better the results that can be achieved in linear regression analysis. We need to define what is meant by better results. This topic will be given considerable attention later. For now we can say that distances of the dots from the regression line through the scatter diagram become progressively smaller as r approaches either $+1.00$ or -1.00.

Before turning our attention to the problem of making inferences regarding correlation coefficients, we provide one additional interpretation of the correlation coefficient r. This interpretation is related to the previous chapter on analysis of variance, in which the variability of a set of scores was described in terms of sums of squares between groups and sums of squares within groups.

Suppose that in a one-way ANOVA problem the measurements of the dependent variable are considered measurements for Y. The designation as to which experimental group a subject belongs to can be considered to be measurements on a variable, X. We would then have X and Y scores for everyone in the study, with X having a range corresponding to the number of groups. In the ANOVA, the finding of a significant F value indicates that there are real differences for the means of Y for different values of X. We test for this significant difference by comparing that portion of the variance of Y scores which is related to differences in values of X ($SS_{bgroups}$) with that component which is not related to differences in values of X ($SS_{wgroups}$).

In a similar fashion, it is possible to think of correlation problems in terms of an attempt to account for the variability in Y scores (or X scores). The total variability of Y scores can be divided into that component which can be explained on the basis of differences in observed X scores and that component which is independent of differences in X. All individuals sharing a common X score can be thought of as members of a common group. If r is either $+1.00$ or -1.00, they will also share a common Y score, and there will be no variability within groups. In this case, the variability of Y scores arises solely from the variability between groups.

In most instances, r will be a value other than $+1.00$ or -1.00, and persons sharing the same X score will display some variability in Y scores. In this case, the total variability of Y scores arises partly from differences in X scores and partly for other reasons that cannot be attributed to differences in X. Exact prediction of Y from X scores will not be possible in this case. For the instance in which there is no linear relationship between X and Y, so that r equals 0.00 none of the variability of Y scores can be attributed to a linear relationship between X and Y.

Note that we have limited our discussion to the linear model. It is possible for r to be 0.00, indicating that no linear relationship exists between X and Y, and yet to be able to predict Y quite accurately from X by use of a curvilinear model. The use of such models can be quite complex and will generally require the use of computer programs to carry out the required calculations. The topic of curvilinear models will not be covered in this book. You will, however, learn about an index of association that includes curvilinear relationships. For now, we continue to restrict our discussion to the linear model.

The discussion of correlation in terms of partitioning the variance of Y into component parts leads to a useful concept called the *coefficient of determination*. This coefficient is a fraction in which the total variance of Y scores is the denominator and the part of the variance that arises from the linear relationship between X and Y is the numerator. In essence, we are saying that the meaning is as follows.

$$\text{Coefficient of determination} = \frac{\text{``variance between groups''}}{\text{total variance}}$$

Note that variance between groups has been placed in quotation marks. Because we are limiting the discussion to linear models, the concept has a somewhat different meaning than it does in ANOVA problems, and it would have to be calculated differently. Fortunately, the problem does not arise. Although the equation expresses the meaning of the coefficient of determination, it does not indicate its usual method of calculation. For once, a useful concept is calculated by a very simple formula.

$$\text{Coefficient of determination} = r^2 \qquad \text{12.1-1}$$

The formula for the coefficient of determination provides additional insight into the meaning of a correlation coefficient. It says that if the correlation coefficient is squared, the result will indicate, by use of a linear regression model, what portion of the total variance of one score can be explained on the basis of differences in the other score. Any remaining variance is unexplained.

Interpreting correlation coefficients in terms of the coefficient of determination can be a rather humbling experience for persons given the responsibility of making personnel decisions. Examples are the selection of persons for advanced education or for management training. Examinations used to help in making these decisions generally have modest correlations with later measures of performance. The usual range is 0.2 to 0.4, and seldom does the correlation exceed 0.6. Squaring correlation coefficients of 0.2 and 0.6 gives coefficients of determination of 0.04 and 0.36. In the optimal case, only a little more than one third of the variability in that which we wish to predict is related to our examination. In the worst case, the portion of the variance that can be explained is negligible, compared with the total variability of performance. In a later lesson, in addition to learning how to find the formula that best estimates Y from X, you will also learn how to improve this estimation by using multiple predictors. For both the single-predictor case and the multiple-predictor case, you will learn how to use a concept similar to the standard error of a statistic as a means of describing the precision of the estimates that result. Before turning to these regression topics, we need to complete our discussion of correlation coefficients. In the remainder of this lesson, I describe the process of making inferences about r. In the following lesson, we examine some other measures of relationship and their interpretation and use.

INFERENCES INVOLVING r

Your first introduction to the correlation coefficient r was contained in an early section of the book involving other descriptive statistics, such as the mean and the variance. Since then you have learned how to make inferences about the corresponding parameters, such as μ and σ^2. As with these parameters, it is also possible to make inferences concerning the population correlation coefficient. We may wish to know whether there is a relationship between two variables, X and Y, in some defined population. Unable to obtain scores for X and Y for all members of the population, we choose a random sample from the population and compute the correlation coefficient r. Based on this sample value, r, we wish to make a statement about the value of the corresponding parameter computed for the entire population.

To distinguish between sample correlation coefficients and population correlation coefficients, it is again necessary to use a different symbol for each. Since we have used r for a sample correlation coefficient, we should be consistent and use the Greek letter for r to represent the population value. Unfortunately, the Greek letter "rho" is very similar in appearance to a printed Arabic p and would lead to a lot of confusion. Accordingly, although the symbol for rho is used in many statistics books, we will instead use the capital letter R to stand for the population correlation. This should cause no particular confusion when other books are used as references.

Learning to make inferences about R will be easy. Everything you have learned about confidence intervals and hypothesis testing is applicable once the sampling distribution of r is known. We begin, therefore, with a general discussion of the sampling distribution of r. For all of the following discussion of sampling distributions and inference making, it is assumed that X and Y represent continuous, normally distributed variables.

Sampling distribution of r

Starting with the easiest part first, we state that r is an unbiased statistic. This is another way of saying that the central tendency of the sampling distribution of r is the parameter value R. This is consistent with our earlier statements regarding the central tendency of the sampling distributions of p, $\overline{X}$, and s^2.

The standard error of the sampling distribution of r (SE_r) is found from the following formula.

$$SE_r = \frac{1 - R^2}{\sqrt{n - 1}} \qquad \textbf{12.1-2}$$

Although the exact form of the formula could not have been anticipated, the inclusion of n in the denominator of the formula should not be a surprise. Once again, the formula for the standard error indicates that increasing sample size reduces sampling variability. It is important to note that the formula involves R, the population value. We will have more to say about this later.

Finally, the shape of the sampling distribution is approximately normal when R equals zero and n is large (say 30 or more). When R is not zero, the sampling distribution of r is skewed, with the degree of skewness increasing as R approaches either $+1.00$ or -1.00. This last statement should make sense, because sample values are limited by ± 1.00. For a population value of R equal to $+0.90$, sample values can only approach $+1.00$ in one direction, but they can extend down to -1.00 in the other. The result will be a skewed distribution. You will learn a method for dealing with this skewness in one of the sections that follow.

Test that R = 0

Based on your knowledge of hypothesis testing and the information just given, you should be able to figure out how to test a null hypothesis that $R = 0$ for large samples. You know the sampling distribution will be normal, with a mean of zero and a standard error equal to $1 - R^2/\sqrt{n - 1}$. Therefore, to test whether R in a population is zero, we need only take a large sample, compute the statistic r, convert this r to a z score, and look up the z score on the table of the normal curve.

Suppose that we wish to know whether the weights of mothers and the weights of their newborn babies are correlated. We expect a positive correlation. Our null hypothesis is that, for the entire population, $R \leq 0.00$. We take a random sample of 65 mothers and find that $r = 0.24$. Is this evidence that $R > 0$? To answer the question, we need to see where our sample r would fall on a sampling distribution from the standard population in which $R = 0.00$.

With an n of 65, sample r's from a population in which R is zero will be approximately normally distributed, with a mean of zero and a standard deviation found by Formula 12.1-2.

$$SE_r = \frac{1 - R^2}{\sqrt{n - 1}}$$

Since our standard population has an $R = 0$, $SE_r = 1/\sqrt{n - 1}$, or $1/\sqrt{64}$, or 0.125. The z score for an r of 0.24 would therefore be:

$$z = \frac{r - R}{SE_r}$$

$$z = \frac{0.24 - 0}{0.125} = 1.92$$

Since we are making a one-tail test, we look at the one-tail section of the table of the normal curve. We find that if $R = 0$ and $n = 65$, the probability of a sample r of 0.24 or higher is only 0.0274. We will probably conclude that there is indeed a positive correlation between the mother's weight and the weight of the newborn baby. Of course, as in any hypothesis-testing situation, the possibility exists that we have made a Type I error.

How would this problem change if n had been small, say six? In this case, the sampling distribution could not be expected to be normal. We could still calculate a z score for the observed r, but the result would not be a standard *normal* score and could not be evaluated in terms of the probabilities from the table of the normal curve. It has been shown that for small sample sizes, a ratio that contains the sample r both in the numerator and the denominator is distributed approximately as the t distribution with $(n - 2)\, df$. The statistic t is calculated, using Formula 12.1-3.

$$t = \frac{r\sqrt{n - 2}}{\sqrt{1 - r^2}} \qquad\qquad \textbf{12.1-3}$$

If, for the previous problem, n had been six and the same value of r had been found, we would have carried out the following calculations:

$$t = \frac{r\sqrt{n - 2}}{\sqrt{1 - r^2}} = \frac{0.24\sqrt{4}}{\sqrt{1 - (0.24)^2}} = \frac{0.48}{0.97} = 0.49$$

To be significant even at the 0.20 level with 4 df and a one-tail test, the value of t would have to be 0.941 or higher. We would decide that an r of 0.24 with an n of 6 does not provide evidence against the null hypothesis that $R = 0$.

Tests that R equals some nonzero value

Suppose that we are interested in testing the hypothesis that the correlation between X and Y is greater than 0.80 in some population. The null hypothesis is that $R \leqslant 0.80$, and the alternative is that $R > 0.80$. Unfortunately, unless samples are extremely large, sample r's from a population in which $R = 0.8$ will not be normally distributed. Knowing that the mean will be R and that SE_r will be $(1 - R^2)/\sqrt{n - 1}$ will be of little use, since once again the z scores calculated for sample r's will not be standard *normal* scores. A solution to the problem has been provided by Fisher. It involves transforming both the sample r from the test population and the parameter R from the standard population into values that are approximately normally distributed, with a standard error equal to $1/\sqrt{n-3}$. We will use the symbol r_F to designate the transformed sample r and R_F to designate the transformed parameter R. The test statistic will therefore be as shown in Formula 12.1-4.

$$z = \frac{r_F - R_F}{1/\sqrt{n-3}}$$ **12.1-4**

Our only remaining problem is to learn the transformation required. The formula involves a logarithmic transformation.

$$r_F = \frac{log_e\left(\dfrac{1 + r}{1 - r}\right)}{2}$$

The formula says that r_F is found by obtaining the natural log of $(1 + r)/(1 - r)$ and then dividing by 2. The calculations are not difficult with the use of a calculator with log functions. However, a transformation table makes even this step unnecessary. In Table 12-1, entries in the body of the table give Fisher's transformation for any desired r. The column on the left contains the first digit of r, and the row at the top contains the second digit.

Notice that the effect of the transformation is very small when r is near 0.00 and that it increases as r approaches 1.00. For $r = 0.15$, $r_F = 0.151$, whereas for $r = 0.95$, $r_F = 1.832$. For negative values of r, the value of r_F is found from the table and given a negative sign. Finally, it should be noted that R_F is obtained in exactly the same manner as r_F, by substituting R for r either in the formula involving a natural log or when referring to the table.

Table 12-1. Fisher's transformation for r

r	.00	.01	.02	.03	.04	.05	.06	.07	.08	.09
.0	.000	.010	.020	.030	.040	.050	.060	.070	.080	.090
.1	.100	.110	.121	.131	.141	.151	.161	.172	.182	.192
.2	.203	.213	.224	.234	.245	.255	.266	.277	.288	.299
.3	.310	.321	.332	.343	.354	.365	.377	.388	.400	.412
.4	.424	.436	.448	.460	.472	.485	.497	.510	.523	.536
.5	.549	.563	.576	.590	.604	.618	.633	.648	.662	.678
.6	.693	.709	.725	.741	.758	.775	.793	.811	.829	.848
.7	.867	.887	.908	.929	.950	.973	.996	1.020	1.045	1.071
.8	1.099	1.127	1.157	1.188	1.221	1.256	1.293	1.333	1.376	1.422
.9	1.472	1.528	1.589	1.658	1.738	1.832	1.946	2.092	2.298	2.647

From Dixon, W. J., and Massey, F. J., Jr.: Introduction to statistical analysis, ed. 3, New York, 1969, McGraw-Hill Book Co.

As an example of a one-sample test of a hypothesis that R is greater than some nonzero value, let us examine the problem of weather forecasting. Suppose that for several years a radio station with its own meteorologist has kept records of its ability to predict the next day's high temperature. The correlation between predicted highs and actual highs has been found to be 0.80. Then, the purchase of a sophisticated instrument, which is said to aid in weather forecasting, is proposed. The question is whether R will be greater than 0.80 once the equipment is available. A rival radio station possesses such an instrument and has been using it during the same period of time. It is difficult to go back and obtain the entire data set for the second station, but it is possible, with some effort, to discover the data for any individual day. Eighty-four days are chosen at random from the past several years, and the data for predicted and actual highs for those days are obtained. The sample r is found to be 0.87. Is this sufficient evidence to conclude that the R for the second station is greater than the R of 0.80 for the first station?

Our first step is to transform both the sample r from the test population (0.87) and the parameter R from the standard population (0.80) into Fisher's transformed correlation coefficients. Reference to Table 12-1 provides the following transformations.

$$\text{For } r = 0.87, \; r_F = 1.333$$
$$\text{For } R = 0.80, \; R_F = 1.099$$

We now have transformed correlation coefficients that can be expected to be normally distributed, with $SE = 1/\sqrt{n - 3}$. We can therefore make use of the normal curve. To do so, we convert r_F to a z score.

$$z = \frac{r_F - R_F}{1/\sqrt{n - 3}} = \frac{1.333 - 1.099}{1/\sqrt{81}} = \frac{0.234}{0.111} = 2.11$$

The logic of our problem indicates that a one-tail test is appropriate. Reference to the table of the normal curve shows that a value as great as 2.11 will occur less than 2% of the time by chance. We will probably conclude that R for the second station is indeed greater than 0.80.

Two-sample tests of R

Fisher's transformation of correlation coefficients can also be used in the case where the problem is to compare two unknown parameters on the basis of *two independent* samples. If you think back to the method of two-sample tests of means or proportions, you should be able to apply the method to the present case. For example, to test for the equality of the *means* of two normally distributed populations, A and B, you take samples from each, calculate $\overline{X}_A$ and $\overline{X}_B$ and then find the test statistic as follows.

$$z = \frac{(\overline{X}_A - \overline{X}_B)}{SE_{\overline{X}_A - \overline{X}_B}}$$

You learned that the standard error of a difference between two statistics, which appears in the denominator, is equal to the square root of the sum of squares of the individual standard errors. For sample *means,* the formula you learned to use is:

$$SE_{\bar{X}_A - \bar{X}_B} = \sqrt{\frac{\sigma_A^2}{n_A} + \frac{\sigma_B^2}{n_B}}$$

To compare R_A and R_B, we combine the method for a two-sample test of means with the information we have just acquired concerning a one-sample test of R. The calculations are described by Formula 12.1-5.

$$z = \frac{r_{F_A} - r_{F_B}}{\sqrt{\frac{1}{n_A - 3} + \frac{1}{n_B - 3}}} \qquad \textbf{12.1-5}$$

Notice that, as with means or proportions, when the sampling distribution is normal, we compute a z score. The numerator is once more the difference between the two observed statistics, and the denominator is again the standard error of the difference between the statistics. Notice also that to obtain the standard error of the difference between the transformed r's, we again take the square root of the sum of the squares of the individual standard errors.

As an example of a two-sample test of R, suppose that two tests are available for predicting success in medical school. Success is defined as performance on a first-year comprehensive examination. The question is whether the tests are equal in predictive ability. A sample of 39 students is *independently chosen* to take test A. Another 75 students are *independently chosen* to take test B. The correlations with the comprehensive exam are 0.52 and 0.38 respectively. Is there evidence of a difference in R_A and R_B? Our calculations are as follows.

$$\text{For } r_A = 0.52, \ r_{F_A} = 0.576$$
$$\text{For } r_B = 0.38, \ r_{F_B} = 0.400$$

$$z = \frac{0.576 - 0.400}{\sqrt{\frac{1}{(39 - 3)} + \frac{1}{(75 - 3)}}} = \frac{0.176}{\sqrt{\frac{3}{72}}} = \frac{0.176}{0.2} = 0.88$$

In this problem, where the question is simply whether a difference exists, the test is a two-tail, or nondirectional, test. Looking at the table of the normal curve, we find that a value as great as 0.88 in either direction occurs more than 37% of the time. We will probably conclude that there is not sufficient evidence to claim that one test has more predictive ability than the other.

Confidence intervals for R

You should recall that the approach used to answer the question "What is the parameter value?" is to set up a confidence interval as follows.

Parameter = observed statistic ± (appropriate tabled value)(standard error of the statistic)

Suppose that our goal is to answer the question "What is R for variables X and Y in some defined population?" We begin by taking a random sample from that population. Next, we calculate the statistic r as our best estimate of R. We then arbitrarily choose a level for our confidence interval. What problems will we now encounter? First we must decide which table is appropriate; then we must calculate the standard error of r. Unfortunately, we cannot expect the sampling distribution to be normal. Furthermore, the calculation of SE_r involves the unknown parameter value R. What would you guess the solution to be? The answer is to once more make use of Fisher's transformation.

We began by changing the observed sample r to r_F. A distribution of such sample r_F's is expected to be normally distributed, with a standard error of $1/\sqrt{n-3}$. If we choose a 0.90 confidence interval, our calculations will be based on Formula 12.1-6.

$$R_F = r_F \pm (1.645)(1/\sqrt{n-3}) \qquad \textbf{12.1-6}$$

We will then have boundaries for a 90% confidence interval for a Fisher's transformation of R. To find the interval for R itself we will have to transform the two boundaries for R_F back to ordinary correlation coefficients. To do so we again refer to the transformation table (Table 12-1), this time finding from the body of the table the R_F corresponding to each *boundary* and obtaining the corresponding R values from the margins.

Suppose that a 90% confidence interval for R is desired and that for a sample of size 67, r is found to be 0.55. First the r of 0.55 is changed to an r_F of 0.618. The standard error is $1/\sqrt{64}$, or 0.125. The 90% confidence interval in terms of transformed correlation coefficients is as follows.

$$R_F = 0.618 \pm (1.645)(0.125) = 0.618 \pm 0.205$$

The boundaries for R_F are 0.413 and 0.823. Each of these boundaries must now be changed back to ordinary correlation coefficients. Referring to Table 12-1, we cannot find the exact values of 0.413 and 0.823 in the body of the table. For example, the two values closest to the lower boundary of 0.413 are 0.412 and 0.424. Since we are looking for the lower boundary of our interval for R_F, a conservative approach would be to use the lower value, 0.412. The corresponding R value from the margins is 0.39. Similarly, for the value of R_F equal to 0.823, we choose as our upper boundary an R of 0.68, which is the value for an R_F of 0.829. Our 90% confidence interval for R is therefore the interval 0.39 to 0.68. It should be noted that the interval is not symmetrical about the observed r of 0.55. This should not come as a surprise, since we expect the sampling distribution of r to be skewed.

SUMMARY

This first lesson on correlation and regression was devoted to a discussion of the correlation coefficient r. You were first shown the logic behind the formula for calculating r. We began by showing that the magnitude of ΣXY is a function of the relationship between X and Y. We then showed how three other factors influencing the magnitude of ΣXY could be removed, resulting in the definitional formula.

Following this discussion of the derivation of the formula for r, you were given information about its interpretation. First, it was explained that r is an index of the *linear* relationship between X and Y. This is equivalent to saying that r provides a measure of how closely the dots in a scatter diagram will be to a *straight* line drawn through the diagram. To provide further meaning for the coefficient r, an analogy to ANOVA was provided. The total variance of the Y scores can be thought of as being made up of two components; the variance of Y scores within groups sharing the same X score, and the variance between groups. When r is either ± 1.00, there is no variance within groups, since equal X scores have identical Y scores. In this case, the total observed variance of Y scores can be attributed to differences in X scores. When r is 0.00, there is no relationship between X and Y; none of the variability in Y scores can be attributed to differences in X. The idea of dividing the total variance of Y into that part attributable to X (variance between groups) and that not attributable to X (variance within groups) leads to a useful concept called the coefficient of determination. The coefficient of determination can be thought of as the proportion of the total variance of Y that can be attributed to differences in X when a linear model is used. It is calculated by squaring the correlation coefficient.

In the remainder of the lesson, you learned to make inferences regarding the correlation coefficient r. The symbol R was used to represent the parameter value for r. You learned how to test that $R = 0$, using the normal distribution when n was large and the t distribution when n was small. You learned to use Fisher's transformation to test that R equals a specified nonzero value, to test that two R's are equal, and to set up confidence intervals for R.

In the review exercises that follow, you are given a chance to test your understanding of these concepts before proceeding to the next lesson, involving other measures of relationships that can be used when r is not appropriate.

LESSON 1 EXERCISES

1. Consider the following three groups of three scores each.

Group number	Score	Group number	Score	Group number	Score
1	0	2	1	3	2
1	1	2	2	3	3
1	2	2	3	3	4

a. Find SS_{total}, $SS_{bgroups}$, and $SS_{wgroups}$.
b. What is the ratio of $SS_{bgroups}$ to SS_{total} and what does this mean?

Now consider each score to be a score for variable Y and the number identifying each group to be a score for variable X, resulting in 9 pairs of scores for X and Y.

c. Find r_{XY}.
d. Find the coefficient of determination. What is the meaning of this value?

Now in place of each Y score, substitute the group mean, resulting again in 9 pairs of scores. Notice that there is now a perfect linear correlation of group means, with the numbers identifying the groups.

e. Compare your answer to b with your answer to d. When the means have a perfect linear relationship from one group to the next, what is the relationship between the coefficient of determination and the ratio of $SS_{bgroups}$ to SS_{total}?

2. Test to see whether the correlation computed for answer c in question 1 is significantly greater than zero at the 0.01 level.

3. If an r of 0.707 had been found for a sample of size 37, would you reject the null hypothesis at the 0.01 level?

4. Two independent samples of size 21 are chosen. For the first sample, r_{ZX} is found to be 0.47. For the second sample, r_{ZY} is found to be 0.77. Is there evidence that $r_{ZX} \neq r_{ZY}$?

5. For a sample of size 28, the correlation between X and Y is found to be 0.68. Find a 95% confidence interval for R_{XY}.

Lesson 2 □ **Measures of relationship for discrete variables**

In the last lesson, you were provided information about the correlation coefficient r, which represents the linear relationship between two continuous variables. There are many situations in which a measure of relationship would be useful but the variables involved are not continuous. In other situations, the underlying variables are continuous but the measurements are discrete. In this lesson, you will learn about a number of indexes used to indicate the extent of the relationship for these situations. The first group includes indexes to be used when one or both variables are dichotomies (only two classes). The second group includes indexes to be used when the data are a set of ranks. Then, at the end of the lesson, you will learn about an index of correlation that can be used when the relationship is thought to be nonlinear.

The material in this chapter will not be used nearly as much as the material in other lessons in the book, and it need not be studied as carefully. Since other lessons will not depend on concepts described in this lesson, you can merely read the lesson to learn what it contains. At a later time, the material will provide a resource to be called upon as needed.

INDEXES OF RELATIONSHIP FOR DICHOTOMIES

Quite frequently, one or both variables to be studied result in data involving only two classes. The two classes may be inherent in the variable (a real dichotomy), or they may arise as the result of the measurement procedure (an artificial dichotomy). Sex is an example of a real dichotomy in which only the two classes, male and female, are relevant. An example of an artificial dichotomy would occur in a study in which age is recorded in only two classes; for example, under 21 or 21 and over.

If variables are classified as either continuous, artificial dichotomies, or real dichotomies, several combinations are possible for any two variables X and Y. The various combinations can be represented in a 3×3 table as follows:

		Y		
		Continuous	*Artificial dichotomy*	*Real dichotomy*
	Continuous	r	r_{bis}	r_{pb}
X	*Artificial dichotomy*		r_t	
	Real dichotomy			ϕ

Within the cells of the table are symbols that are used to represent the measure of relationship appropriate for the corresponding conditions. The three cells in the lower left of the diagram represent duplications of conditions that appear in the upper right portion. The table therefore indicates that an index has been devised for each of the possible combinations of conditions except one, an artificial dichotomy paired with a real dichotomy. The first cell contains an r, representing the correlation coefficient with which you are already familiar. The other measures of relationship will now be described. When appropriate, the method of making inferential tests involving the index will also be presented.

Biserial correlation. The symbol r_{bis} stands for the biserial correlation coefficient, which is appropriate when one of the variables is continuous and the other is an artificial dichotomy. This index, which was developed by Pearson, appears rather frequently in the literature in spite of serious drawbacks. Perhaps most serious is the difficulty of interpreting its meaning, since r_{bis} is not limited by the values $+1.00$ and -1.00. A second drawback is that the sampling distribution of r_{bis} is not known, making direct inferential tests impossible. A third problem is that the calculations for r_{bis} are tedious and the logic underlying the calculations is difficult to explain. In my judgment, these drawbacks are so serious as to make r_{bis} of limited usefulness. Accordingly, its calculation and a worked-out example will not be presented.

Point biserial correlation. The symbol r_{pb} is used to represent the point biserial correlation coefficient. Unlike r_{bis}, it is not necessary to assume that the variable that is recorded as a dichotomy is in reality a continuous, normally distributed variable. Instead, if X is the dichotomy and Y the continuous variable, it is only assumed that the continuous variable Y is normally distributed, with equal variances within each of the two classes for X.

The point biserial also does not suffer from the drawbacks mentioned for r_{bis}. The range of r_{pb} is $+1.00$ to -1.00, tests of inference using the t distribution are possible, and the familiar formula used to compute r can be used to compute r_{pb}. One class of the dichotomous variables is recorded as 0 and the other as 1.

As an example, the question might be asked whether there is a relationship between the sex of an individual and performance on a standardized test. Suppose that X is used to represent sex and Y is used to represent performance on the test. A man might be recorded as a 0 on variable X and a woman as a 1. The usual formula for the correlation r can then be used to calculate the index r_{pb}. Most statistics books provide another formula, which makes calculating easier. The fact that all X scores are either 0 or 1 makes this modification of the ordinary formula for r possible. However, since students now have ready access to calculators, learning this formula is unnecessary.

Suppose that for 100 individuals, half men and half women, the summary data are as follows:

$$n = 100$$

$$\Sigma X = 50 \qquad\qquad \Sigma Y = 10{,}200$$
$$\Sigma X^2 = 50 \qquad\qquad \Sigma Y^2 = 1{,}060{,}000$$
$$\Sigma XY = 5500$$

These values are the ones needed to make use of the computational formula for r (Formula 3.3-5). We carry out exactly the same computations, as though both variables were continuous. We call the result r_{pb}, since one variable is a dichotomy recorded as 0's and 1's.

$$r_{pb} = \frac{\Sigma XY - \frac{(\Sigma X)(\Sigma Y)}{n}}{\sqrt{\left[\Sigma X^2 - \frac{(\Sigma X)^2}{n}\right]\left[\Sigma Y^2 - \frac{(\Sigma Y)^2}{n}\right]}}$$

$$r_{pb} = \frac{5500 - \frac{(50)(10,200)}{100}}{\sqrt{\left[50 - \frac{(50)^2}{100}\right]\left[1,060,000 - \frac{(10,200)^2}{100}\right]}}$$

$$r_{pb} = 0.57$$

A test of the null hypothesis that the point biserial correlation in the population is 0.00 is exactly the same as the small sample test that the product moment correlation R is 0.00. Substitute r_{pb} for r in Formula 12.1-3. The test statistic t is calculated as follows:

$$t = \frac{r_{pb}\sqrt{n - 2}}{\sqrt{1 - r_{pb}^2}}$$

This value is evaluated with $n - 2$ *df*. For our example, we would calculate as follows:

$$t = \frac{0.57\sqrt{98}}{\sqrt{1 - 0.57^2}} = \frac{5.64}{0.82} = 6.88$$

Evaluated with 98 *df,* the t value of 6.88 would represent very strong evidence of a relationship between sex and performance on the test.

As an aside, did you notice something peculiar about the summary data for this example? Both ΣX and ΣX^2 equal 50. Can you figure out why? The values of X were 0's for men and 1's for women. With 50 men and 50 women, ΣX would have to equal 50. Since the square of 0 is also 0, and the square of 1 is also 1, the same would be true of ΣX^2. This observation is the basis for the alternate formula that makes the work of calculating r_{pb} a bit easier when a sophisticated calculator is unavailable.

Tetrachoric correlation. The symbol r_t is used to represent the tetrachoric correlation coefficient, calculated when both variables are artificial dichotomies. The dichotomies are formed by dividing the distribution for each variable in half at the median. Scores below the median are recorded as zero and scores above the median are one. If an association exists between the two variables, there will be a tendency for scores to be either above the median on both measures or below the median on both measures.

Before the widespread use of inexpensive calculators, r_t was often used as a quick estimate of r. Tables available in statistic books allowed the calculation of r_t by finding the number of cases for which both the X and the Y score were above the median.

The use of tetrachoric correlation coefficients has now been largely replaced by either *r* computed from the actual *X* and *Y* scores or, in the case of two real dichotomies, the phi (usually pronounced "fee" by statisticians) correlation coefficient.

Phi correlations coefficients. When both variables are real dichotomies, the appropriate correlation coefficient is the phi coefficient. The symbol used is ϕ. When calculating ϕ, it is convenient to summarize the data by recording frequencies in a 2×2 table with cells labeled as follows:

		Y	
		First class	*Second class*
X	*First class*	(a)	(b)
	Second class	(c)	(d)

The coefficient ϕ is then found, based on the frequencies of scores in each cell. Formula 12.2-1 is used.

$$\phi = \frac{ad - bc}{\sqrt{(a + b)(a + c)(c + d)(b + d)}} \qquad \textbf{12.2-1}$$

As an example, suppose that the problem is to determine whether left-handedness is associated with order of birth in a group of 100 individuals. For variable *X*, firstborns are classified as 1's, all others as 0's. For variable *Y*, left-handed individuals are classified as 1's and right-handed individuals as 0's. Suppose that the frequencies are as follows:

		Y		
		1	*0*	
X	*1*	8 (a)	52 (b)	60
	0	2 (c)	38 (d)	40
		10	90	

We can then use the formula just given to calculate ϕ as follows:

$$\phi = \frac{(8)(38) - (2)(52)}{\sqrt{(60)(10)(40)(90)}} = 0.136$$

An alternative method can be used to get the same result. Just as with the point biserial coefficient, the ordinary formula for calculating *r* can be used. This time the scores for both variables are 0's and 1's. The reason I have chosen to provide the new formula for ϕ is twofold. First, it is easy to use, and second, it provides a clue as to the method that is used to make significance tests involving ϕ.

The formula for ϕ should remind you of a formula presented earlier. Remember the alternate formula for computing X_p^2 for the 2 × 2 table (Formula 9.3-1)?

$$X_p^2 = \frac{[|ad - bc| - n/2]^2 n}{(a + b)(a + c)(c + d)(b + d)}$$

Without the correction for continuity, the formula would be:

$$X_p^2 = \frac{(ad - bc)^2 n}{(a + b)(a + c)(c + d)(b + d)}$$

If you were to divide both sides of the equation by n, you would have:

$$X_p^2/n = \frac{(ad - bc)^2}{(a + b)(a + c)(c + d)(b + d)}$$

But the right side of the equation is the square of the formula for ϕ. From this we find that:

$$X_p^2/n = \phi^2 \quad \text{or} \quad X_p^2 = n\phi^2$$

What implication does this last statement have regarding inferential tests involving ϕ? The answer is that when the data are a random sample from a larger population, Pearson's chi-square statistic provides a means of testing to see whether the sample ϕ coefficient is sufficiently large to indicate that a relationship exists in the population. If we did not have to worry about correcting for continuity, we could square ϕ and multiply by n to obtain X_p^2. This value of X_p^2 could then be evaluated by reference to a table of χ^2 with 1 *df*.

Because we must correct for continuity in calculating X_p^2 for a 2 × 2 table, we calculate X_p^2 using either of the two formulas learned earlier. Let us use the general formula. For the data for which we calculated ϕ to be 0.136, we get the following:

8		52	
	(6)		(54)
2		38	
	(4)		(36)

$$X_p^2 = \sum \frac{(|f_o - f_e| - 0.5)^2}{f_e}$$

$$X_p^2 = \frac{(1.5)^2}{6} + \frac{(1.5)^2}{54} + \frac{(1.5)^2}{4} + \frac{(1.5)^2}{36}$$

$$X_p^2 = 1.042$$

We could evaluate the result by referring to a table of χ^2 with 1 *df*. However, remembering that with 1 *df*, $\sqrt{\chi^2} = z$, we take the square root of 1.04 and get a value 1.02. Since z values as extreme as 1.02 in either direction have a probability greater than 0.30, we will probably conclude that these data do not provide evidence against the null hypothesis that hand preference and order of birth are *unrelated* in the population from which this sample was chosen.

INDEXES OF ASSOCIATION FOR RANKED DATA

In the previous section, you learned about a number of indexes of association that can be used when one or both of the variables appear as dichotomies. For each of the indexes, other than the biserial correlation coefficient, the coefficient turned out to be a special case of the correlation coefficient r. In fact, the new coefficients can be calculated using the ordinary formula for r by coding one class of the dichotomous variables as 0 and the other as 1. In this section, you will learn of measures of association that are appropriate for *ranked* data. The first one to be discussed is also a special case of the correlation coefficient r.

Spearman's rank correlation coefficient. Suppose that a study is to be made of the relationship between infant mortality and the number of physicians per 1000 in the population. The study is to include 50 countries. The data available for the study include rankings for the 50 countries on each of the two variables. The variables underlying the rankings may or may not be continuous, normally distributed variables. The measure of association commonly used for this situation is called Spearman's rank correlation coefficient. The symbol often used is r_S.

As with the point biserial correlation coefficient, r_S can be calculated using the ordinary formula used to compute r. In this case, the rankings are treated as X and Y scores. It can also be computed by a formula that makes the work easier when a calculator is unavailable. Again, there is now little reason for learning the alternate formula.

When we calculate r_S for a set of data, a question often arises over what to do about ties. The solution is to assign an average rank. Suppose, for example, that two countries have exactly the same infant mortality and that there are six countries with lower rates. The two countries would share ranks 7 and 8, and each would be assigned a rank of 7.5. Once we have a rank for each variable, we compute r in the usual manner. Because the scores are actually ranks, we call the result r_S.

Inference tests regarding Spearman's rank correlation coefficient are similar to those for R. Let us use the symbol R_S to designate Spearman's rank correlation coefficient for a population. The limits for R_S are $+1.00$ and -1.00, and R_S will equal 0.00 if there is no relationship between the variables underlying the ranks. In the case where R_S equals 0.00, the sampling distribution of r_S will approach the normal curve as n increases. However, quite large samples are required for the normal curve to be a satisfactory approximation. Therefore, the approach used for small sample tests of $R = 0$ is used. The formula is the same, merely substituting r_S for r.

$$t = \frac{r_s \sqrt{n-2}}{\sqrt{1 - r_s^2}} \text{ with } n - 2 \ df$$

Because r_s is based on ranks that are not continuous or normally distributed, some caution must be used when we interpret the meaning of r_S. The value of r_S cannot be used as an indication of the linear relationship between the underlying variables. For example, suppose that for a set of scores every Y score is exactly equal to X^2. If X and Y scores are ranked, the ranks will be identical, and r_S will be $+1.00$. However, the relationship between the actual X and Y scores cannot be represented by a straight line, and r will not be $+1.00$ for the actual X and Y scores.

Kendall's tau coefficient. Another index sometimes used to measure the relationship between two sets of ranks is Kendall's tau coefficient. This coefficient has some advantage in terms of the size of n necessary for its sampling distribution to be normal. Unfortunately, for even moderate-size samples, the nature of the computations required make the use of a calculator of little value. Since Spearman's rank correlation coefficient provides a satisfactory measure of the relationship between two ranked variables, the reader need not learn the computations and interpretations of Kendall's tau coefficient.

Kendall's coefficient of concordance. Suppose that instead of having two rankings for each thing being ranked, there are a number of rankings. We might ask whether, in general, there is agreement (concordance) among the rankings. To answer the question, it would be possible to find r_S for each possible pairing of the several rankings. Suppose that five supervisors are asked to rank the work performance of ten employees. The result would be a set of five rankings, which could be compared two at a time. In all, there would be $_5C_2$, or 10, different r_S coefficients to compute. These could then be averaged to find an average r_S.

Kendall has developed an index that is directly related to the average r_S and that requires much less work to calculate. The index is called Kendall's coefficient of concordance and is given the symbol w. The idea behind this index is to find the sum of the ranks for each thing being ranked and then to examine the variability of this sum. If the rankings are in *perfect agreement,* the variability among these sums will be a *maximum.* For the case where the work of ten employees is being ranked by five supervisors, perfect agreement would result in one employee's being ranked number one by all five supervisors. The sum of ranks for this person would be five. Similarly, one person would be ranked number ten by all judges, for a sum of ranks equal to 50. The sum of ranks would range from 5 to 50. Notice that this large a range in the sum of ranks can only be achieved when agreement is perfect. Any disagreement among the ranks will reduce the range of the sum of ranks. Based on this concept, the coefficient of concordance is the ratio of the observed variance of sum of ranks to the maximum possible variance of sum of ranks.

$$w = \frac{\text{observed variance of sum of ranks}}{\text{maximum variance of sum of ranks}} \qquad \text{12.2-2}$$

Of course, the limits for w cannot exceed 1.00 and cannot be negative.

If we let T represent the sum of ranks for each thing being ranked, the variance of the sum of ranks is found by the usual formula.

$$Var_T = \frac{\Sigma T^2 - \dfrac{(\Sigma T)^2}{n}}{n}$$

It can be shown that the maximum variance of T is $m^2(n^2 - 1)/12$, where m is the number of sets of rankings and n is the number of things being ranked. The formula for w then becomes:

$$w = \frac{\dfrac{\Sigma T^2 - \dfrac{(\Sigma T)^2}{n}}{n}}{\dfrac{m^2(n^2 - 1)}{12}}$$

This simplifies to the computational formula for w:

$$w = \frac{12\left[\Sigma T^2 - \dfrac{(\Sigma T)^2}{n}\right]}{nm^2(n^2 - 1)} \qquad \text{12.2-3}$$

where T = sum of ranks for each thing being ranked,
m = number of rankings, and
n = number of things being ranked.

For the example of five supervisors' ranking of ten employees, the results and calculations might be as shown in Table 12-2. From the data in this table we get a w of 0.90, which indicates a great amount of agreement. To interpret its meaning, we need to relate it to the average r_S that would be computed for the same data. There is a direct relationship.

$$\text{Average } r_S = \frac{mw - 1}{m - 1} \qquad \text{12.2-4}$$

Table 12-2. Rankings of ten employees by five supervisors

Employee	Supervisor					T	T²
	a	b	c	d	e		
1	2	3	1	2	1	9	81
2	1	2	3	1	3	10	100
3	4	1	4	4	2	15	225
4	3	5	2	3	4	17	289
5	5	6	6	5	6	28	784
6	7	4	5	6	5	27	729
7	6	7	8	7	8	36	1296
8	9	10	7	8	7	41	1681
9	10	8	9	10	9	46	2116
10	8	9	10	9	10	46	2116
						275	9417

$$w = \frac{12\left[9417 - \dfrac{(275)^2}{10}\right]}{10(5)^2(99)} = 0.90$$

For our problem:

$$\text{Average } r_S = \frac{5(0.90) - 1}{5 - 1} = \frac{3.5}{4.0} = 0.875$$

Our conclusion is that the average Spearman rank correlation coefficient computed for the ten possible pairings of ranks would be 0.875.

A NONLINEAR MEASURE OF RELATIONSHIP

To complete this lesson on measures of relationship, we describe a measure of relationship that is related both to the coefficient of concordance and to the earlier discussion of the coefficient of determination. It is the *correlation ratio*, which is often called eta (pronounced āta). The coefficient w was described as a ratio of the observed variance of the sum of ranks compared with the maximum possible variance of the sum of ranks. The coefficient of determination was described as the ratio of the variance of Y that can be attributed to a linear relationship with X, compared with the total variance of Y scores. The correlation ratio eta is similar to both of these concepts in that it is also based on a ratio of variances.

The square of eta has a meaning very similar to that of the coefficient of determination (the square of r). Both are ratios in which the denominator is the total variance of Y and the numerator is the variance of Y that can be attributed to the relationship between Y and X. The difference is that for eta^2, the relationship between Y and X is not restricted to a linear relationship. The variance for the numerator is therefore a true variance between groups, based on groupings of scores of variable X. We can therefore use the methods we have already learned for ANOVA to compute eta^2.

$$\text{Eta}^2 = \frac{SS_{bgroups}}{SS_{total}}$$

and

$$\text{Eta} = \sqrt{\frac{SS_{bgroups}}{SS_{total}}} \qquad \textbf{12.2-5}$$

As an example of the use of eta, consider an experiment involving the learning of a psychomotor task. The dependent variable is the number of trials required before the task is learned. The independent variable is the age of the learner, with ten subjects at each age level from 2 years old to 7 years old. You expect that there will be a relationship between age and the number of trials required to learn the task. However, the relationship may be nonlinear. The scores of the children are reported in Table 12-3.

To complete an ANOVA for the data of Table 12-3, you first find SS_{total}, then $SS_{bgroups}$. To find SS_{total}, we treat all 60 scores as one large group and find the sum of squared deviations. To use the computational formula, we need ΣX (call this T) and ΣX^2. For our data, the values are 1272 and 30,648. We note that the correction factor equals $(1272)^2/60$, or 26,966.4.

Table 12-3. Number of trials until success on a psychomotor task, by age

i	Age (in years)					
	2	**3**	**4**	**5**	**6**	**7**
1	36	27	32	14	12	17
2	32	15	18	15	15	9
3	30	25	17	19	16	23
4	37	25	26	33	15	7
5	35	24	20	18	26	20
6	37	25	18	19	16	11
7	34	22	20	20	17	20
8	36	25	19	15	17	17
9	23	30	12	14	9	12
10	33	28	22	19	11	13
TOTAL	333	246	204	186	154	149

$$SS_{total} = \Sigma X^2 - \frac{T}{n} = 30{,}648 - \frac{(1272)^2}{60} = 30{,}648 - 26{,}966.4 = 3681.6$$

$$SS_{bgroups} = \frac{333^2 + 246^2 + 204^2 + 186^2 + 154^2 + 149^2}{10} - C = 29{,}353.4 - 26{,}966.4 = 2387$$

$$SS_{wgroups} = SS_{total} - SS_{bgroups} = 3681.6 - 2387 = 1294.6$$

To find eta, divide $SS_{bgroups}$ by SS_{total} and then take the square root.

$$\text{Eta} = \sqrt{\frac{2387}{3681.6}} = \sqrt{0.648} = 0.805$$

A test for the significance of eta is identical to the test for a significant difference between group means in the one-way ANOVA. We first obtain mean squares for between and within by dividing the appropriate sums of squares by *df*.

$$ms_{bgroups} = 2387/5 = 477.4$$
$$ms_{wgroups} = 1294.6/54 = 23.97$$

The *F* ratio is then $ms_{bgroups}/ms_{wgroups}$.

$$F = 477.4/23.97 = 19.9 \text{ with 5 and 54 } df$$

The result is highly significant.

The high value for eta indicates a fairly strong relationship between the ages of the children and the number of trials required to learn the task. The relationship is not necessarily a linear one, however. If you plot the scores on a scatter diagram, you will see that the swarm of dots shows some indication that the relationship is curvilinear.

We might compare eta for these data with *r* computed for the same data. Since *r* is a measure of linear relationship between age and number of trials, we might expect it to be less than eta. We choose to make age the *X* variable, with six values ranging from two to seven. We calculate *r*, using Formula 3.3-5.

$$r = \frac{\Sigma XY - \dfrac{(\Sigma X)(\Sigma Y)}{n}}{\sqrt{\left[\Sigma X^2 - \dfrac{(\Sigma X)^2}{n}\right]\left[\Sigma Y^2 - \dfrac{(\Sigma Y)^2}{n}\right]}}$$

$$r = \frac{5117 - \dfrac{(270)(1272)}{60}}{\sqrt{\left[1390 - \dfrac{(270)^2}{60}\right]\left[30,648 - \dfrac{(1272)^2}{60}\right]}}$$

$$r = \frac{-607}{\sqrt{(175)(3681.6)}} = \frac{-607}{802} = -0.757$$

Notice that although eta must always be positive, r calculated for the same data is negative in this case. The negative value indicates that there is an inverse relationship; the older the child, the fewer the trials.

If we square both eta and r, we obtain two indications of the relative proportion of the variance of Y accounted for by the relationship of Y with X. Eta squared is 0.648, indicating that if we do not restrict the relationship to one that is linear, about 65% of the variance of Y can be explained by differences in X. If we square r, we get a coefficient of determination of 0.573, indicating that if we consider only a linear relationship, about 57% of the variance of Y can be explained by differences in X. The closeness of these two values is an indication that a linear model will work almost as well as a more complex curvilinear one. If we wish to do so, we can test the null hypothesis that a curvilinear model is no better than a linear model by comparing eta^2 with r^2. We again use an F ratio.

$$F = \frac{(\text{eta}^2 - r^2)/(k - 2)}{(1 - \text{eta}^2)/(n - k)} \text{ with } k - 2 \text{ and } n - k \, df \qquad \textbf{12.2-6}$$

where k is the number of groups and
       n is the total number of scores.

For our data, we obtain the following:

$$F = \frac{(0.648 - 0.573)/4}{(1 - 0.648)/54} = 2.876 \text{ with 4 and 54 } df$$

Referring to the table of F, we find that the value we obtained would be a rather rare event when the relationship is actually linear within the population. We conclude that there is sufficient evidence to reject the null hypothesis of a linear relationship. In other words, there is evidence that a curvilinear model represents the relationship between X and Y better than does a linear one. In terms of the calculations to be made to find the formula for predicting Y from X, this is an unwelcome finding. In the next lesson, you will learn how to find the optimal formula, based on the assumption that a linear model adequately represents the relationship. These methods of regression analysis become much more complex when a curvilinear model is assumed.

SUMMARY

This lesson contained a description of some commonly used measures of relationship other than r. It included a set of indexes used when the data represent dichotomies, another set for ranked data, and, finally, a measure that can be used for describing nonlinear relationships. For dichotomies, the point biserial correlation coefficient and the ϕ coefficient were found to be useful indexes, the first for the case where one of the variables is a dichotomy, and the second when both are. For both indexes, if the classes of the dichotomous variable are scored as 0 and 1, the usual formula for r can be used to compute the coefficient. Information was provided about how to make inferences regarding each coefficient.

For ranked data involving two variables, Spearman's rank correlation coefficient was described. It also can be calculated using the ordinary formula for r, with the rankings representing the scores for X and Y. The test that the Spearman rank correlation coefficient is zero in a population is made using the small sample procedure for testing that R is 0.00. Kendall's coefficient of concordance was described as a method of estimating the average Spearman rank correlation coefficient in a situation involving each of several judges' ranking the same set of things.

Finally, the eta coefficient was presented as a means of indicating the extent of the relationship between X and Y for the case where the relationship may be nonlinear. The comparison between the coefficient of determination and the square of eta was explored. A significance test was then explained, which provides a test of nonlinearity. For the example given, the result was significant, providing evidence in favor of a curvilinear model. It was explained that this will be an unwelcome finding in most studies, since the steps involved in linear regression analysis are much simpler than those of curvilinear regression analysis. In the two remaining lessons of this chapter, the process of regression analysis will be described only for the linear model.

LESSON 2 EXERCISES

The first five exercises all make use of the following set of scores.

i	X	Y	Z
1	67	78	56
2	60	74	57
3	112	99	84
4	72	54	59
5	64	68	74
6	79	90	77
7	87	86	90
8	76	84	83
9	76	71	62
10	56	84	57

1. Find the correlation r between X and Y. Test for significance at an alpha level of 0.05.
2. Change variable X to a dichotomy. If $X \geq 75$, assign a 1; otherwise, assign a 0. Compute r_{pb}. Test for significance at an alpha level of 0.05.

3. Change variable Y so it is also a dichotomy. If $Y \geq 80$, assign a 1; otherwise, assign a 0. Compute ϕ. Test for significance at an alpha level of 0.05.

4. Rank the X and Y scores. The highest score in each group is ranked as 1. In case of ties, assign an average rank. Compute r_S. Test for significance at an alpha level of 0.05.

5. Rank the scores for Z. Compute w for the three sets of ranks. Calculate the estimated average r_S.

6. Chapter 11, Lesson 1 begins with an example of four samples of medical students, one sample from each class (see Table 11-1). A measure of altruism for the students in each sample is provided. Calculate eta for these data to measure the extent of the relationship between altruism and class in school. Is there evidence that the relationship is curvilinear?

Lesson 3 □ One-predictor variable linear regression analysis

In the first two lessons of this chapter, you learned to measure and to make inferences about the relationship between any two variables X and Y. This information can be used to answer questions like the following. For a *sample* of individuals, what is the magnitude of the relationship between smoking history and blood pressure? For the *population* from which the sample was chosen, is there a relationship? For a *sample* of persons, is age or amount of regular exercise more highly correlated with pulse rate at rest? For the *population* from which the sample was chosen, is there a difference in the correlations of these two "predictors" with the criterion pulse rate at rest?

Answering questions such as these is usually only the first step in an investigation. Once it is learned that an association exists, the investigator then seeks a way to make use of this knowledge. This takes the form of attempting to predict or to understand one variable (the criterion) by studying its relationships with other variables (the predictors). The process used to find a formula for predicting a criterion is called *regression analysis*. If the prediction is based on a linear relationship, the process is called *linear* regression analysis. If more than one predictor is used, the process becomes *multiple* regression analysis.

STEPS IN A REGRESSION STUDY

Before describing the calculations carried out as part of a regression analysis, I will describe the steps involved in a typical study in which a regression analysis is used. The first step is to identify a criterion and group of "persons" for whom the criterion is to be predicted. An example might be an attempt to predict the performance of potential students in a graduate nursing program. The word *persons* was placed in quotation marks above because there are many regression studies in which the criterion is to be predicted not for persons but for things. Economists attempt to predict the unemployment rate for each *month;* agronomists attempt to predict crop yields for each *county;* marine biologists attempt to predict the survival rate for hatchery-raised fish for each *lake*.

Once a criterion and a population of interest have been chosen, the next step is to identify variables that might be used to predict the criterion. At one time it was necessary to limit the number of predictor variables, because each additional predictor greatly increased the complexity of the calculations required for a regression analysis. Today, most regression analyses are done on computers, using programs designed to handle virtually any desired number of predictors. Essentially, the researcher is now able to include as many predictor variables as desired, so long as each can be measured and recorded. We will return later to a discussion of how to decide which predictor variables should be included in a study. First, let us complete the description of the steps in the study.

Having identified the criterion and a set of predictor variables for some population of interest, the next step is the selection of a sample from the population. The sample must be one for which measurements of both criterion and predictor variables are available. Suppose that the study is one designed to predict performance in a graduate nursing program. The criterion is to be first-year grade point average, and the predictors are to be undergraduate grade point average and the verbal and quantitative scores of the Graduate Record Examination. The sample might be a recent class for whom all of these data can be obtained, or, if such a class is not available, an incoming class for which data will be gathered.

In either case, whether from past data or as the result of data gathered as part of the study, once predictor scores and criterion scores are available for each member of the sample, the regression analysis can be carried out. We will delay a description of the process for the last section of this lesson. For now, we will merely say that the result of the regression analysis will be a single formula that tells how each predictor variable is to be weighted to come up with an estimate of the criterion. When two predictors, X_1 and X_2 are used to predict the criterion Y, the formula might appear as follows:

$$\hat{Y}_i = 6.2 + 1.5\,(X_{1_i}) + 0.5\,(X_{2_i})$$

The formula says that to predict any individual's score on variable Y, multiply that individual's X_1 score by 1.5 and his X_2 score by 0.5. Then add these two results to the constant 6.2.

Once the formula has been obtained through regression analysis, it can be used to go back and "predict" the criterion score for each member of the sample. You already know the actual criterion score, so in each case you have an observed and a predicted criterion score. The nearness of the predicted scores to the actual scores will be an indication of how well the regression formula works. You will learn later to use a concept called the *standard error of estimate* as an index of the nearness of predicted scores to actual scores.

The process of going back and predicting criterion scores for the members of the sample is a post hoc process. The real goal of the study is to be able to use the formula that has been obtained to make predictions of the criterion in advance for new individuals. As an example, having completed a regression study for this year's first-year students, we wish now to apply what we have learned to make predictions about the performance of applicants for next year's class. To the extent that the applicants to next year's class are like those in this year's class, the *real* predictions for these individuals can be expected to be approximately as good as the *post hoc* predictions for the sample used in the regression study. If there are significant differences in what is being measured, either for the criterion or for any of the predictor variables, the usefulness of the formula obtained from the regression analysis will be reduced. For this reason, the regression formula, once obtained, cannot be used blindly forever. Instead, there needs to be a periodic check to see how well predictions for the criterion match with actual observations.

CHOICE OF A PREDICTION BATTERY

If there is virtually no limit to the number of predictor variables that can be included in a study, what makes up a satisfactory battery of predictors? In the ideal situation, it would be possible to identify all of the factors contributing to performance on the criterion and then to obtain a ''pure'' measure of each of these factors for use as predictors. For example, the performance of individual students in a graduate nursing program will be affected by prior education, motivation, ability to solve problems, ability to manage one's time, and many other factors. If a pure measure of each of these factors could be obtained, it would be possible to make very accurate predictions about how each prospective student would perform.

Unfortunately, pure measures of all of the factors affecting the criterion are rarely available. Instead, each of the predictors is also a complex variable, which shares components not only with the criterion but also with other predictors. In addition to components shared in common, each measure (both criterion and predictors) also has unique components not shared by any other variable included in the study.

When one plans a regression study, the general rule is to try to include predictor variables that share components (are correlated) with the criterion, but not with other predictor variables. Each is then contributing something unique to the prediction of the criterion. It does no harm to include two predictor variables that are highly correlated with each other; it is just that the two scores will represent duplicate information. The results of the regression analysis will not be improved much by the inclusion of the second variable. Similarly, it does no harm to include a variable that is unrelated to the criterion. In the regression analysis, that variable will end up with a weighting of zero. The rule of thumb would therefore be: If you are in doubt about the usefulness of a predictor, include it.

There is one case in which a variable with no relationship to the criterion is useful as a part of the predictor battery. As an example, suppose that a criterion to be predicted is made up of only two components, A and B. A predictor variable is available that is made up of components A and C. Since the criterion and the predictor share component A in common, they will be correlated, and the predictor variable will indeed be useful in the prediction of the criterion. Predictions will be less than perfect, however, because the criterion has a component B, which is not measured at all by the predictor. Also, the predictor is not a pure measure of A, but it includes a component C, which is of no use in predicting the criterion. The presence of component C in the predictor will function as an ''error'' component, or ''noise,'' and will distract from the prediction of the criterion. Now suppose that another predictor were available that had C as its only component. This new variable would have no correlation with the criterion. Its inclusion in the test battery would serve a useful purpose, however. If this second predictor variable were to be subtracted from the first, the result would be a pure measure of A. Component C, which represents noise, or error, would be subtracted out or suppressed. This second predictor variable, which is unrelated to the criterion but is nevertheless of value in the battery of predictors, is called a *suppressor* variable. You will see a concrete example later.

It has now been explained that the data for a regression analysis will ordinarily represent a criterion variable and a number of predictor variables, each of which is made up of a number of components. Usually there will be a complex set of interrelationships among the components of the criterion and those of the predictors. Attempting to sort out all of this information in some systematic way sounds hopelessly complex. Actually, the calculations range from very complex, in the case of multiple predictors, to simple, in the case of one predictor variable. Fortunately, the logic of the process and the interpretation of the results are the same for either case. We will therefore start in this lesson with the one-predictor case, showing the calculations, the logic, and the interpretations. In the next lesson, we will use the two-predictor case to present additional concepts relevant to multipredictor problems. Since the calculations are still rather straightforward, they will also be included. The chapter will then conclude with a description of the rationale and output from a typical computer program used for the case of a large battery of predictor variables.

A SAMPLE DATA SET

To illustrate the calculations involved in a regression analysis, we need a data set of limited size that contains a criterion and one or more possible predictor variables. We could choose actual data from some health-related study. This would have the advantage of providing a sense of realism and inherent meaning. A second alternative is to construct an artificial data set. Constructing an artificial data set allows us to make up variables while knowing exactly what factors are present in each variable. Either method of providing data for an example will be equally good for showing calculations. Since the latter method will make the meaning and results of regression analysis more understandable, we will use this alternative.

Table 12-4 contains a set of artificially constructed data that will be used as examples both in this lesson and the one to follow. There are 10 variables, with 25 scores for each variable. It is important to understand how each variable was constructed in order to gain the maximum insight into the process of regression analysis in later examples. The first five variables—A, B, C, D, and E—were generated by a computer. Each has a distribution that is approximately normal, a mean of 500, and a standard deviation very nearly equal to 100. These variables have one other important characteristic. Each has essentially a zero correlation with each of the others. Variables A through E may therefore be thought of as pure measures, or pure factors. The remaining variables represent combinations of variables A through E and will be used either as criterion variables or predictor variables in examples in this lesson and the next. The advantage is that in each case we will know the exact composition of each of the composite variables. For example, variable AB is formed by an equal weighting of variables (factors) A and B.

The means at the bottom of Table 12-4 for the composite variables are logical, because of how each variable was formed. The standard deviations are not as readily understood. Can you figure out why variables AB, AC, and BD should have a standard deviation of approximately 141, why variable ABD should have a standard deviation of approximately 173, and why variable $ABCE$ has a standard deviation of approximately 200? Remember that if a composite is formed by adding two uncorrelated variables, the variance of the composite will be equal to the sum of the variances. For variables *AB*, *AC*, and *BD*, each composite is the sum of two uncorrelated variables, each having a variance of approximately 10,000. The variance of the composite should be approximately 20,000. Its square root, the standard deviation, should be approximately 141. The only difference for variables *ABD* and *ABCE* is the number of components in the composite.

We have now: (1) summarized the steps in a regression study, (2) described the basis for selection of variables to be included in a test battery, and (3) provided a data set that includes five pure variables and five composite variables of known composition. We can now proceed to a discussion of the calculations of regression analysis for the case of a single predictor. We begin our discussion with a review of some concepts probably first encountered in an eighth- or ninth-grade math class.

GRAPHING AND THE X, Y COORDINATE SYSTEM

At one time or another, almost all readers of this book have probably learned to draw a straight line to represent a formula on an X, Y coordinate system. For those who have not done so and for those needing a review, the basic information about such graphic representation of an equation will be summarized.

An *X*, *Y* coordinate system is traditionally drawn with *Y* as a vertical line and *X* as a horizontal line. The point at which the lines intersect is called the origin and represents the

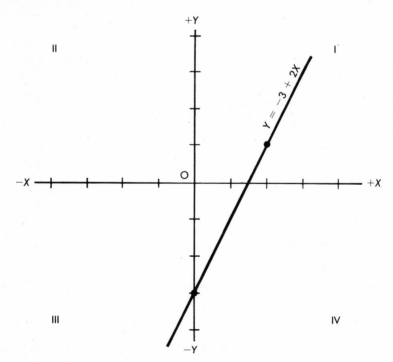

Fig. 12-1. Sample *X*, *Y* coordinate systems.

value zero for both X and Y. Vertical distances for Y are considered positive if above the origin and negative if below the origin. Horizontal distances for X are considered positive if to the right of the origin and negative if to the left of the origin. Fig. 12-1 contains an X, Y coordinate system with a line representing the equation $Y = -3 + 2X$.

It is traditional to call that part of the diagram which is above and to the right of the origin "quadrant I," that part which is above and to the left of the origin "quadrant II," that part which is below and to the left of the origin "quadrant III," and that part which is below and to the right of the origin "quadrant IV." Any point on the diagram can be identified by telling how far it is from the origin horizontally and vertically and labeling the direction of the distance with a + or − sign. These signed distances are called the X and Y *coordinates* of the point. It has become customary to list the X coordinate first and then the Y coordinate, separated by a comma. In Fig. 12-1, two dots have been placed on

Table 12-4. Artificial data to be used in regression examples

	Pure variables					Composite variables				
i	A	B	C	D	E	AB	AC	BD	ABD	ABCE
1	539	360	555	538	624	899	1094	898	1437	2078
2	417	575	605	474	369	992	1022	1049	1466	1966
3	721	473	411	538	548	1194	1132	1011	1732	2153
4	468	472	554	505	701	940	1022	977	1445	2195
5	489	442	452	433	510	931	941	875	1364	1893
6	569	390	390	432	395	959	959	822	1391	1744
7	388	669	483	464	319	1057	871	1133	1521	1859
8	437	472	308	558	396	909	745	1030	1467	1613
9	469	545	565	612	498	1014	1034	1157	1626	2077
10	346	503	369	421	511	849	715	924	1270	1729
11	469	349	667	538	383	818	1136	887	1356	1868
12	427	554	451	305	523	981	878	859	1286	1955
13	398	442	432	443	676	840	830	885	1283	1948
14	518	595	390	569	472	1113	908	1164	1682	1975
15	600	339	370	464	383	939	970	803	1403	1692
16	376	503	605	316	523	879	981	819	1195	2007
17	539	576	432	580	651	1115	971	1156	1695	2198
18	447	595	431	642	536	1042	878	1237	1684	2009
19	560	442	503	612	396	1002	1063	1054	1614	1901
20	457	338	707	563	421	795	1164	991	1448	1923
21	590	473	564	306	523	1063	1154	779	1369	2150
22	710	523	574	484	511	1233	1284	1007	1717	2318
23	408	565	513	664	536	973	921	1229	1637	2022
24	680	740	595	432	471	1420	1275	1172	1852	2486
25	478	565	574	517	624	1043	1052	1082	1560	2241
n	25.0	25.0	25.0	25.0	25.0	25.0	25.0	25.0	25.0	25.0
Mean	500.0	500.0	500.0	500.0	500.0	1000.0	1000.0	1000.0	1500.0	2000.0
SD	100.0	99.8	99.8	100.1	99.9	138.1	141.9	137.8	168.5	198.5

the line for the equation. One is on the Y axis and the other is in quadrant I. The coordinates for the two dots are $(0, -3)$ and $(2, 1)$ respectively.

The slanted line on the diagram represents the formula $Y = -3 + 2X$. If the formula is solved for Y for any value of X, the resulting X and Y values will be coordinates of points falling on the line. For example, if $X = 0$, then $Y = -3$, and if $X = 2$, then $Y = 1$. These are the dots identified on the figure. Suppose $X = 3$ in the formula, what would be the value of $Y?$ We can solve the equation as follows:

$$Y = -3 + 2(3) = -3 + 6 = +3$$

We could also find the value of Y corresponding to any X value directly from the graph. For example, to find the Y value for $X = 3$, we find the point on the horizontal axis three units to the right of the origin. The corresponding Y value is then the vertical distance from this point to the slanted line (in this case 3). The value is negative if the slanted line is below the X axis at this point. Use this process to find what Y equals when $X = 1$. Whether solving the equation or examining the figure, you should get a value of -1.

The general formula for any straight line on the X, Y coordinate system is shown in Formula 12.3-1.

$$Y = a + bX \qquad \qquad \textbf{12.3-1}$$

A different line will result from any different combination of values for a and b. The values of a and b provide immediate information as to how the line will appear when plotted. The value of a always indicates where the line will cross the Y axis and is therefore called the Y *intercept*. The value of b tells how much of a change there is in Y for any change in X. For example, in Fig. 12-1, b equals $+2$. This indicates that for any increase of one unit in X there will be a corresponding two-unit increase in Y. The magnitude of b therefore indicates how sharply the line slants either up or down. If b equals 1, the line will slant upward at a 45° angle. As b gets larger and larger, the line becomes more nearly vertical. As b approaches 0, the line becomes more nearly horizontal. The sign of b indicates whether the line slants upward from left to right (positive sign) or downward from left to right (negative sign). Because of its relationship with the slant of the line, b is often referred to as the *slope of the line*.

In math classes, one of the assignments involving the X, Y coordinate system is to draw the line and discover the formula for the line represented by any two data points. As a test of your understanding of the material just presented, see if you can draw the line and give the formula for the line corresponding to these two data points: $(-4, -1)$ and $(2, 2)$. Try to answer before reading on.

Since these two points determine a line, we need only to mark them on an X, Y coordinate system and then draw a line through them. The resulting line crosses the Y axis at $+1.00$, so we know that $a = +1.00$. It slants upward from left to right, so we know that b is a positive value. For every increase of one unit in X, Y increases by one-half unit.

We therefore know that $b = 0.5$. Substituting $+1.00$ for a and 0.5 for b in the general formula, we see that $Y = 1 + 0.5X$.

THE LEAST SQUARES CRITERION

Now suppose that instead of having only two data points, you were given four data points and again asked to draw the line and find the formula. Try this for the following four data points: $(1,1)$, $(2,3)$, $(3,2)$, and $(4,4)$. We begin by locating these points on a coordinate system (Fig. 12-2).

Now what problem do you encounter when attempting to draw the line? The difficulty is that the four data points do not fall on a straight line. This is more representative of what actually happens when two variables are represented on an X, Y coordinate system. Earlier you learned to call the result a scatter diagram. Even though the dots do not fall on a straight line, because the correlation between X and Y is not perfect, we still wish to find an equation for predicting Y from X or X from Y. That equation will correspond to a line drawn through the scatter of dots. As stated earlier, in this discussion we are limiting our consideration to a straight line. The goal is to find the "best" equation or the "best" line. The formula for the line again involves a and b. If we are trying to predict Y from X, we write the formula as follows:

$$\hat{Y}_i = a + bX_i \qquad \qquad \textbf{12.3-2}$$

where X_i is any individual score and
          $\hat{Y}_i$ is the predicted value of Y for that X score.

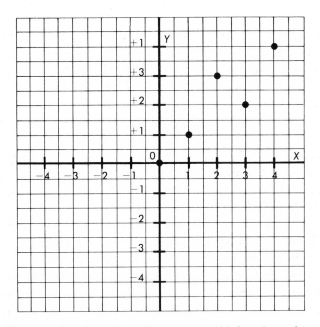

Fig. 12-2. Sample X, Y coordinate system with four data points.

We could also find a similar formula for predicting X from Y. The process of finding the equation of the line that best predicts either Y from X or X from Y is called *regression analysis*.

If you were asked to draw a line through the dots on Fig. 12-2, where would you draw it? If several persons are asked to do this task, a line through points (1,1) and (4,4) would probably be the most common solution. This line has the equation:

$$\hat{Y}_i = 0 + (1)(X_i)$$

The formula provides exact predictions for Y for two of the points. How far from the line are the other two points? Since we are trying to predict Y from X, the vertical distances represent the amount of error. For both of the points not on the line, the error of prediction is one unit. Is this the best result possible? To answer the question, we first have to define the word *best*.

In regression analysis, the criterion used to define the best line is the *least squares criterion*. To explain the meaning of this criterion, we specify that for a particular problem we are interested in predicting Y from X. Then the best line, according to the least squares criterion, is the line that minimizes the sum of the *squares* of the *vertical* distances from each point to the line. Since these vertical distances represent errors of prediction, the use of the least squares criterion also means that the sum of the squares of the errors of prediction will be a minimum. In symbols, when the least squares criterion is used, $\Sigma (Y_i - \hat{Y}_i)^2$ is a minimum. For reasons that will become apparent later, this last expression is often referred to as the "residual sums of squares."

For the line we have discussed, the sum of squared errors of prediction is $0^2 + 1^2 + (-1)^2 + 0^2$, or 2. Would it be possible to draw a line for which $\Sigma (Y_i - \hat{Y}_i)^2$ is less than 2? Try formulas for other lines and see what happens. In each case, let a and b take on a different value for the formula $Y = a + bX$. You would have to be very fortunate to stumble onto the formula $Y = 0.5 + 0.8X$. However, if you did, you would get the following results when you applied the formula to predict each Y score:

X	Y	$\hat{Y}$	$Y - \hat{Y}$	$(Y - \hat{Y})^2$
1	1	1.3	−0.3	0.09
2	3	2.1	0.9	0.81
3	2	2.9	−0.9	0.81
4	4	3.7	0.3	0.09
				1.80

In this case, the sum of squared errors of prediction is 1.8 units, compared with 2 units for the line previously discussed. The formula $Y = 0.5 + 0.8X$ is, in fact, the best formula possible for predicting Y from X, if "best" is defined by the least squares criterion. Plot this line to see where it is located in relation to the dots on Fig. 12-2.

You are not expected to find the optimal values for a and b, either by trial and error or by inspecting the scatter diagram. Instead, these values for the formula are found by using concepts you are familiar with from the earlier discussion of correlation. The conceptual and computational formulas for *b* are as follows:

$$b = \frac{\Sigma xy}{\Sigma x^2}$$

12.3-3

and

$$b = \frac{\Sigma XY - \dfrac{(\Sigma X)(\Sigma Y)}{n}}{\Sigma X^2 - \dfrac{(\Sigma X)^2}{n}}$$

12.3-4

For the data of Fig. 12-2, we have the following:

X	Y	X²	Y²	XY
1	1	1	1	1
2	3	4	9	6
3	2	9	4	6
4	4	16	16	16
10	10	30	30	29

Using Formula 12.3-4 (the computational formula), we calculate *b* as follows:

$$b = \frac{29 - \dfrac{(10)(10)}{4}}{30 - \dfrac{(10)(10)}{4}} = \frac{4}{5} = 0.8$$

To find *a*, the formula is Formula 12.3-5.

$$a = \text{mean}_Y - b(\text{mean}_X)$$

12.3-5

For our data, the means are both 2.5.

$$a = 2.5 - 0.8\,(2.5) = 0.5$$

The formula is therefore:

$$\hat{Y}_i = 0.5 + 0.8\,(X_i)$$

In summary, to find the best linear formula for predicting Y from X, first find b, using the formula $b = \Sigma xy / \Sigma x^2$ or its computational form. Then find a, using the formula $a = \text{mean}_Y - b(\text{mean}_X)$. Finally, substitute these values in the general formula $\hat{Y}_i = a + bX_i$. Once the formula is determined, it is then possible to go back and predict Y for each X score. These predictions will be the best possible in the sense that the sum of the squared errors of prediction will be a minimum. In a graphic presentation, b represents the slope, and a represents the Y intercept of the line for which the sum of the squares of vertical distances to the line is a minimum.

At this point, you should be able to carry out a least squares linear regression analysis for any set of paired scores. Although no proof has been given, it has been stated that the result is the best one possible, so long as "best" is defined by the least squares criterion. It is worth noting that a different criterion would not necessarily result in the same "best" line. For example, suppose that we had decided to minimize the sum of the *absolute* distances from the line rather than the sum of the *squared* distances. The line originally proposed, which goes through two of the points and misses the other two by one unit each, would be superior to that resulting from the regression analysis. The sum of absolute differences is 2 for the first line and 2.4 for the least squares regression line. The choice of the least squares criterion is therefore an important one. In effect, it is the decision that relatively *small* errors of prediction for *many* scores are preferable to *large* errors of prediction for a *few* scores. This is because squaring magnifies the effect of the large errors.

A WORKED EXAMPLE

Let us now look at an example of the use of regression analysis for the case of a single variable. It will be possible to show more clearly the relationship between regression analysis and correlation if we choose a pure measure as the predictor variable. We will, therefore, use pure variable A from Table 12-4 as the predictor for this first example. We will choose as the criterion variable the first composite, variable AB, which is formed by adding variables A and B. We can therefore anticipate in advance that approximately half of the variability of our criterion will be attributable to the variability of the predictor.

Table 12-5 contains information summarizing the regression analysis for our example. As is rather common in the one-predictor case, I have used the symbol X to represent the predictor variable and the symbol Y to represent the criterion. There are three sections for the table. Some of the information represents concepts you have already used many times. Other parts of the information contained in the table represent concepts you have just learned in this lesson. Let us examine the table carefully.

In the first section of the table, there are nine columns. The first six should be familiar. They are the same columns used in calculating a correlation coefficient. Notice that columns 2 and 3 are the columns taken from Table 12-4. The totals for columns 2 through 6 are used in finding the deviations sums, the means, and the standard deviations contained in the second section of the table. Columns 7 through 9 relate to the completion of the regression analysis and will be discussed last.

Notice two things about the information in the second section. First, the three entries at the upper left part of section 2 refer to deviation scores, not raw scores. Second, when standard deviations are calculated, N is used in the equation, not $n - 1$. The use of N rather then $n - 1$, along with the use of the symbol σ rather than s, indicates that the data are being treated as a population of scores. More will be said about this later.

The third section of the table is the one that is of primary interest. First, the correlation between X and Y was calculated and found to be 0.691. This represents a fairly strong correlation, indicating that X will be quite effective as a basis for predicting Y. The square of the correlation coefficient is 0.477, which is the coefficient of determination. Recall

Table 12-5. Example of a single-variable regression analysis

Col 1 i	Col 2 X	Col 3 Y	Col 4 X^2	Col 5 Y^2	Col 6 XY	Col 7 $\hat{Y}$	Col 8 $Y - \hat{Y}$	Col 9 $(Y - \hat{Y})^2$
1	539	899	290,521	808,201	484,561	1037.25	−138.25	19,113.06
2	417	992	173,889	984,064	413,664	920.73	71.27	5079.41
3	721	1194	519,841	1,425,636	860,874	1211.08	− 17.08	291.72
4	468	940	219,024	883,600	439,920	969.44	− 29.44	866.71
5	489	931	239,121	866,761	455,259	989.49	− 58.49	3421.08
6	569	959	323,761	919,681	545,671	1065.90	−106.90	11,427.61
7	388	1057	150,544	1,117,249	410,116	893.03	163.97	26,886.16
8	437	909	190,969	826,281	397,233	939.83	− 30.83	950.48
9	469	1014	219,961	1,028,196	475,566	970.39	43.61	1901.83
10	346	849	119,716	720,801	293,754	852.92	− 3.92	15.36
11	469	818	219,961	669,124	383,642	970.39	−152.39	23,222.71
12	427	981	182,329	962,361	418,887	930.28	50.72	2572.51
13	398	840	158,404	705,600	334,320	902.58	− 62.58	3916.25
14	518	1113	268,324	1,238,769	576,534	1017.19	95.81	9179.55
15	600	939	360,000	881,721	563,400	1095.51	−156.51	24,495.38
16	376	879	141,376	772,641	330,504	881.57	− 2.57	6.60
17	539	1115	290,521	1,243,225	600,985	1037.25	77.75	6045.06
18	447	1042	199,809	1,085,764	465,774	949.38	92.62	8578.46
19	560	1002	313,600	1,004,004	561,120	1057.31	− 55.31	3059.19
20	457	795	208,849	632,025	363,315	958.93	−163.93	26,873.04
21	590	1063	348,100	1,129,969	627,170	1085.96	− 22.96	527.16
22	710	1233	504,100	1,520,289	875,430	1200.57	32.43	1051.70
23	408	973	166,464	946,729	396,984	912.13	60.87	3705.15
24	680	1420	462,400	2,016,400	965,600	1171.92	248.08	61,543.68
25	478	1043	228,484	1,087,849	498,554	978.99	64.01	4097.28
TOTAL	12,500	25,000	6,500,068	25,476,940	12,738,837	25,000.02	− 0.02	248,827.14

$$\Sigma x^2 = 250,068 \qquad \sigma_X^2 = \frac{\Sigma x^2}{N} = \frac{250,068}{25} = 10,002.72$$

$$\Sigma y^2 = 476,940 \qquad \sigma_X = \sqrt{\sigma_X^2} = \sqrt{10,002.72} = 100.01$$

$$\Sigma xy = 238,837 \qquad \sigma_Y^2 = \frac{\Sigma y^2}{N} = \frac{476,940}{25} = 19,077.6$$

$$\mu_X = 500$$

$$\mu_Y = 1000 \qquad \sigma_Y = \sqrt{\sigma_Y^2} = \sqrt{19,077.6} = 138.1$$

$$R = \frac{\Sigma xy}{\sqrt{\Sigma x^2 \Sigma y^2}} = \frac{238,837}{\sqrt{(250,068)(476,940)}} = 0.691$$

$$b = \frac{\Sigma xy}{\Sigma x^2} = \frac{238,837}{250,068} = 0.9551$$

$$a = \mu_Y - b(\mu_X) = 1000 - 0.9551(500) = 522.456$$

$$\hat{Y}_i = 522.456 + 0.9551 \, (X_i)$$

that this coefficient indicates what portion of the total variability of one variable can be explained on the basis of the other. The value 0.477 is consistent with our expectation that approximately half of the variability of Y scores could be attributed to differences on X.

The second entry in the third section of the table is the calculation of b, the slope of the regression line. The sign of b is positive, indicating that the regression line will slant upward from left to right. Since the value (0.9551) is very nearly 1.00, we know that the line will slant upward at approximately a 45° angle to the X axis. The value of 0.9551 also tells us that for every increase of one unit in X we would expect an increase of very nearly one unit in Y.

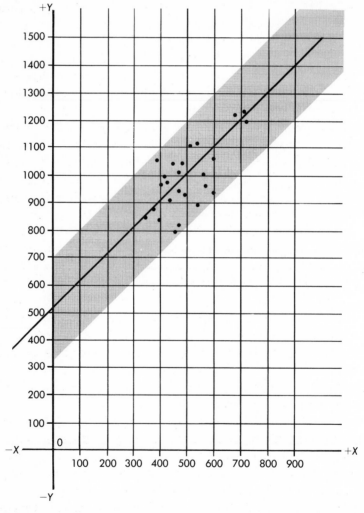

Fig. 12-3. Regression of composite variable AB and pure variable A from Table 12-4.

The third entry in section 3 of the table is the calculation of *a,* the *Y* intercept. The value is found by multiplying the mean of the predictor variable by *b* and subtracting the result from the mean of the criterion variable. The value of 522.456 indicates that the regression line will intercept the *Y* axis at a point approximately 522 units above the origin.

The last entry in section three provides the result of our regression analysis. The formula indicates that to predict the *Y* score for any individual, you first multiply that person's score on *X* by 0.9551 and then add the constant 522.456.

Fig. 12-3 shows the scatter diagram of the *X* and *Y* scores, along with a line representing the equation we have found. Rather than using the formula to predict any score for *X,* we could also use the information in Fig. 12-3 to get an approximate prediction. For example, to predict the *Y* score for a person who has an *X* score of 400, we find 400 on the horizontal scale and then follow a vertical line up from that point to the regression line. We find the height of the regression line at this point to be approximately 900. Our prediction for *Y* for an *X* score of 400 would therefore be just over 900. By comparison, use of the formula from the bottom of Table 12-5 would yield a value of 904.5.

Notice that the scores for *X* in this example extend only from about 350 to 750. The regression line was calculated for scores within these limits for *X.* Visual inspection of the scatter diagram in Fig. 12-3 indicates that the relationship between *X* and *Y* scores is indeed linear for *X* scores between 350 and 750. It would be very risky to conclude that the relationship remains linear beyond these limits. Use of the formula to make predictions when the *X* score is outside these bounds is unwarranted.

Let us now return to columns 7 through 9 of the first section of the table. Column 7 contains the post hoc predictions of *Y* for each individual. The values in the column result from substituting each individual's *X* score in the formula at the bottom of Table 12-5. Notice that, except for rounding, the total of the predictions for *Y* is equal to the total of the actual *Y* scores.

Column 8 contains the differences between the actual *Y* scores and the predicted *Y* scores. Since we have subtracted the predicted score from the actual score, a negative value indicates that the prediction was too high. The total of the negative values equals the total of the positive values, resulting in a sum equal to zero. This is a desirable characteristic of our process of regression analysis, since a bias in favor of either overprediction or underprediction would reduce the value of the results.

On the graph of Fig. 12-3, the values in column 8 are represented by vertical distances from the dots to the regression line. The line represents the prediction made in each case. Therefore, if the value in column 8 is a plus value, it indicates that the dot is above the line and that the *Y* score was underpredicted. A negative value results when the dot is below the line, meaning that the *Y* score was overpredicted.

Column 9 is the one that tells us how well our formula works. It contains the squares of the differences between actual and predicted values. You have been told that the sum of these squares of the errors of prediction will be a minimum. In other words, the total at the bottom of this column is less than it will be if any other linear formula is used to predict Y from X. Just knowing that the total 248,827.14 is as small as possible is not very informative, however. What we would like to know is how to use this number to indicate the precision of our estimates. The concept we will use is called the standard error of estimation.

THE STANDARD ERROR OF ESTIMATION

Every one of us has at one time waited impatiently for a car to be repaired or for a product to be delivered. We ask, ''How much longer will it be?'' only to be told, ''I'm sorry, we're doing the best we can.'' Such answers will provide little comfort when what you want is some indication of how good ''the best'' will be.

So far you are in an analogous position regarding regression analysis. You have learned a procedure that you have been told will provide the best possible estimates of Y from X. The logical next question is, ''How good is the best?'' To answer the question, the statistician provides an index called the standard error of estimation, SE_{est}. The concept is similar to the idea of a standard error of a statistic and is used in a similar manner. Just as $SE_{\bar{x}}$ can be used to set a 90% confidence interval for the mean, so SE_{est} can be used to set a 90% confidence interval for the Y score on the basis of a given X score.

Like the standard error of a statistic, the standard error of estimation can be thought of as a standard deviation of sorts. Visualize a *horizontal* line drawn through Fig. 12-3 *exactly at the height of the mean for Y*. Now consider the vertical distances of the dots to this line. These distances represent what we have learned to call deviation scores. The sum of squares of these distances, when divided by N, provides a measure of variability we have learned to call the variance. Its square root is the standard deviation of Y scores. If we perform exactly the same calculations using *distances to the regression line*, rather than distances to the line representing the mean of Y, we will have SE_{est}. The SE_{est} therefore represents the standard deviation of scores about the regression line, or the standard deviation of Y scores with the effect of variable X removed.

Let us now calculate SE_{est}. The definitional formula is Formula 12.3-6.

$$SE_{est} = \sqrt{\frac{\Sigma (Y_i - \hat{Y}_i)^2}{N}} \qquad \textbf{12.3-6}$$

For our example, the numerator of the fraction is the sum at the bottom of column 9. By substituting in the formula, we find the standard error of estimation to be 99.76.

For a large data set, each X score will occur more than once. Unless the correlation between X and Y is perfect, the corresponding Y scores will not be identical. Now suppose that for each observed X score, the observed Y scores have a mean at the regression line. Suppose also that distribution of Y scores about this point is approximately normal. We can then set up confidence intervals for the estimates of Y. For example, suppose that we want a 95% confidence interval for the predicted value of Y when X is 400. Earlier, using the formula at the bottom of Table 12-5, we found that when X is 400, the best prediction of Y is the value 904.5. The 95% confidence interval for this prediction is as follows:

$$904.5 \pm (1.96)(99.76) \quad \text{or} \quad 904.5 \pm 195.53 \quad \text{or} \quad 708.97 \text{ to } 1100.03$$

We have incorporated the concept of the SE_{est} into the drawing of Fig. 12-3 by placing lines parallel to the line of regression. The lines are drawn $1.96\ SE_{est}$ above and below the regression line. We expect approximately 95% of all the dots to be within the resulting band. This area is shaded in the figure. For this example, the 95% confidence band includes all of the observed scores. It is essential to keep in mind that for all of this discussion we are assuming that we are dealing with the entire population of scores. Later in this lesson you will learn of modifications required when dealing with samples.

Relationship between SE$_{est}$ and R^2. Let us now look at our example to see if the values obtained are consistent with what we know to be true of the scores used as the predictor and the criterion. Since the criterion was made up of equal parts of A and B, we expect that about half of the variability of the Y scores can be attributed to variability in X. Indeed, we found that the correlation coefficient was 0.691, resulting in a coefficient of determination of 0.477. We can express this in an equation as follows:

$$R^2 = 0.477 = \frac{\text{variance of } Y \text{ related to } X}{\text{total variance of } Y}$$

We are now in a position to verify this meaning for the coefficient of determination. You have learned that SE_{est} is a form of *standard deviation* that is a measure of the variability of Y that is independent of X. We computed SE_{est} by taking the square root of the average of the squares of the errors of prediction. If we do not take the square root, but simply take the average of the squares of the errors of prediction, we will have a measure of the *variance* of Y that is independent of X. For our data, we divide the sum at the bottom of column 9 of Table 12-5 by 25 and obtain the value 9953.0856.

The meaning of R^2 is stated in terms of the variance of Y *related* to X, compared with the total variance of Y. We now have a measure of the variance of Y that is *unrelated* to X (9953.0856). We can get the total variance of Y from the second section of Table 12-5. It is 19,077.6. Now we subtract 9953.0856 from 19,077.6 to obtain a measure of the variance of Y that is *related* to X. The result is 9124.5144. According to the meaning we have stated for the coefficient of determination, dividing 9124.5144 by 19,077.6 should result in the same value that we obtained by squaring R. We see that, except for rounding error, the two values are the same (0.477 compared with 0.478).

Computational formula for SE$_{est}$. In addition to confirming the meaning of the coefficient of determination, the steps we have just completed also serve as the basis for a computational formula for SE_{est}. If you track through the steps just taken and some algebraic manipulation, you will see that what has been said can be expressed as Formula 12.3-7.

$$R^2 = \frac{Var_Y - SE_{est}^2}{Var_Y}$$

$$R^2 = 1 - \frac{SE_{est}^2}{Var_Y}$$

$$1 - R^2 = \frac{SE_{est}^2}{Var_Y}$$

$$Var_Y(1 - R^2) = SE_{est}^2$$

or

$$SE_{est} = \sigma_Y\sqrt{1 - R^2} \qquad \textbf{12.3-7}$$

This last formula says that to find SE_{est} it is not necessary to: (1) make post hoc predictions for each Y score, (2) find the differences between actual and predicted scores, (3) square these differences, (4) sum these squares, (5) divide by n, and then (6) take the square root. Instead, we need the standard deviation of Y and the coefficient of determination for the computational formula. For our data, we have the following:

$$SE_{est} = \sigma_Y\sqrt{1 - R^2}$$
$$SE_{est} = 138.1\sqrt{1 - (0.691)^2}$$
$$SE_{est} = 99.83$$

Except for rounding errors, this is the same value we obtained earlier by the definitional formula (99.83 compared with 99.76).

Computational formula for $\Sigma(Y - \hat{Y})^2$. Manipulation of the computational formula for SE_{est} leads to a computational formula for the sum of squared errors of prediction, $\Sigma(Y - \hat{Y})^2$.

$$SE_{est} = \sigma_Y\sqrt{1 - R^2}$$

$$\sqrt{\frac{\Sigma(Y - \hat{Y})^2}{N}} = \sigma_Y\sqrt{1 - R^2}$$

$$\frac{\Sigma(Y - \hat{Y})^2}{N} = \sigma_Y^2(1 - R^2)$$

$$\frac{\Sigma(Y - \hat{Y})^2}{N} = \frac{\Sigma y^2}{N}\left[1 - \frac{(\Sigma xy)^2}{\Sigma x^2 \Sigma y^2}\right]$$

$$\Sigma(Y - \hat{Y})^2 = \Sigma y^2\left[1 - \frac{(\Sigma xy)^2}{\Sigma x^2 \Sigma y^2}\right]$$

or

$$\Sigma(Y - \hat{Y})^2 = \Sigma y^2 - \frac{(\Sigma xy)^2}{\Sigma x^2} \qquad \textbf{12.3-8}$$

For the example of Table 12-5, we can get the value at the bottom of column 9 without ever making any predictions for Y. We use Formula 12.3-8 instead.

$$\Sigma\,(Y - \hat{Y})^2 = \Sigma y^2 - \frac{(\Sigma xy)^2}{\Sigma x^2}$$

$$\Sigma\,(Y - \hat{Y})^2 = 476{,}940 - \frac{(238{,}837)^2}{250{,}068}$$

$$\Sigma\,(Y - \hat{Y})^2 = 248{,}829.59$$

Except for rounding errors, the value is the same as that obtained earlier.

The expression $\Sigma\,(Y - \hat{Y})^2$ represents the variability of Y scores that remains after varibility related to X has been removed. It is often referred to as *residual variance* or simply *residuals*.

INFERENTIAL STATISTICS AND REGRESSION ANALYSIS

It was specified earlier that, in the example given, the data were being treated as a population of scores. In the usual regression problem, the data represent a sample of scores. When the data are considered to be a sample, the concept of df must be considered. You have already learned that the formula for calculating standard deviations is modified by dividing by $n - 1$ rather than n. Both the conceptual formula (12.3-6) and computational formula (12.3-7) for SE_{est} must be modified in a similar fashion. The comparisons for these formulas for the SE_{est} for *samples versus populations* are shown in Table 12-6. The symbol k in the formulas for samples in Table 12-6 (Formulas 12.3-9 and 12.3-10) represents the number of predictor variables, making the formulas applicable to the case of multiple predictors.

For the sake of clarity, the examples in this first lesson have been treated as populations; thus, we have used the formulas in the first column. In practice, the formulas in the second column are almost always used, except in the case of large samples. In that case, the modifications have little effect.

Whether we are dealing with populations or samples, the methods we use to calculate the value b for the regression formula remains unchanged. The only change in the formula for calculating a is the substitution of $\overline{X}$ for μ_x. Since both b and a are usually statistics, it is possible to set confidence intervals for each or to make inferential tests for each. Some

Table 12-6. Formulas for SE_{est} for populations and samples

	Populations	Samples	
Conceptual	$SE_{est} = \sqrt{\dfrac{\Sigma\,(Y_i - Y)^2}{n}}$	$SE_{est} = \sqrt{\dfrac{\Sigma\,(Y_i - \hat{Y})^2}{n - 1 - k}}$	**12.3-9**
Computational	$SE_{est} = \sigma_Y \sqrt{1 - R^2}$	$SE_{est} = \sqrt{\dfrac{n - 1}{n - 1 - k}}\, s_Y \sqrt{1 - r^2}$	**12.3-10**

statistics books describe the appropriate procedures. You do not need to learn these procedures. If you wish to know whether there is a statistically significant relationship between X and Y, you already know how to make the test that $R = 0$. This test gives the same result as the test that the slope of the regression line in the population is zero. Once it is determined that a relationship exists, you know how to find the regression equation for predicting Y from X. You also know how to calculate SE_{est}.

There is one further complication when a sample is used to find the regression formula. As stated before, both a and b are sample statistics. When the formula involving these sample statistics is used to make predictions, the predictions will not be as good as they would be if the population regression formula were used. The calculation of confidence intervals for actual scores will have to be modified to allow for the use of the sample statistics in the regression formula.

Recall that when a population regression formula was used, we said that confidence intervals could be calculated based on a value from the table of the normal curve and the value of SE_{est}. When a sample regression formula is used, we will instead use a value from the t distribution and a weighted SE_{est}. The df for choosing a value of t is the number in the sample minus 2. The weighting for SE_{est} is one that depends on sample size, sample standard deviation, and distance of the observed X score from the observed mean of X. It is $\sqrt{1 + (1/n) + [(X_i - \overline{X})^2/(n - 1)s_X^2]}$. The confidence interval for the actual scores, based on a sample regression formula, is therefore as shown in Formula 12.3-11.

$$\text{Confidence interval} = \hat{Y}_i \pm (\text{value of } t)\left[SE_{est}\sqrt{1 + \frac{1}{n} + \frac{(X_i - \overline{X})^2}{(n - 1)s_X^2}}\right] \qquad \textbf{12.3-11}$$

Notice that the weighting of SE_{est} gets greater the farther an observed X score is from the sample mean for X. This means that instead of having a confidence interval of equal width for all observed values of X, the confidence interval becomes increasingly large as the observed value of X becomes more extreme. Instead of a band bounded by two parallel lines, as in Fig. 12-3, we would have a confidence band bound by lines that are nearest the regression line at the mean of X and that then curve away from the regression line at each extreme of the distribution.

SUMMARY

This lesson began with a description of the steps in a regression study. They are as follows:

1. Selection of a criterion of interest for a population of interest
2. Selection of a battery of predictors
3. Selection of a sample from the population of interest
4. Recording of measurements for all variables for each member of the sample
5. Calculation of the regression formula
6. Post hoc predictions of the criterion
7. Calculation of SE_{est} as an indication of how well the regression formula works

8. Predictions of the criterion for members of the population not included in the study.

The procedures for graphing linear equations on an X, Y coordinate system were presented next. It was explained that the formula for such graphing would be $Y = a + bX$. The concepts for graphing linear equations were then applied to the case where data are not perfectly correlated; such data produce scatter diagrams. The least squares criterion for choice of the best line for this situation was then explained.

Next, you were shown the calculations involved in a regression analysis for a worked example. With Y as the criterion and X as the predictor, you learned to obtain b by the formula $b = \Sigma xy / \Sigma x^2$. You then learned that a could be obtained by the formula $a = \text{mean}_Y - b(\text{mean}_X)$. The regression equation then became $\hat{Y}_i = a + bX_i$. For the worked example, you were shown how to calculate SE_{est} using either a definitional formula or a computational formula. You were also shown how to use this concept to set a confidence interval for the predictions made using the regression formula.

Finally, you learned how the formulas for regression analysis are modified for the usual case in which the regression analysis is for a sample.

The exercise that follows provides an opportunity to test your knowledge of the concepts from this lesson before we proceed to the two-predictor case.

LESSON 3 EXERCISES

From Table 12-4, use variable AB as the criterion variable and variable AC as the predictor variable. Consider the data to be a sample.
1. Find r.
2. Find the regression formula.
3. Find SE_{est} by both methods, and compare the results.
4. Find the coefficient of determination. What is the interpretation of this coefficient?
5. Compare your findings with those of the example in the lesson. Are the comparisons reasonable in view of what you know about the composition of the criterion and the predictor variables?

Lesson 4 □ Multiple regression analysis

An admissions committee for a college of medicine has one remaining place to fill to complete the school's first-year class. Of the two top candidates for the position, one has an outstanding academic record from a good undergraduate program but only an average score on the Medical College Admission Test (MCAT). The other candidate has a good, but not outstanding, academic record and MCAT scores well above average. In all other respects the two candidates are equal. The committee has decided to select the candidate who can be expected to perform better in the first year of medical school.

Past data show that both undergraduate grade point average (GPA) and MCAT performance are correlated with first-year medical school performance, with GPA having a slightly higher correlation coefficient. Using the methods of the last lesson, it would be possible to predict each candidate's performance by using each of the predictors separately. However, the results would not be very helpful, since the first candidate would receive a higher prediction when GPA was used as the predictor and the second candidate would receive a higher prediction when the MCAT was used. Neither would it be satisfactory to simply take the candidate who had performed better on the variable that correlates highest with first-year performance. This procedure would ignore differences on the second variable, no matter how great those differences might be.

SOLUTION FOR THE CASE OF TWO-PREDICTOR VARIABLES

The problem just posed is a common one faced by admissions committees at various levels of higher education. The criterion may be GPA, class rank, or a comprehensive examination. The predictors almost always include: (1) some type of measure of prior academic performance and (2) performance on some form of a standardized test. Usually it is assumed that the best prediction of the criterion can be achieved by weighting the applicant's score on each predictor variable and then adding the two weighted scores. Multiple regression analysis provides a systematic method for finding the best weights to use. Once more, it is assumed that "best" is defined by the least squares criterion.

There is one additional assumption for the multiple predictor case. It can be referred to as the *assumption of additivity*. When the methods of linear multiple regression are used, it is assumed that the performance on the criterion task is a function of the *linear addition* of performance on the predictor variables. This means that a low score on one variable can be offset by a high score on another. An applicant with a low score on one variable and a high score on another would have the same predicted criterion score as a person with average scores on both.

Use of an additive model is in contrast to the decision to use cutoff points for one or both predictor variables. When a nonadditive model involving a cutoff point for GPA is used, the decision is made that below a certain GPA even outstanding performance on the standardized test is of no consequence. In this text, we will assume that an additive model is to be used throughout the observed range of the predictor variables.

A graphic model for the case of two-predictor variables

In the last lesson you learned how to determine a formula of the form $\hat{Y}_i = a + bX_i$ to make the best predictions of Y by using variable X. Graphically, the solution was portrayed as a straight line through a scatter diagram of dots on an X, Y coordinate system. For each "individual" for whom predictions were to be made, the scores on the predictor and criterion variables represented coordinates for a dot representing the individual. With Y as the criterion, the goal was to draw a line in such a way as to minimize the sum of the squared vertical distances from the dots to the line.

For the two-predictor case, the coordinate system must be thought of as one with three dimensions. The easiest way for you to visualize this is to look at one of the corners of the room in which you are reading. The two walls and the floor intersect, resulting in three lines, each perpendicular to the others. Let us think of the vertical line formed by the intersection of the two walls as representing the criterion. Each of the lines formed by the intersection of a wall with the floor represents a direction for one of the predictors. The corner where the three lines come together is the origin. For each individual, there are three scores, representing three coordinates. The score on the criterion provides the height of the dot above the floor. The score for each predictor variable tells how far the dot is from each wall. The three coordinates determine the location of a dot for each individual, the dot being suspended somewhere in the three-dimensional space represented by the room. The set of scores for a group of individuals represents the coordinates for a swarm of dots suspended within the room.

When the predictor variables are positively related to the criterion, higher scores on the predictors are associated with higher scores on the criterion. In other words, the swarm of dots will slant upward, out into the room. If neither predictor were related to the criterion, the swarm of dots would be horizontal to the floor at an average height equal to the mean of the criterion.

For the *one-predictor* case, the goal is to find the formula for a straight line through a cluster of dots. For the *two-predictor* case, the goal is to find an equation representing a plane passing through a swarm of dots. The goal is to locate the plane in such a way as to minimize the vertical distances of dots to the plane. To visualize this, think of trying to position a large pane of glass through a swarm of dots slanting outward into the room.

The formula for the regression plane will be similar to the formula for a regression line. The difference is that now there are two predictors and two b values to determine. We need to be able to designate each separately. In most statistics books, when the two-predictor case is discussed, the predictors are represented by X and Y and the criterion by Z. We will follow a similar practice. We will refer to the two b values as b_X and b_Y. The formula we seek has the following form:

$$\hat{Z}_i = a + b_X X_i + b_Y Y_i \qquad\qquad \textbf{12.4-1}$$

Just as in the one-predictor case, we will first find the values for b and then use these in a formula with the means to find a. From the discussion you have just read, can you describe the meaning of *a* in the two-variable case? In the description we have used it is the height of the regression plane at the point of intersection of the two walls.

Calculations for the case of two-predictor variables

In this section, the formulas used to make calculations for the two-predictor case are presented, and an example is provided. It is assumed throughout that samples, rather than populations, are involved. It is not worthwhile to spend a lot of time trying to understand the basis for the formulas. They are presented as a reference for later use, in case you wish to solve a regression problem involving two-predictor variables. The calculations can easily be accomplished using any calculator that makes computing correlations a reasonable task.

The example again uses data from Table 12-4 of the last lesson. As the criterion, Z, we will use composite variable $ABCE$, which is formed by adding pure factors A, B, C, and E. For X, we will use composite variable ABD. For Y, we will use composite variable AC. The scores are listed in the top section of Table 12-7.

Knowing the composition of the criterion variable and of each predictor allows us to guess in advance what our regression solution should look like. Our predictions will be less than perfect, since the criterion has a factor, E, not represented by either of the predictors. We expect that the predictor variables will be correlated with each other, since they share factor A. Can you guess which predictor will be most useful in predicting the criterion? Each shares two components with the criterion. However, predictor X also has a component that is unrelated to the criterion. Predictor X is therefore not as "purely" related to the criterion. As a result, the second predictor, Y, should be more highly correlated with the criterion and therefore of greater use in its prediction.

The interim statistics needed for a two-predictor variable solution include means, standard deviations, and all correlation coefficients. The second section of Table 12-7 contains these statistics. As an aside, note that these statistics correspond to our knowledge of the composition of the variables used. We have treated the scores as samples and computed s, using $n - 1$ for the denominator.

Once all means, standard deviations, and correlations have been calculated, we are ready to find the b weights. The formulas are as shown:

$$b_X = \frac{s_Z}{s_X}\left[\frac{r_{ZX} - r_{ZY}r_{XY}}{1 - r_{XY}^2}\right] \qquad \text{12.4-2}$$

$$b_Y = \frac{s_Z}{s_Y}\left[\frac{r_{ZY} - r_{ZX}r_{XY}}{1 - r_{XY}^2}\right] \qquad \text{12.4-3}$$

The expressions inside the brackets are often referred to as beta (β) weights. When all scores are changed to standard scores before computations begin, all standard deviations equal 1.00, so b values equal beta weights. All standard scores also have a mean of zero, resulting in a zero value for the Z intercept. As a result, to predict the standard score for Z, the formula is simplified as follows:

$$\hat{z}_i = \beta_x(x_i) + \beta_y(y_i)$$

Since the researcher is generally interested in predicting the actual score for the criterion (rather than the standard score) the examples given here will use b values. The results of the calculations for the values of b for our example appear in the third section of Table 12-7.

Table 12-7. Example of a two-predictor regression solution

i	X	Y	Z	$\hat{Z}$	$Z - \hat{Z}$	$(Z - \hat{Z})^2$
1	1437	1094	2078	2045.77	32.23	1038.77
2	1466	1022	1966	2002.84	− 36.84	1357.18
3	1732	1132	2153	2195.34	− 42.34	1792.68
4	1445	1022	2195	1994.25	200.75	40,300.56
5	1364	941	1893	1899.48	− 6.48	41.99
6	1391	959	1744	1924.22	− 180.22	32,479.25
7	1521	871	1859	1910.42	− 51.42	2644.02
8	1467	745	1613	1792.45	− 179.45	32,202.30
9	1626	1034	2077	2077.41	− .41	0.1681
10	1270	715	1729	1689.05	39.95	1596.00
11	1356	1136	1868	2044.60	− 176.60	31,187.56
12	1286	878	1955	1819.63	135.37	18,325.04
13	1283	830	1948	1781.88	166.12	27,595.85
14	1682	908	1975	2004.43	− 29.43	866.12
15	1403	970	1692	1937.50	− 245.5	60,270.25
16	1195	981	2007	1860.80	146.2	21,374.44
17	1695	971	2198	2057.69	140.31	19,686.90
18	1684	878	2009	1982.41	26.59	707.03
19	1614	1063	1901	2094.57	− 193.57	37,469.34
20	1448	1164	1923	2103.54	− 180.54	32,594.69
21	1369	1154	2150	2063.62	86.38	7461.50
22	1717	1284	2318	2304.88	13.12	172.13
23	1637	921	2022	1995.91	26.09	680.69
24	1852	1275	2486	2353.24	132.76	17,625.22
25	1560	1052	2241	2064.11	176.89	31,290.07
TOTAL	37,500	25,000	50,000	50,000.04	− 0.004	420,759.77

Summary statistics

Deviation totals	Means	Standard deviations	Correlations
$\Sigma x^2 = 709{,}940$	$\bar{X} = 1500$	$s_X = 171.99$	$r_{XY} = 0.413$
$\Sigma y^2 = 503{,}974$	$\bar{Y} = 1000$	$s_Y = 144.91$	$r_{XZ} = 0.572$
$\Sigma z^2 = 985{,}770$	$\bar{Z} = 2000$	$s_Z = 202.67$	$r_{YZ} = 0.687$
$\Sigma xy = 247{,}254$			
$\Sigma xz = 479{,}310$			
$\Sigma yz = 484{,}542$			

$$b_X = \frac{s_Z}{s_X}\left[\frac{r_{ZX} - r_{ZY}r_{XY}}{1 - r_{XY}^2}\right] = \frac{202.67}{171.99}\left[\frac{0.572 - (0.687)(0.413)}{1 - (0.413)^2}\right] = 0.409$$

$$b_Y = \frac{s_Z}{s_Y}\left[\frac{r_{ZY} - r_{ZX}r_{XY}}{1 - r_{XY}^2}\right] = \frac{202.67}{144.91}\left[\frac{0.687 - (0.572)(0.413)}{1 - (0.413)^2}\right] = 0.76$$

$$a = \bar{Z} - b_X\bar{X} - b_Y\bar{Y} = 2000 - 0.409(1500) - 0.761(1000) = 625.5$$

$$\hat{Z}_i = 625.5 + 0.409(X_i) + 0.761(Y_i)$$

Once the b values are known, the Z intercept can be calculated by the following formula:

$$a = \overline{Z} - b_X \overline{X}_i - b_Y \overline{Y}_i \qquad \text{12.4-4}$$

These calculations are also included in section 3 of Table 12-7.

The best formula for predicting Z using predictors X and Y is as follows:

$$\hat{Z}_i = 625.5 + 0.409(X_i) + 0.761(Y_i)$$

Returning to the first section of Table 12-7, we see that this formula has been used to predict each Z score for the data of the example. The results are in column 5. The errors and the squares of the errors of prediction appear in columns 6 and 7. As in the one-variable case, we can calculate SE_{est} using either the sum of the values in column 7 or a computational formula. Using the values from column 7, we calculate SE_{est} by using the formula from Table 12-6 of the last lesson.

$$SE_{est} = \sqrt{\frac{\Sigma (Z - \hat{Z})^2}{n - 1 - k}}$$

$$SE_{est} = \sqrt{\frac{420,759.77}{22}} = 138.3$$

To find SE_{est} using the computational formula, we again will use the formula from Table 12-6 of the last lesson.

$$SE_{est} = \sqrt{\frac{n - 1}{n - 3}} s_Z \sqrt{1 - r_{Z \cdot XY}{}^2}$$

The formula contains the symbol $r_{Z \cdot XY}$, which is an index referred to as a multiple correlation coefficient. It is the correlation between Z and the best prediction of Z that can be made using predictors X and Y. In other words, it is the correlation between Z and $\hat{Z}$. You can use the usual formula to find the correlation between Z and $\hat{Z}$ in Table 12-7. Doing so will yield a value of 0.757. This same value can be obtained computationally, using Formula 12.4-5.

$$r_{Z \cdot XY} = \sqrt{\frac{r_{ZX}{}^2 + r_{ZY}{}^2 - 2r_{ZX}r_{ZY}r_{XY}}{1 - r_{XY}{}^2}} \qquad \text{12.4-5}$$

We have already found all the necessary correlation coefficients and listed them in Table 12-7. Substituting these in the formula, we get the same result that is obtained by finding the correlation between Z and $\hat{Z}$.

$$r_{Z \cdot XY} = \sqrt{\frac{(0.572)^2 + (0.687)^2 - 2(0.572)(0.687)(0.413)}{1 - (0.413)^2}} = \sqrt{\frac{0.474}{0.829}} = 0.756$$

Having calculated $r_{Z \cdot XY}$ computationally, we can now go back and substitute this value in the computational formula for SE_{est}.

$$SE_{est} = \sqrt{\frac{n-1}{n-3}} s_Z \sqrt{1 - r_{Z \cdot XY}^2}$$

$$SE_{est} = \sqrt{\frac{24}{22}}(202.67)(\sqrt{1 - 0.756^2}) = 138.56$$

Except for rounding error, this is the same value we obtained using the sum of the squares of $(Z - \hat{Z})$. Just as in the two-variable case, SE_{est} can be used to set confidence intervals for predicted Z scores. When the regression formula has been calculated from a sample, the confidence interval will have to be based on the t distribution and a weighted SE_{est}.

PARTIAL CORRELATION COEFFICIENTS

Correlation coefficients between two variables can often be misleading if the relationship of each with another closely related variable is ignored. For example, it may be surprising to find a high correlation between shoe size and achievement scores in elementary school. This correlation results because of a high correlation of each variable with a third variable, maturation. It is often relevant to ask what the relationship would be between two variables if the effect of a third variable were removed. This is analogous to asking what the correlation would be between shoe size and achievement scores if children were all at the same level of maturity. The question can be answered through the use of a measure of relationship often calculated as part of a multiple regression analysis. The index is called a partial correlation coefficient.

Conceptualizing the meaning of a partial correlation coefficient is difficult. We want to know the correlation between Z and X with the effect of Y removed from each. Suppose we were to do two regression analyses, one using Y to predict Z and another using Y to predict X. As the result of the first solution, we could obtain $Z - \hat{Z}$ for each individual. The resulting difference would represent Z with the effect of Y removed. Similarly, we could obtain $(X - \hat{X})$ for each individual, representing X with the effect of Y removed. The correlation between these sets of differences is what is being measured by the partial correlation coefficient $r_{ZX \cdot Y}$. The meaning of a partial correlation can be expressed as follows:

$$r_{ZX \cdot Y} = r_{(Z - \hat{Z})(X - \hat{X})}$$

where   in each case the prediction is based on linear regression using variable Y as the predictor.

To further clarify the meaning of this formula, think of the case of achievement scores, shoe size, and maturation. Achievement score is the criterion Z, shoe size is predictor X, and age is predictor Y. Instead of finding the correlation between achievement level and shoe size, we first adjust both scores based on the age of the child. The 10-year-old child's achievement scores is adjusted by subtracting the achievement level expected for all 10-year-olds. Similarly, the child's shoe size is adjusted by subtracting the shoe size expected for all 10-year-olds. The same procedure is carried out for each age level. The high correlation originally observed will probably all but disappear when the correlation is calculated for these adjusted scores. In other words, the correlation between achievement and shoe size will be near zero when the effect of maturation is "partialed out."

To obtain partial correlation coefficients by the method just described would involve a lot of calculations. Computational formulas will provide an identical result much more easily. These formulas, which make use of the correlation coefficients calculated as part of the regression analysis, are as follows:

$$r_{ZX \cdot Y} = \frac{r_{ZX} - r_{ZY}r_{XY}}{\sqrt{1 - r_{ZY}^2}\sqrt{1 - r_{XY}^2}}$$ **12.4-6**

and

$$r_{ZY \cdot X} = \frac{r_{ZY} - r_{ZX}r_{XY}}{\sqrt{1 - r_{ZX}^2}\sqrt{1 - r_{XY}^2}}$$ **12.4-7**

For the example for which we have done the regression analysis, we get the following values:

$$r_{ZX \cdot Y} = \frac{0.572 - (0.687)(0.413)}{\sqrt{1 - (0.687)^2}\sqrt{1 - (0.413)^2}} = \frac{0.288}{0.662} = 0.435$$

$$r_{ZY \cdot X} = \frac{0.687 - (0.572)(0.413)}{\sqrt{1 - (0.572)^2}\sqrt{1 - (0.413)^2}} = \frac{0.451}{0.747} = 0.604$$

Notice that in each case the partial correlation is somewhat less than the original correlation: 0.435 compared with 0.572, and 0.604 compared with 0.687. This does not come as a surprise to us, since we knew at the outset that all three variables included component A.

REGRESSION ANALYSIS WITH MORE THAN TWO PREDICTORS

This lesson began by posing a problem in which a committee on admissions for a college of medicine was faced with the decision of how to weight two variables: prior academic performance and performance on a standardized test. The goal was to predict first-year medical college performance. You can now help the committee by doing a two-predictor variable regression solution and arriving at a formula that provides ideal weights. You can also calculate SE_{est} for the predictions resulting from the application of the formula. You can further calculate the multiple r, which, when squared, provides a coefficient of determination. This index indicates what portion of the variability in first-year performance is predictable from performance on the two predictor variables. You can even calculate partial correlations to indicate to the committee the extent of the independent relationship between each predictor and the criterion.

You are now all prepared to collect a big consultant's fee for your help when you learn that the admissions committee separates prior academic performance into two parts—science GPA and nonscience GPA. You also learn that the MCAT results in several part scores to be considered. Instead of having two predictor variables, you have seven or eight.

A solution to the problem is available and is described in many statistics texts. It involves a solution for a set of simultaneous equations. The amount of work increases rapidly as the number of predictor variables increases. As indicated earlier, the reader who needs to obtain a solution to a multiple regression problem involving several predictor variables will almost certainly make use of a computer program to solve the problem. As a result, the calculations involved in a solution for the multiple predictor case will not be presented in this text. Instead, a procedure will be described that, as carried out by a popular computer program, is called *stepwise multiple regression*. With the many predictor variables possible in such a procedure, we would soon run out of letters to use in designating each one. To get around this problem, predictor variables are often designated X_1, X_2, X_3, X_4, and so forth, and the criterion is designated Y. We will follow this practice for the example that is used in the next section.

Stepwise multiple regression

In a regression problem involving a large number of predictor variables, there is usually a great amount of overlap among the predictors. As a result, in a battery that includes 20 predictor variables, the prediction of the criterion will often be nearly as good when only five of the variables are represented in the regression equation as it is when all 20 are included. This observation has led to a procedure called stepwise multiple regression. When this procedure is used, the analysis involves a series of steps, with a new predictor variable being added at each step. For step 1, the single variable with the highest correlation with the criterion is selected. A one-predictor variable regression solution is obtained, including calculations of the residual sum of squares, $\Sigma(Y - \hat{Y})^2$. At each subsequent step, the variable added is the one that, together with those already chosen, will result in the smallest $\Sigma(Y - \hat{Y})^2$. When no remaining variable will result in a significant reduction in the sum of $(Y - \hat{Y})^2$, the process is terminated.

Most computer programs designed to carry out a stepwise regression solution allow the user to specify a level of significance for inclusion of a new variable. Some also allow the user to specify a maximum number of steps and a level for exclusion of variables. The level of exclusion is used to determine, at the end of each step, whether any variable selected earlier is no longer contributing significantly to the prediction of the criterion. This can happen when a predictor variable selected in one of the early steps overlaps almost totally with a combination of variables selected in later steps. For each step, and at the end of the analysis, a complete regression solution is carried out.

The remainder of this lesson provides an example of the process of stepwise regression, using the program that is part of the biomedical (BMD) package developed at the University of California at Los Angeles. Many readers who seek to do multiple regression solutions will have access to a computer facility that makes use of this software package. Familiarity with the output of the BMD stepwise regression program will therefore be advantageous.

An example using BMD. The data chosen for the following example are again selected from the variables included in Table 12-4 of the last lesson. This makes it possible to compare the results of the computer analysis with the result expected based on our knowledge of the composition of variables. Let us use the same variables as those that were just used in the example for the two-predictor case; we will merely add two more predictors. One that is added is pure variable D, and the other is composite variable ABD. Our battery and criterion have the following composition:

$$Y = ABCE$$
$$X_1 = ABD$$
$$X_2 = AC$$
$$X_3 = D$$
$$X_4 = BD$$

Can you anticipate in advance some of the results of the stepwise regression analysis? Which variable do you expect to have the highest correlation with the criterion and, therefore, to be selected in step 1? Both X_1 and X_2 share two components with Y; however, X_1 also has a component made up of pure factor D, which is not in the criterion. Since X_2 is made up of only two components, both of which are included in Y, it should be selected at step 1. Can you anticipate what the correlation will be between Y and X_2? The criterion Y is made up of pure factors A, B, C, and E, each with an equal weighting. Half of these factors overlap with X_2. The coefficient of determination should therefore be approximately 0.5. The square root of 0.5 is 0.707. We expect the analysis to show that X_2 is selected at step 1 and that it has a correlation of about 0.70 with the criterion.

Can you estimate in advance the expected SE_{est} at the end of step 1? To do so, we use the computational formula $(SE_{est} = \sqrt{n-1/n-1-k}\ s_Y\sqrt{1-r^2})$. We have already concluded that r^2 will be about 0.5, but what about the standard deviation of Y? Since the factors making up the composite Y are essentially uncorrelated, the variance of Y can be expected to be nearly equal to the sum of the variances of its components. Each component has a σ near 100, or a variance near 10,000. The variance of the composite will therefore be near 40,000, and the standard deviation will be near 200. Substituting in the formula, we get the following for the expected SE_{est} at the end of step 1:

$$SE_{est} = \sqrt{\frac{n-1}{n-1-k}}\ s_Y\sqrt{1-r^2}$$

$$SE_{est} = \sqrt{\frac{24}{23}}\ 200\sqrt{1-0.5}$$

$$SE_{est} = 144.5$$

Can you anticipate which variable will be selected at step 2, keeping in mind that this variable is to be used along with variable X_2? We can eliminate X_3 from consideration at this step, since it shares no components with either the criterion or with X_2. What about

X_1, as compared with X_4? Variable X_1 has two components in common with Y, whereas X_4 has only one. Nevertheless, X_4 should be chosen at step 2. Can you understand why? Part of the information in X_1 (component A) is already being measured in X_2. Including X_1 in the regression formula in order to include component B will result in the inclusion of an extra component of A, as well as an unwanted component, D. Inclusion of X_4 will provide a measure of component B while adding only one unwanted component, D. We therefore expect the regression formula at the end of step 2 to include variables X_2 and X_4. These two include all of the components of Y, except pure factor E, plus one unwanted component D.

At step 3, only two variables are left. Will either add significantly to the ability to predict Y when used in conjunction with X_2 and X_4? This time, X_3, which is a measure of pure factor D, cannot be disregarded. By itself, X_3 would be of no use in predicting Y. It would also have been of no use at step 2, when it would have been used in conjunction with X_2. It is only when X_4 is added to the battery that X_3 becomes useful. Its use at this point has been described previously as that of a *suppressor* variable. Its use provides a pure measure of component B by suppressing component D in variable X_4. Since it accomplishes this by subtraction, we expect variable X_3 to have a negative weight when added to the regression equation at step 3.

At the end of step 3, the battery of predictors will include all components of Y except E. Since the one remaining variable, X_1, does not contain E, it cannot be expected to add further to the prediction of Y. We therefore expect the procedure to terminate at the end of step 3.

Having decided in advance what should happen when a stepwise multiple regression solution is sought for our data, we now look at the results of using the BMD program, as shown in Table 12-8. The table contains excerpts from the printout provided by the computer. Those parts of the results to which I wish to call particular attention have been boxed. At the top of the table is an intercorrelation matrix. The last line of the matrix contains the correlations of each of the predictors with the criterion. As expected, X_2 has the highest correlation, which is near 0.7. Also as expected, X_3 has the lowest correlation, essentially 0.00. Looking at step 1, we find that indeed variable X_2 was selected and that the standard error of estimate using this single predictor was 150. Remember we had projected a value near 144.5. The value of $\Sigma(Y - \hat{Y})^2$ at the end of step 1 is called the residual sum of squares and is listed as 519,910.19. The part of the variability of Y that can be accounted for by using regression based on X_2 is called a regression sum of squares and is listed as 465,861.94. Adding these two values provides the total sum of squares for Y, or 985,772.13. Can you remember what the ratio of 465,861.94 to 985,772.13 indicates? Division results in a value of 0.4726, which is the value given for the multiple r^2 (in this case $r_{YX_2}^2$). We have again confimed that the coefficient of determination indicates what portion of the total variability of Y is attributable to variability on the predictor(s). The table for step 1 also provides the regression formula for predicting Y from X_2. It is as shown at the top of p. 320.

Table 12-8. Example of stepwise regressor using BMD

INTERCORRELATION MATRIX

		1	2	3	4	5
1.	X_1	1.0000				
2.	X_2	0.4134	1.0000			
3.	X_3	0.5733	0.0428	1.0000		
4.	X_4	0.8049	−0.0105	0.6904	1.0000	
5.	Y	0.5730	0.6874	−0.0052	0.3466	1.0000

STEP 1

Variable entered X_2

Multiple r 0.6874
Multiple r^2 0.4726
SE_{est} 150.34

Analysis of variance

	Sum of squares	df	Mean square	F ratio
Regression	465,861.94	1	465,861.9	20.61
Residual	519,910.19	23	22,604.79	

Variables in equation

Variable	Coefficient	F to remove
(Y intercept 1038.556)		
X_2	0.961	20.61

Variables not in equation

Variable	Partial correlation	F to enter
X_1	0.43671	5.18
X_3	−0.04780	0.05
X_4	0.48721	6.85

STEP 2

Variable entered X_4

Multiple r 0.7732
Multiple r^2 0.5978
SE_{est} 134.2487

Analysis of variance

	Sum of squares	df	Mean square	F ratio
Regression	589,272.63	2	294,636.3	16.35
Residual	396,499.50	22	18,022.70	

Variables in equation

Variable	Coefficient	F to remove
(Y intercept 523.756)		
X_2	0.967	26.13
X_4	0.510	6.85

Variables not in equation

Variable	Partial correlation	F to enter
X_1	0.00917	0.00
X_3	−0.60990	12.44

Table 12-8. Example of stepwise regressor using BMD—cont'd

STEP 3

Variable entered X_3

Multiple r	0.8645
Multiple r^2	0.7474
SE_{est}	108.8925

Analysis of variance

	Sum of squares	df	Mean square	F ratio
Regression	736,763.06	3	245,587.7	20.71
Residual	299,009.06	21	11,857.57	

Variables in equation

Variable	Coefficient	F to remove
(Y intercept	483.949)	
X_2	1.004	42.65
X_3	−1.062	12.44
X_4	1.043	22.75

Variable not in equation

Variable	Partial correlation	F to enter
X_1	0.00237	0.00

F-levels (0.05, 0.050) or tolerance insufficient for further stepping

Lists of predicted values, residuals, and variables

Case number	Predicted	Residual	Y	X_1	X_2	X_3	X_4
1	1947.6460	130.3540	2078	1437	1094	538	898
2	2100.8159	−134.8159	1966	1466	1022	474	1049
3	2103.6614	49.3386	2153	1732	1132	538	1011
4	1992.7930	202.2070	2195	1445	1022	505	977
5	1881.5500	11.4500	1893	1364	941	433	875
6	1845.4099	−101.4099	1744	1391	959	432	822
7	2047.4167	−183.4167	1859	1521	871	464	1133
8	1713.6147	−100.6147	1613	1467	745	558	1030
9	2078.9221	− 1.9221	2077	1626	1034	612	1157
10	1718.4573	10.5427	1729	1270	715	421	924
11	1978.3489	−110.3489	1868	1356	1136	538	887
12	1937.5627	17.4373	1955	1286	878	305	859
13	1769.8936	178.1064	1948	1283	830	443	885
14	2005.3711	− 30.3711	1975	1682	908	569	1164
15	1802.6484	−110.6484	1692	1403	970	464	803
16	1987.5901	19.4099	2007	1195	981	316	819
17	2048.6062	149.3938	2198	1695	971	580	1156
18	1973.8411	35.1589	2009	1684	878	642	1237
19	2000.6169	− 99.6169	1901	1614	1063	612	1054
20	1992.7805	− 69.7805	1923	1448	1164	653	991
21	2130.2170	19.7830	2150	1369	1154	306	779
22	2309.4849	8.5151	2318	1717	1284	484	1007
23	1985.3086	36.6914	2022	1637	921	664	1229
24	2527.7739	− 41.7739	2486	1852	1275	432	1172
25	2119.6846	121.3154	2241	1560	1052	517	1082

$$\hat{Y}_i = 1038.556 + (0.96)(X_{2_i})$$

Looking at step 2, we find that the program did indeed select variable X_4 as the second variable to be included in the regression formula. Notice that as a result of the inclusion of variable X_4, the multiple r increases to 0.77 and SE_{est} decreases to 134.2. The ratio of regression sum of squares to total sum of squares for Y has increased to 0.598. The regression formula is now:

$$\hat{Y}_i = 523.756 + (0.967)(X_{2_i}) + (0.510)(X_{4_i})$$

At step 3, the variable entered was variable X_3, which can be considered a suppressor variable. The prediction of Y has been improved, because X_3 serves as a means of suppressing the noise contributed as component D of X_4. The multiple r becomes 0.864, and the standard error of estimate decreases to 108.9. With X_2, X_4, and X_3 all in the regression equation, the proportion of variability in Y that can be attributed to the regression is 0.747. Notice that, as expected, the weight for variable X_3 is negative.

$$\hat{Y}_i = 483.949 + (1.004)(X_{2_i}) + (-1.062)(X_{3_i}) + (1.043)(X_{4_i})$$

As anticipated, the program has terminated the process at the end of step 3, indicating that the addition of variable X_1 would not significantly add to the prediction of Y. Component E is not measured by any of the predictor variables and is therefore unique to the criterion. Since it makes up about one fourth of the criterion, approximately one fourth of the variability of Y is not predictable from the battery. This is consistent with the conclusion we reach when we look at the square of the multiple r (0.747) at the end of step 3.

The remainder of Table 12-8 contains the actual and predicted values of Y. The sum of the errors of prediction squared is the value called "residual" in the results of step 3. It should be noted that the computer program has treated the data as a sample.

As you can see by this example, doing a stepwise multiple regression using a computer program such as that provided in BMD is very informative. It provides information concerning the results at each step of the process. Even large data sets for large predictor batteries are handled very easily. Generally, it will be found that the result at the end of step 4 or 5 will be almost as good as the result obtained by using all of the predictor variables. Usually, the result of a stepwise multiple regression solution will be very similar to an ordinary regression analysis, which involves the simultaneous solution of a large set of equations.

SUMMARY

This lesson described the extension of linear regression to the case of two or more predictor variables. An assumption of additivity for the case of multiple predictors was explained. The least squares criterion still serves as the basis for selecting the optimal formula. Graphically, the process was described for the two-predictor case as an attempt to find the best plane fitted through a swarm of dots suspended in three-dimensional space. The formula for the plane is:

$$\hat{Y}_i = a + b_X(X_i) + b_Y(Y_i)$$

The formulas for finding the values of a and b were provided for the two-predictor case. A worked-out example was presented, and the concepts of multiple correlation and partial correlation were explained and related to this example.

For a case involving more than two predictors, a procedure called stepwise multiple regression was described. The procedure was illustrated using a data set of known composition. A computer program from the BMD package was used to solve the problem. The output from the program was examined and found to conform to the results expected on the basis of the known composition of variables.

LESSON 4 EXERCISE

Students in a school of public health generally take a course in biometry early in the program. Subsequently or concurrently, they take a basic course in epidemiology. The chairman of the epidemiology department wishes to predict in advance which students are likely to have difficulty in the basic epidemiology course. A preadmission GPA, along with scores on the biometry and epidemiology finals, is available for 20 students who have taken both courses. The scores for the students are contained in the following table.

i	GPA (X)	Biometry final (Y)	Epidemiology final (Z)
1	4.57	101	80
2	3.86	58	64
3	3.96	48	63
4	4.14	86	82
5	4.06	95	74
6	3.52	63	61
7	3.98	56	54
8	4.16	123	87
9	3.58	65	64
10	3.14	53	41
11	4.03	55	72
12	4.61	108	78
13	4.56	113	81
14	3.90	40	54
15	4.17	98	74
16	4.80	79	65
17	3.30	86	75
18	4.31	82	63
19	4.47	62	81
20	4.16	77	66

Consider the scores to be a sample.
1. Find $\bar{X}$, $\bar{Y}$, and $\bar{Z}$.
2. Find s_X, s_Y, and s_Z.
3. Find r_{ZX}, r_{ZY}, and r_{XY}.
4. Find the optimal formula for predicting Z by using X and Y.
5. Find $r_{Z \cdot XY}$ and $r_{Z \cdot XY}^2$.
6. Find SE_{est}.
7. Find partial correlations $r_{ZX \cdot Y}$ and $r_{ZY \cdot X}$.

CHAPTER 13

Analysis of covariance

Lesson 1

In Chapter 11, you learned how to partition the total variance of scores for a dependent variable into variance *between* experimental groups and variance *within* experimental groups. The process was called ANOVA and was used as a method for comparing the means of several groups. As part of Chapter 12, you learned to partition the total variance of scores for a criterion variable into variance *related* to predictor variables and variance *unrelated* to predictor variables. The process was called regression analysis and resulted in a formula for predicting criterion scores. In this chapter you will learn about a technique of data analysis that is based on a combination of the concepts of ANOVA and regression analysis. The process is called *analysis of covariance,* which we will abbreviate as ANCOVA.

Many statistics books written at the level of this one do not include a discussion of ANCOVA. As originally planned, this book was intended to devote only a few paragraphs to a description of the purpose of the procedure. The reluctance to include the topic arises from its complexity. A large number of readers are thoroughly confused after reading a description of the process. However, ANCOVA has such wide applicability in research involving human subjects that it was decided to devote a lesson to an elementary discussion of the topic. This lesson will describe: (1) the circumstances in which ANCOVA is applicable, (2) the general rationale for the procedure, (3) the calculations for an uncomplicated example, and (4) the assumptions and requirements for using the procedure. After studying this lesson, *all* students should be able to recognize when ANCOVA is applicable and be able to carry out the calculations for what you will learn to call the one-covariate case. It is hoped that *most* students will also have an understanding of the general rationale for the procedure.

THE USE OF ANCOVA

Suppose that, as in the introduction to Chapter 8, we are interested in investigating methods for treating patients with a certain type of damage to knee ligaments. This time, let us consider the case in which three different methods of treatment are to be compared. The *dependent* variable used to compare the effects of the three methods is the length of

Table 13-1. Rehabilitation periods for three methods of treating damaged knee ligaments

	Method A	Method B	Method C
	50	75	20
	76	51	42
	42	43	50
	13	17	80
	29	41	68
	42	29	
		38	
		66	

Summary statistics

	A	B	C	Combined
n	6	8	5	19
ΣY	252	360	260	872
ΣY^2	12,814	18,686	15,688	47,188
$\bar{Y}$	42	45	52	45.89

ANOVA table

Source	SS	df	ms	F
Between	283.79	2	141.89	0.33
Within	6884	16	430.25	
TOTAL	7167.79	18		

time until the rehabilitated knee reaches 90% of the strength and mobility of the undamaged knee. If the required assumptions of normality and homogeneity of variance can be met, we will probably analyze the data by using a one-way ANOVA.

Table 13-1 contains a listing of scores that might result from such an experiment. Summary statistics and an ANOVA table describing the analysis of the data are also included. Although treatments *B* and *C* have means that are higher than that of treatment *A*, the difference is not found to be significant.

When the researcher writes a report of the investigation, he or she will attempt to explain the results. The researcher will examine two questions. First, in light of the results of this experiment, is he or she prepared to believe that population means are really equal for the three methods of treatment? Second, if the researcher remains convinced that a difference exists among population means, why were the results of the experiment insignificant? The most obvious answer for the second question is that small sample sizes were used in the experiment. However, another, less obvious, explanation may be the real answer.

In an ideal experimental design, a group of patients with ligament damage would be assigned at random to the methods of treatment. Random assignment ensures that, except for chance variability, the groups will be similar for characteristics such as age, sex, or physical condition. If there is a variable known in advance to be related to the dependent variable, subjects may be matched on that variable first, before being randomly assigned to one of the experimental groups. For two groups, the analysis then becomes a matched pair *t* test. For several groups, the analysis becomes a randomized-block ANOVA.

In experiments involving human subjects, random assignment of subjects to experimental conditions is seldom carried out. In the field of medicine, ethical considerations require that each physician be allowed to choose the method of treatment he or she feels is best for each patient. Even when ethical considerations are not a factor, random assignment may not be feasible. Instead, experimental groups are formed from *intact* groups, such as patients in each of a set of hospitals, students in each of a number of classes, or residents in each of a group of communities.

When subjects are not randomly assigned to experimental conditions, it is quite likely that the groups will differ on variables such as age, sex, or physical condition. If one of these variables is related to the dependent variable, the results of the experiment will be affected. Differences may be found and attributed to the experimental conditions when in fact they are due to differences in the related variable. Or, in an experiment with results similar to those reported in Table 13-1, real differences in population means may not be found, because they are obscured by differences in the related variable.

When it is not possible to design the experiment in such a way as to control for related variables, the procedures of ANCOVA provide a way to *statistically* remove the effect of these variables. This is accomplished through an adjustment of the means of the dependent variable, based on differences in the means of related variables. The related variable or variables are called *covariates*. We will confine our examples to the case of a single covariate. As with regression analysis, if there are several covariates, the analysis will usually be done using a computer program. The logic of the analysis will remain the same as for the simple case.

THE LOGIC OF ANCOVA

Suppose that the researcher who has carried out the experiment resulting in the data of Table 13-1 remains convinced that the three methods of treatment do not result in equal rehabilitation periods. Perhaps the failure to find significant differences is even further complicated by the direction of the differences observed. Method *A* is the traditional method of treatment. This method has been pretty much replaced, first by method *B*, and more recently by method *C*. Practitioners believe that the average rehabilitation times should be just in the reverse order, with the highest average occurring when method *A* is used.

While the experimenter is examining the data to try to figure out what happened, it is discovered that ages of patients were different for the three groups. Fig. 13-1 shows the means of the three groups, with the dependent variable plotted as *Y* and the covariate, age, plotted as *X*. The importance of the discovery that the three groups differ on the variable age depends on the relationship between age and length of time required for rehabilitation. If there is no relationship, there will be no significance to the findings. If a relationship

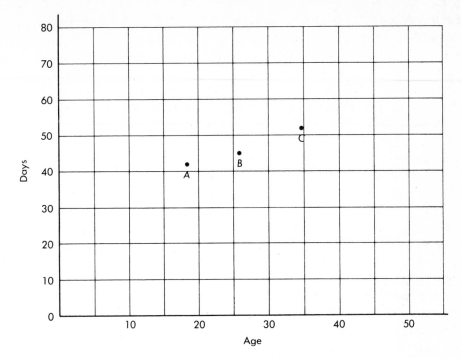

Fig. 13-1. Means for the data of Table 13-1 plotted with dependent variable as Y and age as X.

exists but differs from one group to another, the finding will be interesting, but it will not be very helpful in explaining the results of the research project. It is when there is a consistent relationship between age and period of rehabilitation that we will want to look closely at the effects of age differences on the results of the experiment.

Figs. 13-2, 13-3, and 13-4 represent three possible situations involving the relationship of the two variables age and rehabilitation period (X and Y). In each case, the scatter diagram of the scores appears as an oval. The line drawn through each oval represents a least squares regression line.

In Fig. 13-2, the relationship between X and Y varies from one group to another, as indicated by the different slopes of the regression lines. The lack of a consistent relationship between X and Y makes it unreasonable to attempt to adjust group means for Y with the goal of eliminating the effects of differences in X.

Notice the difference in Fig. 13-3. Here, the regression lines all have the same slope. It would now make sense to use this common slope to adjust each group mean of Y to eliminate the effects of differences in X. For example, we would find the *overall* group mean for X, 25.63, and then ask: "What would $\overline{Y}_A$ be if $\overline{X}_A$ had been 25.63?" and "What would $\overline{Y}_B$ be if $\overline{X}_B$ had been 25.63?" and so forth. Can you estimate the answers just by looking at Fig. 13-3? In each case, you find the height of the regression line at a point

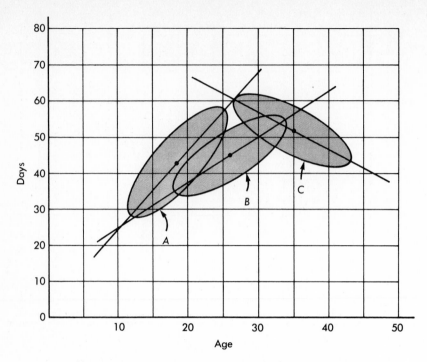

Fig. 13-2. Three scatterplots of X and $Y;$ slopes unequal.

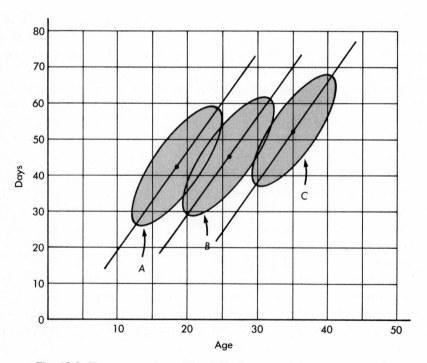

Fig. 13-3. Three scatterplots of X and $Y;$ slopes equal, intercepts unequal.

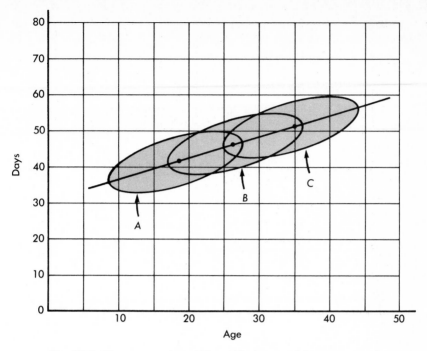

Fig. 13-4. Three scatterplots of X and Y; slopes and intercepts equal.

directly above 25.63 on the X axis. If the relationship between X and Y is that shown in Fig. 13-3, *adjusted* means for Y differ greatly from *observed* means. Notice that the order of the adjusted means would be reversed from the original order and would then correspond to expected results.

Finally, look at Fig. 13-4. In this case, the regression lines again have identical slopes. They also have identical *intercepts with the Y axis*. What does it mean if our scatter plots for the three groups turn out to be like those of this third figure? We can still adjust each mean to a value expected when $\overline{X}$ equals 25.63, but what would be the result of the adjustment? In this case, the adjusted means would all be equal.

The examination we have just made of Figs. 13-2, 13-3, and 13-4 is essentially the reasoning behind ANCOVA. We begin by asking whether the data indicate that Fig. 13-2 describes the relationship between X and Y. In other words, we first test to see whether the slopes of the individual regression lines differ. If the answer is "yes," we conclude that age is not a useful covariate and discontinue looking at differences in age as a possible explanation for our unexpected results.

If the first test leads to the conclusion that it is reasonable to assume that the regression lines have equal slopes, we then make a second test, this time to determine whether the situation is that of Fig. 13-3 or of Fig. 13-4. In other words, having determined that the slopes are equal, we now test to see whether the lines also have the same Y

intercept. If the answer is "yes," we conclude that the means of *Y* would be equal if mean ages were equal. If the answer is "no," we conclude that the rehabilitation periods would differ for the three methods of treatment if only we could control for age.

THE TWO MAJOR TESTS OF ANCOVA

All that remains is to learn how to make the two tests just described. The procedure involves three methods of finding sums of errors of prediction (residuals). For method 1, a separate least squares regression line is calculated independently for each group. The sum of errors of prediction is calculated separately within each group. These sums are then added to obtain a total of residuals for method 1. For method 2, a regression line is once more determined separately for each group, this time with the restriction that each line must have the same slope. Residuals are again calculated for each group and then summed to obtain a total of residuals for method 2. Finally, for method 3, all groups are combined, and a single regression line is found for the entire set of pooled scores. Residuals about this line are calculated to obtain a total of residuals for method 3.

A comparison of the residuals obtained for methods *1 and 2* provides a way of determining whether the slopes of the regression lines are equal in the populations represented by the samples. To understand the test, consider the residuals for methods 1 and 2. Which method will result in a smaller sum of residuals? A least squares criterion has been used to determine the three regression lines for method 1. The residuals for method 1 must therefore be the smallest possible. Under what circumstance would the residuals for method 2 equal the residuals from method 1? Can you see that this will happen only when the slopes of the regression lines calculated for method 1 are all equal? A comparison of residuals from methods 1 and 2 therefore provides the test we seek. Since the residuals represent a type of variance, the comparison will be based on an *F* ratio. The formula will be described in the next section, which shows the calculations.

A comparison of the residuals from methods *2 and 3* provides a way of determining whether the lines have not only a common slope, but also a common *Y* intercept. Under what circumstance would the residuals of method 3 equal the residuals of method 2? Method 2 residuals are based on lines with a common slope. What will happen if the three lines of method 2 also have a common intercept? The lines will then coincide and correspond to the single line of method 3. A comparison of residuals from methods 2 and 3 therefore provides the test we seek. Again, the test is based on an *F* ratio to be described in the following section.

When it is determined that the regression lines have a common slope but different *Y* intercepts, the researcher concludes that the means for the criterion would be different if it had been possible to control for the covariates. He or she then calculates what these adjusted means would have been. Pairwise comparisons can then be made to determine which adjusted means are significantly different from the others. The methods for doing this, along with the calculations, will be described later.

Before we go to a description of the calculation of ANCOVA, it should be pointed out that even when samples are chosen from a single population, the residuals of methods 1, 2, and 3 will not be exactly equal. Sampling variability will affect both the slopes and intercepts of the lines of methods 1, 2, and 3. The F tests comparing the residuals will be used to determine whether the differences are greater than those that would be expected by chance when sampling is performed from populations with equal slopes and equal Y intercepts.

THE CALCULATIONS OF ANCOVA

Based on what you have learned about the logic of ANCOVA, you have no doubt concluded that the calculations will be lengthy and involved. Determining all of the regression lines and then finding the residuals about each line is not an easy task. Fortunately, the calculations are not quite as difficult as you might anticipate. This is because it is possible to make use of the computational formula for residuals. Do you remember Formula 12.3-8 from the last lesson?

$$\Sigma (Y - \hat{Y})^2 = \Sigma y^2 - \frac{(\Sigma xy)^2}{\Sigma x^2}$$

By using this formula, we can find the residuals without ever finding the regression formulas. The values in the right-hand side of the formula will be calculated differently for each of methods 1, 2, and 3.

The calculations of ANCOVA can be confusing to talk about without an example. Let us therefore complete the data for the study described earlier. So far, you have seen all of the scores for Y in Table 13-1. You have also seen the means for both X and Y plotted in Fig. 13-1. Table 13-2 provides the pairs of scores for X and Y.

When the calculations of ANCOVA are being done, it is difficult to keep all of the results in order. Making up a table similar to Table 13-3 will be helpful. All of the contents of the upper part of the table are familiar and easy to obtain by using any

Table 13-2. Ages and rehabilitation periods for three methods of treating damaged knee ligaments

Method A		Method B		Method C	
Age (X)	Rehabilitation period (Y)	Age (X)	Rehabilitation period (Y)	Age (X)	Rehabilitation period (Y)
20	50	35	75	27	20
30	76	27	51	34	42
19	42	24	43	31	50
8	13	18	17	37	80
19	29	32	41	43	68
13	42	20	29		
		23	38		
		27	66		

Table 13-3. Calculations of ANCOVA for the data of Table 13-2

Statistic	Separate group data			Summary data
	Group A	Group B	Group C	
ΣX	109	206	172	487
ΣX^2	2255	5536	6064	13,855
ΣY	252	360	260	872
ΣY^2	12,814	18,686	15,688	47,188
ΣXY	5279	9888	9402	24,569
n	6	8	5	19
$\overline{X}$	18.167	25.75	34.4	25.63
$\overline{Y}$	42.0	45.0	52.0	45.89

Deviation sums	Separate for A	Separate for B	Separate for C	$SS_{wgroups}$	SS_T
Σx^2	274.8	231.5	147.2	653.5	1372.42
Σy^2	2230	2486	2168	6884	7167.79
Σxy	701	618	458	1777	2218.26
$\Sigma(Y - \hat{Y})^2$	441.79	836.22	742.97	2051.97	3583.23

Method 1 residuals $= 441.79 + 836.22 + 742.97 = 2020.98$
Method 2 residuals $= 2051.97$
Method 3 residuals $= 3583.23$

calculator that will perform correlations. The contents of the second section of the table are the values needed for Formula 12.3-8. For the first three columns, the values are readily obtained from the entries in the upper section of the table. For example, for the first column, we calculate as follows:

$$\Sigma x^2 = \Sigma X^2 - \frac{(\Sigma X)^2}{n} = 2255 - \frac{(109)^2}{6} = 274.8$$

$$\Sigma y^2 = \Sigma Y^2 - \frac{(\Sigma Y)^2}{n} = 12814 - \frac{(252)^2}{6} = 2230$$

$$\Sigma xy = \Sigma XY - \frac{(\Sigma X)(\Sigma Y)}{n} = 5279 - \frac{(109)(252)}{6} = 701$$

The values for groups B and C are obtained in a similar fashion.

For the column headed $SS_{wgroups}$, the values are obtained by adding the entries for the individual groups. For example:

$$274.8 + 231.5 + 147.2 = 653.5$$

For the column headed SS_T, the values are obtained using the summary data in the upper section of the table. For example:

$$\Sigma x^2 = \Sigma X^2 - \frac{(\Sigma X)^2}{n} = 13,855 - \frac{(487)^2}{19} = 1372.42$$

For the ANCOVA, the row of the table in which we are particularly interested is the one for the sum of residuals, $\Sigma(Y - \hat{Y})^2$. In each case, $\Sigma(Y - \hat{Y})^2$ is found using

Formula 12.3-8. The values to substitute in the formula are obtained from the entries in the second section of the table. For example, for group A:

$$\Sigma (Y - \hat{Y})^2 = \Sigma y^2 - \frac{(\Sigma xy)^2}{\Sigma x^2} = 2230 - \frac{(701)^2}{274.8} = 441.79$$

Each of the other values (including those for $SS_{wgroups}$ and SS_T) is obtained in a similar manner.

Note the meaning of each of the values in the row for $\Sigma (Y - \hat{Y})^2$. The number 441.79 is the sum of the squares of errors of prediction when a least squares regression solution is obtained separately for the scores of group A. Adding these values for the three groups gives the sum of residuals for method 1. At the bottom of the table, this addition is shown to result in a value of 2020.98. The number 2051.97 is the sum of residuals that would result if three separate lines were chosen under the restriction that all had to have the same slope. It represents the sum of residuals for method 2. The number 3583.23 is the sum of residuals that results when all groups are combined and a single least squares regression line chosen. It therefore represents the sum of residuals for method 3.

Having obtained the residuals by three different methods, we are now ready to proceed to make the two tests described earlier. We pause before doing so to point out that the columns headed $SS_{wgroups}$ and SS_T have exactly the same meaning as in ANOVA. If you go back to the ANOVA table (Table 13-1) you will see that this is so.

Calculations for the F ratios

We will first provide the formulas for the F ratios to be used in making the two major tests of ANCOVA. Later, we will explain why these formulas are reasonable.

To make the first test, that there is a common slope to the regression lines within the groups, we will use residuals from methods 1 and 2, divided by the proper number of df. The formula is Formula 13.1-1.

$$F_1 = \frac{\dfrac{res_2 - res_1}{k - 1}}{\dfrac{res_1}{n_T - 2k}} \quad \text{with } (k - 1) \text{ and } (n_T - 2k) \text{ df} \qquad \textbf{13.1-1}$$

where n_T is the total number of subjects and
k is the number of groups.

An extreme value for F will lead to the conclusion that the differences in slopes for the individual lines of method 1 are greater than can be expected by chance when slopes are equal in the populations represented by the samples. In this case, the second test will not be performed. What we are hoping will happen is that F will be a value that is a common result when samples are taken from populations in which the slopes are the same.

For our data, the calculations are as follows:

$$F_1 = \frac{\dfrac{2051.97 - 2020.98}{3 - 1}}{\dfrac{2020.98}{19 - 2(3)}} = 0.10 \text{ with 2 and 13 } df$$

For this example, the *F* value is less than 1.00. Only numbers greater than 1.00 represent evidence against the null hypothesis that there is no difference among slopes. We therefore proceed to the second test.

Having determined that it is reasonable to assume a common slope for the regression lines, we now test to see whether they also have a common *Y* intercept. The test compares residuals from methods 2 and 3. If the *F* value is significant, it indicates that the situation is that of Fig. 13-3 rather than that of Fig. 13-4. The formula is Formula 13.1-2.

$$F_2 = \frac{\dfrac{res_3 - res_2}{k-1}}{\dfrac{res_2}{n_T - k - 1}} \quad \text{with } k-1 \text{ and } n_T - k - 1 \ df \qquad \textbf{13.1-2}$$

where n_T is the total number of subjects and
        k is the number of groups.

For our example, the calculations are as follows:

$$F_2 = \frac{\dfrac{3583.23 - 2051.97}{2}}{\dfrac{2051.97}{15}} = 5.59 \text{ with 2 and 15 } df$$

Our resulting *F* value has a probability less than 0.05. We therefore conclude that if we had been able to control for age, there would have been significant differences in average rehabilitation times.

Adjusted means. The finding that means for rehabilitation times would have been different if we could have controlled for age leads us to ask what these means would have been. We could plot the data, as in Fig. 13-3, and find the heights of the regression lines at the point of the overall mean for *X*, 25.63. Instead, we use Formula 13.1-3.

$$\text{adj } \overline{Y}_j = \overline{Y}_j + b_w(\overline{X}_T - \overline{X}_j) \qquad \textbf{13.1-3}$$

where b_w is the slope calculated for method 2 and
        $\overline{X}_T$ is the overall mean for *X*.

We will use Formula 12.3-3 from Chapter 12 to obtain b_w. The values for deviation sums to go into the formula will be taken from the $SS_{wgroups}$ column of Table 13-3. Once we have b_w, we will calculate each of the adjusted means using Formula 13.1-3 and the original means obtained from Table 13-3. For our data, we calculate as follows:

$$b_w = \frac{\Sigma xy}{\Sigma x^2} = \frac{1777}{653.5} = 2.72$$

$$\text{adj } \overline{Y}_A = 42.0 + 2.72(25.63 - 18.167) = 62.3$$
$$\text{adj } \overline{Y}_B = 45.00 + 2.72(25.63 - 25.75) = 44.68$$
$$\text{adj } \overline{Y}_C = 52.0 + 2.72(25.63 - 34.4) = 28.16$$

Individual comparisons. We now ask which of these adjusted means differ from the others. We use a method similar to that described for ANOVA. Confidence intervals are established for *all* possible pairwise comparisons of means, while the overall alpha level is held at a desired value. First we determine *m*, the number of pairwise comparisons possible, by using Formula 13.1-4.

$$m = \frac{k(k - 1)}{2}$$

13.1-4

where   *k* is the number of groups.

In our case, *m* = 3.

Let us say that we decide to use an overall alpha level of 0.15. The confidence level for each individual comparison will be found by using Formula 13.1-5.

$$\text{Individual confidence levels} = 1.00 - \frac{\text{overall alpha level}}{m}$$

13.1-5

For our data, we get the following: $1.00 - (0.15/3)$, or 0.95.

Each of the confidence intervals will be established in the usual way:

Observed difference ± (appropriate tabled value)(standard error of the difference)

The observed difference in this case is the difference between adjusted means. To this difference we add and subtract a constant based on two factors. The first factor is again a tabled value, in this case taken from the *t* distribution with $(n_T - k - 1)$ *df*. For our example, with each confidence level set at 0.95 and with 15 *df*, the value for the first set of parentheses is 2.131.

The standard error for the second set of parentheses is more complicated. It is given by Formula 13.1-6.

$$SE_{(\text{adj } \bar{Y}_A - \text{adj } \bar{Y}_B)} = \sqrt{\left[\frac{1}{n_A} + \frac{1}{n_B} + \frac{(\bar{X}_A - \bar{X}_B)^2}{SS_{w_X}}\right]\left[\frac{res_2}{n_T - k - 1}\right]}$$

13.1-6

For our example, we calculate as follows to find the standard error of the difference between adjusted means *A* and *B*.

$$SE_{(\text{adj } \bar{Y}_A - \text{adj } \bar{Y}_B)} = \sqrt{\left[\frac{1}{6} + \frac{1}{8} + \frac{(18.167 - 25.75)^2}{653.5}\right]\left[\frac{2051.97}{19 - 3 - 1}\right]} = 7.21$$

The confidence interval for the difference between adjusted means for *A* and *B* is: $(62.3 - 44.68) \pm (2.131)(7.21)$, or 2.26 to 32.98. Since the interval does not include zero, we decide that the difference is significant. Using the same procedure, we find that the confidence interval for adjusted means *A* and *C* is 12.26 to 56.01 and that the confidence interval for adjusted means *B* and *C* is -0.004 to 33.04. The results of our experiment, which were very puzzling, now conform to our expectations based on our knowledge of treatments *A*, *B*, and *C*. The average rehabilitation time, when adjusted for age, is significantly lower for treatments *B* and *C* than for *A*.

The basis for the F tests of ANCOVA

The formulas for the F ratios for the two main tests of ANCOVA have been presented without any explanations as to how they are derived. This section will show that they are logically based on concepts of ANOVA and regression analysis. We begin by observing that *when the samples are from a single population* the residuals of methods 1, 2, and 3 all represent error variance. If each sample were a perfect replica of the population, the regression lines of methods 1, 2, and 3 would all coincide. However, since each sample is not a perfect replica of the population, there will be variability in the regression lines of the three methods. Variability will be found both in the slopes of sample regression lines and in the Y intercepts for these lines. Let us see how each of these sources of variability enters into the residuals calculated for methods 1, 2, and 3.

For method 1, the sum of residuals is unaffected by sampling variability in either the slopes of the regression lines or in the Y intercepts of the lines. This is because individual regression lines are calculated for each group. The size of the sum of residuals for method 1 is affected only by the relationship of X to Y within each sample. If the relationship is perfect, there will be no errors of prediction, and the sum of residuals for method 1 will be zero. Can you figure out what the sum of residuals for method 1 will be if there is no relationship between X and Y within each group? In this case, each regression line will be a horizontal line through the mean of Y, and the sum of errors of prediction will be the same as the sum of deviations squared (Σy^2).

Now consider the sum of residuals for method 2. Here again, the extent of the relationship between X and Y within each group will affect the sum of residuals. However, in this case, the residuals are also affected by another factor. The lines are calculated while we are imposing the condition that all lines must have the same slope. Even when samples are from a single population, there will be sampling variability in the slopes of the regression lines calculated separately for each sample. This sampling variability in the slopes of regression lines adds to the residuals for method 2. The size of the sum of residuals for method 2 therefore depends on two factors: (1) the extent of the relationship between X and Y within each group and (2) the sampling variability in the slope of regression lines. Since the residuals for method 1 represent only the first factor, subtracting the residuals for method 1 from the residuals for method 2 provides a measure of sampling variability in the slopes of the regression lines. When divided by *df*, each of the factors represents error variance. These two error variances can be compared by means of an F ratio, as represented by Formula 13.1-1.

$$F_1 = \frac{\dfrac{res_2 - res_1}{k - 1}}{\dfrac{res_1}{n_T - 2k}} \text{ with } (k - 1) \text{ and } (n_T - 2k) \, df$$

When samples are from the same population or from populations with equal slopes, this F ratio can be expected to have a value near 1.00. If samples are from populations with different slopes for the regression lines, the numerator of the F ratio is no longer based on chance variability in the slopes of sample regression lines. Instead, the numerator reflects actual population differences, and the F ratio can be expected to be greater than 1.00.

Now consider the residuals for method 3. In this case a single regression line is found. In other words, both the slopes and the Y intercepts must be the same for the lines drawn for all groups. The sum of residuals is now affected by three factors: (1) the extent of the relationship between X and Y, (2) sampling variability in the slopes of sample regression lines, and (3) sampling variability in the Y intercepts of individual regression lines. Since the residuals for method 2 reflect the first two factors, subtracting these residuals from the residuals for method 3 provides a measure of sampling variability in Y intercepts. When divided by df, each of the factors again represents error variance. The variances can be compared by means of an F ratio, as represented by the formula given earlier as Formula 13.1-2.

$$F_2 = \frac{\dfrac{res_3 - res_2}{k - 1}}{\dfrac{res_2}{n_T - k - 1}} \text{ with } k - 1 \text{ and } n_T - k - 1 \, df.$$

When samples are from the same population, or from populations with equal slopes and equal Y intercepts, this F ratio can be expected to have a value near 1.00. If samples are from populations with equal slopes but different Y intercepts, the numerator is no longer based on chance variability in sample Y intercepts. Instead, the numerator reflects actual population differences, and the F ratio can be expected to be greater than 1.00.

ADDITIONAL APPLICATIONS OF ANCOVA

In this single lesson dealing with ANCOVA, we have described the major use of the technique, comparison of adjusted means. Our example has been one involving only a single covariate. In textbooks devoted entirely to ANOVA, ANCOVA, and regression analysis, the procedures of ANCOVA are shown to be applicable to a variety of additional situations. Most are beyond the scope of this book, including ANCOVA with multiple covariates and ANCOVA for curvilinear regressions. We will, however, present one additional use of ANCOVA that is based directly on what you have already learned.

Suppose that an experiment is conducted in which three treatments are being compared. The means for the dependent variable Y are related in the expected fashion, with $\bar{Y}_C$ highest, followed by $\bar{Y}_B$, followed by $\bar{Y}_A$. However, an analysis of variance results in an insignificant F. The experimenter knows that variable X is related to Y. If the three treatments groups have different means for variable X, you would suggest ANCOVA as a method of adjusting for these differences. But what about the case where the three groups have equal means for variable X? Would ANCOVA accomplish anything in this instance? Suppose that, when carrying out the process of ANCOVA for this problem, you decide that the regression lines have a similar slope. Can you visualize a figure similar to one of those shown earlier that describes this situation? The regression lines will all have the same slope, and the ovals representing the scatter diagrams will be positioned one directly above the other. Since all groups already have the same mean for X, adjustment along the regression lines will have no effect. Nevertheless, the use of ANCOVA will result in a more powerful test than the one-way ANOVA.

Your ability to understand the explanation for this last statement will provide a good test of how well you understand the basis for the computational methods of ANOVA and ANCOVA. In ANOVA, the total variance of Y is partitioned into variance between groups and variance within groups. The variance within groups represents unaccounted-for variance or error variance. This error variance is used as the denominator of the F ratio. Reducing the amount of unaccounted-for variance increases the power of the F test. This is the goal of a randomized-block experimental design.

Consider now the experiment we have just described. Is the variance of Y scores within groups all error variance? When talking about error variance, we are referring to differences on Y that cannot be explained. In an ANOVA, the variance within groups cannot be explained. In an ANCOVA there is, however, an explanation for part of the variance within groups. It is due to the individual differences in variable X. If this portion could be subtracted from the total variability within groups, the resulting difference would be the new error variance. What would this remaining unexplained variance be called? You have just used the concept extensively throughout this lesson. It is what we have called residuals. The greater the correlation between X and Y, the smaller the residuals, compared with $SS_{wgroups}$. Since ANCOVA uses an F ratio with residuals as the error term in the denominator, it will be a more powerful procedure than ANOVA, even when groups have equal means for the covariate. The relative increase in the power of the test will be directly related to the degree of correlation between the covariate and the dependent variable. You will have a chance to see this use of ANCOVA in one of the exercises that follow.

ASSUMPTIONS OF ANCOVA

Now that you have learned the advantages of ANCOVA, it is necessary to state the assumptions required for its use. The process that has been described is a combination of ANOVA and regression analysis, and it rests on the assumptions of these two techniques. These include the assumptions of normality of shape and homogeneity of variance, along with the assumption of a linear relationship between the covariate and the dependent variable. There is another assumption required if the interpretation of differences between adjusted means is to be uncomplicated. This is the assumption that measurements of the covariate have not been affected by the independent variable. Sometimes when ANCOVA is used, the covariate is measured as an afterthought following the experiment. If measurements of the covariate have changed as a result of experimental conditions, it will be difficult to interpret the meaning of the adjusted means.

Suppose that three diets are to be evaluated as methods of weight reduction. At the conclusion of the experiment, it is recognized that the amount of physical activity is an important factor in weight reduction under all three diets. Participants are therefore asked to provide information as to the amount of physical activity they engaged in weekly during the course of the experiment. This measurement is then used as a covariate to adjust average weight losses for the three diet groups. The problem is that physical activity may also be affected by each of the diets. It may not make sense to ask: "What would be the weight loss under diet *A* if the level of physical activity had been as high for those following diet *A* as for those following diet *B?*" The question is meaningless if energy level is so low for those following diet *A* that physical activity is reduced to a point considerably below the level of activity of those following diet *B*. The point to be made is that when ANCOVA is used to adjust means, the experimenter should always ask whether the independent variable has affected the measurements not only of the dependent variable but of the covariate as well. The usual solution is to make measurements of the covariate before or at the outset of the experiment.

One additional caution is necessary. Do you remember that in Chapter 12 you were warned about the danger of using a regression formula to make predictions based on predictor scores outside the range of predictor scores in the original regression study? The same warning applies to ANCOVA. Although means for the covariate differ from one group to another, it is expected that the score distributions for the covariate will overlap. Ideally, each will include the overall mean for the covariate. This is because adjusted means are actually regression predictions using the overall mean of the covariate as the value for the predictor variable. If the score distribution of a group does not extend to the overall mean, the prediction for the adjusted mean will be based on a predictor score outside the range of the predictor scores used to find the regression equation.

SUMMARY

In this lesson, you received an introduction to a complex technique of data analysis called analysis of covariance (ANCOVA). The usual application of the technique is for the case in which a variable that is related to the dependent variable cannot be controlled as part of the experimental design. Differences in this variable, the covariate, may obscure the true relationship among group means for the dependent variable. This situation is one that occurs frequently in research involving human subjects. In such research, groups are seldom based on random assignment. The result is that experimental groups have differing means on the covariate. The process of ANCOVA provides a way of statistically removing the effects of these differences among covariate means.

The process of ANCOVA involves two steps. The first is to determine whether there is a similar relationship between *X*, the covariate, and *Y*, the dependent variable, for each of the experimental groups. The test is to see whether the slopes of the regression lines are the same in each group. If the answer is "no," the process of ANCOVA is discontinued. If a common slope is a reasonable assumption, the second step is to determine whether the regression lines also have a common *Y* intercept. If this proves to be the case, a single regression line through all of the groups adequately represents the data. Adjusted means for Y would all be equal. However, if the data support the hypothesis of a common slope but different Y intercepts, this is evidence of differences among adjusted means for Y. The adjusted means can be thought of as the heights of the regression lines for each group at a point directly above the overall mean for X. The computations for finding adjusted group means and a way of making comparisons among them were described.

The last part of the lesson described a second, less common, application of ANCOVA in which the goal is not to adjust means of Y for differences in means of X. Rather, the goal is to reduce the estimate of error variance by subtracting from $SS_{wgroups}$ that part of the variance that can be explained by the relationship between X and Y. The result is a more powerful test for comparing the means of Y.

Finally, the lesson concluded with a section that described the assumptions of ANCOVA. It was explained that they are the same as those of ANOVA and regression analysis. For the usual application of ANOVA, there is the additional assumption that the covariate is not affected by the experimental conditions.

LESSON 1 EXERCISES

The scores for Y represent weight loss after 6 months for three diet programs. The scores for X represent minutes of weekly strenuous physical activity, estimated at the start of the experiment.

A		B		C	
X	Y	X	Y	X	Y
30	2	100	26	40	12
75	9	100	18	100	17
100	15	50	13	100	14
200	21	70	11	50	26
45	14	80	8	150	34
		40	5	60	14
		130	21	130	32
		150	31		

1. Perform a one-way ANOVA for the Y scores, using an alpha level of 0.10.
2. Perform an ANCOVA for the data, *even though the means of X are equal.* Use an alpha level of 0.10 for the second test, the test of equal Y intercepts. Make all possible pairwise comparisons, using an overall alpha level of 0.30. What do you conclude from a comparison of these results with those of exercise 1?
3. Now add 10 points to every X score for group B and 20 points to every X score for group C. Carry our an ANCOVA for the revised data. What has happened, and what is the explanation for the new result?

CHAPTER 14

Nonparametric statistics

Lesson 1 □ One- and two-sample tests

When first learning about hypothesis testing, you learned to make a one-sample test of the mean. The procedure was used to answer questions such as: "Is the mean for a group undergoing an experimental treatment different from the known mean under a traditional method of treatment?" The usual procedure is called a *one-sample t test of the mean*.

Later you learned to make a two-sample test of means, either for matched or independent samples. An example of a question being tested by this procedure would be: "Are the means for groups undergoing two experimental treatments different?" If samples are independently selected at random for each experimental condition, the usual procedure is called an *independent-sample t test for difference between means*. If subjects are matched before a member of each pair is randomly assigned to one of the experimental conditions, the procedure is called a *matched-pair t test for difference between means*.

For an experiment involving more than two experimental conditions, you learned to compare means by using ANOVA. The least complex case is an extension of the independent-sample *t* test. Subjects are independently assigned at random to three or more groups in a design involving a single variable. The analysis is called a *one-way ANOVA*. If subjects are matched on a variable such as age before being assigned to experimental groups, the design is an extension of the matched-pair *t* test. The analysis becomes a *randomized-block, two-way ANOVA*, with age as the second variable.

In each of the five conditions just described, the test of means depends on assumptions about the populations from which samples are chosen. One assumption is that the populations from which samples are chosen have normal distributions for the variable under consideration. A second assumption, for the case of two or more samples, is that the populations from which the samples are chosen have similar variances. Because the tests are based on these assumptions about population parameters, the tests are referred to as parametric tests. When these assumptions are unwarranted, use of an equivalent nonparametric test is required. This chapter describes at least one such test for each of the five conditions.

Because the tests described in this chapter are numerous, it is easy to become

Table 14-1. Nonparametric equivalents of parametric tests

Condition	One-sample tests	Two-sample tests		Multiple-sample tests	
		Independent samples	Matched samples	Independent samples	Matched samples
Parametric test	One-sample *t* test	Independent-sample *t* test	Matched-pair *t* test	One-way ANOVA	Randomized-block, two-way ANOVA
Nonparametric tests	1. Median test 2. Wilcoxon signed-rank median test	1. Median test 2. Mann-Whitney *U* test or Wilcoxon rank sum test	1. Sign test 2. Wilcoxon matched-pair signed-rank test	1. Median test 2. Kruskal-Wallis ANOVA by ranks test	1. Friedman's ANOVA by ranks test

Table 14-2. Sample pulse rates for nonparametric examples

i	A	B	C	D	E
1	56.3	54.3	51.3	52	44
2	62.3	60	53	49.3	46
3	97	82	86	81	74
4	86	89.3	75.6	77	64
5	81	76.3	69	65.3	54.3
6	77.3	70	62	58.3	53.6
7	92.6	85.6	82	69.6	72
8	94.3	79	83.3	78.3	71
9	83	88	71.3	58.3	61
10	71	66	55.3	57	48.6
11	91	84	79.6	75	67
12	78	74.3	65	72.6	72.3
13	84	78.6	72.6	64	63.6
14	93.3	91	82	78	69
15	80.6	73	67.3	70	68
16	92	80.3	81	76	68.3
17	88	87	77.3	73.6	66.6
18	75.3	78.6	61	50	51.6
19	94	85	85	80	73.6
20	74	70.3	59	55	49

confused as to which test is applicable to which situation. Table 14-1 will help to keep them all straight. The five conditions are listed as headings in the table. Below these, in the first section of the table, are the parametric tests just identified. The second section of the table provides the names of corresponding nonparametric tests. When more than one test is described, the tests are listed in the order of the complexity of the analysis. For example, for the one-sample condition, the median test is easier to carry out than the Wilcoxon signed-rank test. Unfortunately, the tests that are less complex are also general-ly less powerful. The tests will be described in the order and groupings listed in the table. The one- and two-sample tests are described in this first lesson, and the multiple-sample tests are described in the second. The goodness of fit test and the test of association using X_p^2 are also nonparametric tests. Since they have already been described in detail, we need not include them in this chapter.

The nine tests to be described are related in purpose. To make comparisons of the tests easier, a single data set will be used throughout the chapter. Table 14-2 contains five listings of 20 scores each. Each score listed represents an *average* of three measurements of individual pulse rates taken at different measuring stations. At each station, the individ-ual was asked to perform a different form of exercise just before the pulse rate was measured. Because some of the tests to be described are intended for correlated sets of scores, a positive correlation has been built into the data set. There is also a gradual decrease in the magnitude of the scores from column (variable) *A* to column *E*. The overall median of the 100 scores is 73.3. Because the distribution of scores within each listing is negatively skewed, ordinary tests involving means are inappropriate.

ONE-SAMPLE TESTS

Both of the nonparametric, one-sample tests to be described are tests that the median is some designated value in a test population. As an example, it might be known that for a large company, the median pulse rate of employees, measured under the condition described, is 89 beats per minute. The company decides to sponsor and evaluate a jogging program. There are 300 volunteers for the program. Twenty are chosen to participate in a pilot study. Variable A represents their scores for pulse rate before the program gets under way. It is believed that those who volunteer to participate in the program will, in general, be in better physical condition than those who do not volunteer. The null hypothesis to be tested is that the median pulse rate of the volunteers is $\geqslant 89$. Two ways of testing the hypothesis will be described. The first is called the one-sample median test; the second is the Wilcoxon signed-rank test.

The one-sample median test

One of the easiest nonparametric tests to carry out is the one-sample median test. For the example just described, if the null hypothesis is true, approximately half of all values in the sample should be equal to or greater than 89. The other half can be expected to be below 89. Looking at the scores of variable A in Table 14-2, we find that there are seven scores above 89 and 13 scores below 89. Can you anticipate how the goodness of fit application of X_p^2 can be used to test the null hypothesis that the median of volunteers $\geqslant 89$? The data can be placed in a table that includes expected values in parentheses.

Above 89	7 (10)
Below 89	13 (10)

Remember the calculations? Correcting for continuity, we get the following:

$$X_p^2 = \sum \frac{(|f_o - f_e| - 0.5)^2}{f_e}$$

$$X_p^2 = \frac{2.5^2}{10} + \frac{2.5^2}{10}$$

$$X_p^2 = \frac{6.25}{10} + \frac{6.25}{10} = \frac{12.5}{10} = 1.25 \text{ with } 1 \ df$$

Since there is 1 df, we can take the square root of 1.25 and evaluate the results as z from a table of the normal curve. We find that $z = 1.12$. Using the one-tail section of the table of the normal curve, we find that values this extreme in one direction can be expected to occur by chance about 13% of the time when the null hypothesis is true. We will probably decide that we do not have sufficient evidence to reject the null hypothesis that volunteers for the program have pulse rates as high as or higher than those of all employees.

Before we go to a description of the Wilcoxon signed-rank median test, it should be pointed out that the one-sample median test, as described here and in other texts, is a special case of a more general procedure. The choice of the median as the point at which to divide the population is arbitrary. The same procedure can be used to test the null hypothesis that the seventy-fifth percentile, or the ninetieth percentile, is some designated value. It is also arbitrary to divide the distribution of the population into only two parts by identifying the median. The distribution could be divided at each quartile, and a goodness of fit application of χ^2 could be used to test the null hypothesis that the distribution of a test population has quartiles equal to those of some standard population. The only limit is that the expected frequencies within resulting classes must be large enough for X_p^2 to be distributed as χ^2. If expected values are too small, exact probability distributions can be calculated. In the case of two classes, the familiar binomial distribution is used. For more than two classes, a polynomial distribution will be required.

The Wilcoxon signed-rank median test

Another one-sample test that the median equals some specified value has been provided by Wilcoxon. The Wilcoxon signed-rank median test uses more of the data than the median test and will generally be a more powerful test. The first step in the calculations involves subtracting the hypothesized median from each score in the sample. The second step is to rank-order the absolute value of the differences, with a rank of one assigned to the smallest absolute difference. In case of ties, average ranks are assigned using the procedure given in the description of Spearman's rank correlation coefficient in Lesson 2 of Chapter 12. If a score is exactly equal to the hypothesized median, resulting in a difference of zero, that score is dropped from the analysis and n is reduced by one. Once all absolute differences have been ranked, the original signs of the differences are restored to the ranks. If the null hypothesis is true, the *number* of positive- and negative-signed ranks should be equal, except for the effects of random sampling. The same is true for the *sum* of the signed ranks. Extreme values for the sum of signed ranks will constitute evidence against the null hypothesis.

Tables that provide probabilities for extreme values for various-size samples are provided in many textbooks. Students often become confused in attempting to interpret these and the numerous other tables provided for other nonparametric tests. As a result, this text will provide formulas that, for moderate-size samples, result in test statistics that can be evaluated using tables with which you are already familiar. For small samples, you can refer to a nonparametric textbook. In the case of the Wilcoxon signed-rank median test, the table used is that of the normal curve.

Table 14-3. Calculations for the Wilcoxon signed-rank median test

Scores	Score − 89	Rank	Signed rank	Fewer ranks
56.3	− 32.7	20	− 20	
62.3	− 26.7	19	− 19	
97	8	11.5	11.5	+ 11.5
86	− 3	3.5	− 3.5	
81	− 8	11.5	− 11.5	
77.3	− 11.7	15	− 15	
92.6	3.6	5	5	+ 5
94.3	5.3	9	9	+ 9
83	− 6	10	10	
71	− 18	18	− 18	
91	2	2	2	+ 2
78	− 11	14	− 14	
84	− 5	7.5	− 7.5	
93.3	4.3	6	6	+ 6
80.6	− 8.4	13	− 13	
92	3	3.5	3.5	+ 3.5
88	− 1	1	− 1	
75.3	− 13.7	16	− 16	
94	5	7.5	7.5	+ 7.5
74	− 15	17	− 17	
				44.5

$$z = \frac{T - [n(n + 1)/4]}{\sqrt{\dfrac{n(n + 1)(2n + 1)}{24}}}$$

$$z = \frac{44.5 - \dfrac{20(20 + 1)}{4}}{\sqrt{\dfrac{(20)(20 + 1)(40 + 1)}{24}}} = \frac{-60.5}{26.78} = -2.26$$

The sample statistic computed is T. To calculate T, first decide whether there are more positive- or more negative-signed ranks. Then find the sum of whichever set is less frequent. Take the absolute value of this total and call it T. It has been shown that when the null hypothesis is true and the sample has been taken from a population with the designated median, the sampling distribution of the statistic T is approximately normally distributed for samples of even moderate size (say 10 or more). The mean and standard error of the sampling distribution are as follows:

$$\text{mean}_T = n(n + 1)/4 \qquad\qquad \textbf{14.1-1}$$

and

$$SE_T = \sqrt{\frac{n(n + 1)(2n + 1)}{24}} \qquad\qquad \textbf{14.1-2}$$

The test statistic is therefore:

$$z = \frac{T - [n(n + 1)/4]}{\sqrt{\dfrac{n(n + 1)(2n + 1)}{24}}}$$

To test that the scores of variable A in Table 14-2 represent a random sample from a population in which the median $\geqslant 89$, we perform the calculations shown in Table 14-3.

Remember that we have decided to make a one-tail test. We expected the volunteers to have a lower pulse rate. Our null hypothesis is therefore that the median pulse rate for the population of volunteers $\geqslant 89$. Evidence against the null hypothesis that the median of the population $\geqslant 89$ occurs when the sample values are generally less than 89. Since 89 was subtracted from each score, evidence against the null hypothesis occurs when the majority of differences are negative. This was the case for our data, and we continued with calculations. If the majority of differences had been positive, we would not bother to calculate T for this example of a directional test.

The z score for the value of T that we calculated was -2.26. The negative value merely confirms that the sum of the positive ranks was less than expected based on the null hypothesis. Referring to the table of the normal curve, we find the one-tail probability to be 0.0119. At the 0.05 level, we would reject the null hypothesis and conclude that indeed the sample comes from a population in which the median is less than 89.

Notice the increase in the power of the Wilcoxon signed-rank median test over the ordinary median test. The calculations for the test for moderate-size samples are not too difficult. The test provides a good alternative to the one-sample test of the mean in cases where the assumption of normality in the population is questionable.

TWO-SAMPLE TESTS, INDEPENDENT SAMPLES

Information will be presented about two nonparametric tests applicable in the case where two populations are to be compared based on independent samples from each. The first is an extension of the median test; the second is a test generally referred to as the Mann-Whitney U test.

The two-sample median test

The null hypothesis for this test is that two populations have the same median. To test this hypothesis, a random sample is chosen from each population. The samples need not be of the same size, but each n should be at least ten. The two samples are then combined and a median is calculated for the combined samples. If the null hypothesis is true, this sample median will be the best estimate of the true median for the two populations. Except for sampling variations, half of the scores in each sample can be expected to be above this median and half below. A 2×2 table that includes expected frequencies is used to summarize the data. The statistic X_p^2, corrected for continuity, provides a test of the null hypothesis that the two populations have the same median.

To illustrate the two-sample median test, we use variables B and C from Table 14-2. The numbers listed for variable B represent scores of employees who have participated in the jogging program for 1 month. The scores of variable C are for employees who have been jogging for 2 months. The researcher would like to find evidence that the median pulse rate for employees who have jogged for 2 months is lower than that of employees who have jogged for 1 month. The directional null hypothesis is that either the medians are the same or that, if there is a difference, the *higher* median is for those who have jogged for 2 months.

After combining the two samples, we find that 20 scores are 76.3 or less and 20 scores are 77.3 or higher. Choosing a point halfway between these values, we conclude that if the null hypothesis is correct, our best estimate of the common median for the two populations is 76.8. For sample B, 12 scores are above 76.8 and 8 are below. For sample C, 8 are above and 12 are below.

A 2×2 table containing the observed and expected frequencies appears as follows:

Sample

	B	C
Above 76.8	12 (10)	8 (10)
Below 76.8	8 (10)	12 (10)

Since we intend to make a one-tail test, we first examine the table to see if differences are in a direction that represents evidence against the null hypothesis. We expect more scores to be below the common median in group C than in group B. Finding that this is so, we proceed to calculate X_p^2 corrected for continuity.

$$X_p^2 = \frac{1.5^2}{10} + \frac{1.5^2}{10} + \frac{1.5^2}{10} + \frac{1.5^2}{10} = 0.9 \text{ with } 1 \, df$$

Since X_p^2 is less than the *df*, we know that it is not significant even without consulting the table of χ^2. To find the P value for X_p^2, we obtain a z score by finding the square root of 0.9. The result, 0.95, is evaluated using a table of the normal curve. The table of the normal curve shows that the one-tail probability of a z score greater than 0.95 is approximately 0.17. We will probably conclude that there is not sufficient evidence to decide that the median pulse rate is lower for those who have been jogging for 2 months than it is for those who have been jogging for 1 month.

In the description of the one-sample median test, it was pointed out that the same procedure can be used when the distribution is divided at a point other than the median or at several points. This same generalization is appropriate for the procedure described for the two-sample median test. At an extreme, we may decide to test the null hypothesis that two populations have the same decile values. The deciles are found for the combined samples. If the null hypothesis is correct, approximately 10% of each sample can be expected to be below the first decile, another 10% below the second decile, and so forth. Again, the only restriction is that expected values must be large enough for $X_p{}^2$ to be distributed as χ^2. Carried to an extreme, this generalization of the median test becomes a test of the null hypothesis that the two populations have distributions that are the same throughout the entire range of the variable being considered.

The Wilcoxon rank sum test and the Mann-Whitney U test

The test to be described in this section is in fact a test of the null hypothesis that two populations have distributions that are the same. Two tests have been proposed for the case of independent samples: the Wilcoxon rank sum test and the Mann-Whitney U test. The calculations for the two tests differ, but the logic underlying the procedures is the same and the tests themselves are exactly equivalent.

For the procedure proposed by Wilcoxon, the two samples are combined and all scores are then ranked for the combined sample. The ranks identified with each sample are then totaled, resulting in T_1 and T_2. If there is no overlap between the distributions of the samples, T_1 and T_2 will be extreme values, since all of the low ranks will be associated with one sample and all the high ranks with the other. Any overlapping of distributions will result in less extreme values of T_1 and T_2. If the null hypothesis is correct and the two populations have the same distribution, the sample distributions can be expected to overlap. The scores for the two samples will be completely interspersed, and the sum of ranks for each sample will be of moderate size. The sum of ranks is therefore the test statistic, with extreme values constituting evidence against the null hypothesis. Tables containing probabilities for the sum of ranks for samples of varying sizes provide a means of deciding when results are extreme enough to reject the null hypothesis at a designated alpha level. Since the procedure proposed by Wilcoxon is equivalent to the Mann-Whitney U test, an example of the calculations will not be provided.

The Mann-Whitney U test is directly comparable to the Wilcoxon rank sum test and will provide identical results. It is also based on a measure of the extent of overlap between the distributions of two independent samples. Instead of calculating T, the sum of combined ranks for each sample, the test statistic calculated for each sample is U. Suppose that one sample is designated sample 1 and the other, sample 2. Every score of sample 1 is compared with every score of sample 2. The value of U_1 is the number of comparisons in which the score from sample 1 is greater than the score from sample 2. The total number of possible comparisons is $(n_1)(n_2)$. Therefore, once U has been calculated for one sample, the value of U for the other sample can be obtained by subtraction:

$$U_2 = (n_1)(n_2) - U_1$$

Table 14-4. Calculations for the Mann-Whitney U test

i	Score B	Score C	f(C$_i$ > B)*
1	54.6	51.3	0
2	60.0	53.0	0
3	66.0	86.0	16
4	70.0	75.6	7
5	70.3	69.0	3
6	73.0	62.0	2
7	74.3	82.0	12.5
8	76.3	83.3	13
9	78.6	71.3	5
10	78.6	55.3	1
11	79.0	79.6	11
12	80.3	65.0	2
13	82	72.6	5
14	84	82.0	12.5
15	85	67.3	3
16	85.6	81.0	12
17	87	77.3	8
18	88	61.0	2
19	89.3	85.0	14.5
20	91	59.0	1
			$U_C = 130.5$

$$\text{Mean}_U = (n_C)(n_B)/2 = (20)(20)/2 = 200$$

$$SE_U = \sqrt{n_C n_B (n_C + n_B + 1)/12}$$
$$SE_U = \sqrt{(20)(20)(41)/12}$$
$$SE_U = 37.0$$

$$z = \frac{U_C - \dfrac{(n_C)(n_B)}{2}}{\sqrt{n_C n_B (n_C + n_B + 1)/12}}$$

$$z = \frac{130.5 - 200}{37.0}$$

$$z = -69.5/37 = -1.88$$

*f(C$_i$ > B) is the number of scores from sample B that are less than a particular score from sample C.

Like the sum of ranks, the value of U will be extreme when there is little overlap between samples. If there is *no* overlap, so that all measurements for sample 1 are greater than the greatest measurement for sample 2, the measurements for sample 1 will exceed the measurements for sample 2 for all comparisons. In this case, $U_1 = (n_1)(n_2)$, and $U_2 = 0$. When the null hypothesis is true, the scores from the two samples will be interspersed. The measurement of sample 1 should be greater than the measurement of sample 2 in approximately half of the comparisons, and the measurement of sample 2 should be greater than the measurement of sample 1 in the other half of the comparisons. In other words, when the null hypothesis is true, both U_1 and U_2 should be near $(n_1)(n_2)/2$.

When samples are of moderate size (say 10 or more in each sample), the sampling distribution of U will be approximately normal, with a mean and a standard error as shown in Formulas 14.1-3 and 14.1-4.

$$\text{Mean}_U = (n_1)(n_2)/2 \qquad \textbf{14.1-3}$$

and

$$SE_U = \sqrt{n_1 n_2 (n_1 + n_2 + 1)/12} \qquad \textbf{14.1-4}$$

The normal curve can therefore be used to evaluate U.

$$z = \frac{U - (n_1)(n_2)/2}{\sqrt{n_1 n_2 (n_1 + n_2 + 1)/12}}$$

For two-tail tests, U can be calculated for either sample, and the z value can be compared with the two-tail section of the normal curve. The calculation of U will be easier, however, if U is calculated for the sample that has lower scores. For one-tail tests, the direction of the test will indicate which sample should have the smaller value of U. The first step is to calculate U for this sample. If this value of U is greater than $(n_1)(n_2)/2$, no further calculations need be made. If U is less than $(n_1)(n_2)/2$, the z score is again calculated and evaluated using the one-tail section of the normal curve. For the case where sample sizes are small, exact probabilities have been calculated and are given in nonparametric statistics textbooks.

The calculations for the Mann-Whitney U test with variable B as sample 1 and variable C as sample 2 appear in Table 14-4. Since we expect the pulse rates to be lower for those who have jogged for 2 months, we calculate U_C. We wish to make a one-tail test. We first check to see that U_C is less than $(n_B)(n_C)/2$. Finding that this is so, we calculate SE_U and then z. The z value of 1.88 is compared with the one-tail section of the normal curve. We find that values of U_C this extreme in this direction will occur only about 3% of the time when the samples are taken from populations with the same distributions. We will probably reject the null hypothesis and conclude that the pulse rates of men who have been in the jogging program for 2 months are lower than those of men who have been in the program for 1 month.

Notice two things about the calculations in Table 14-4. First, the scores for sample B (the sample with the higher scores) have been listed in order. This makes it easier to count the frequency with which each score for C exceeds a score for B. Notice also that when the comparison is a tie, 0.5 is added to the number of times the scores for C exceed the scores for B. For example, for individual 7, the value of $f(C_i > B)$ is given as 12.5. The score for C for this row is 82. Starting at the top of the column for B, we go down the column of ordered scores until the score is no longer less than 82. We get down to the thirteenth score. The C_i score equal to 82 exceeds 12 B scores. It is tied with one. The count is therefore 12.5.

You have now seen a worked-out example for the Mann-Whitney U test. As stated at the beginning of this section, the Wilcoxon rank sum test is an alternative approach, providing the same test and the same result. Either test provides a powerful alternative to the independent-sample t test for a difference between means. The next section will describe alternatives to the matched-pair t test.

TWO-SAMPLE TESTS, MATCHED SAMPLES

Suppose that the scores for B and C had been scores for the *same* 20 individuals, with B representing measurement made after 1 month and C representing measurements made after 2 months in the jogging program. In this case, a correlation between the scores of B and C would be expected. If the scores were normally distributed, a matched-pair t test would be used to analyze the data. In this section, two alternatives will be presented for nonparametric cases, the sign test and the Wilcoxon matched-pair signed-rank test.

The sign test

One of the nonparametric tests, which is very easy to use and understand, is the sign test. It is appropriate either in cases where subjects have been matched before assignment to experimental groups or in experiments in which there have been two measurements for each individual. The null hypothesis is that both sets of measurements come from the same population or from populations that have the same distribution. If the null hypothesis is true, then for any given pair of measurements, each measurement has a 0.5 chance of being the larger of the two. For the example that has been proposed, a true null hypothesis would mean that for any individual, the measurement of pulse rate at the end of 2 months would have a 0.5 chance of being less than the measurement of pulse rate made at the end of 1 month of jogging.

To test the null hypothesis, measurement C is subtracted from measurement B. If the null hypothesis is true, approximately half of the differences should be positive and half negative. Evidence against the null hypothesis is found when a large proportion of the differences are of the same sign. Either a one-tail or a two-tail test may be used. For small sample sizes, the binomial formula is the basis for the sampling distribution. For samples of ten or more, the goodness of fit application of χ^2 can be used.

For variables B and C, the score for C was less than the score for B 16 times, the two were tied once, and the score for B was higher 3 times. If the tie is ignored, subtracting C_i from B_i results in 16 positive and 3 negative results. We had anticipated that pulse rates would be lower after 2 months than they had been after 1 month. Finding that the majority of differences were in this direction is encouraging. We now need to determine the probability of this result's happening by chance. We could use the binomial formula to find the probability of exactly three, exactly two, exactly one, and exactly zero negative results. Adding these probabilities would provide the one-tail probability desired.

When the sample size is large enough for X_p^2 to be distributed as χ^2, the goodness of fit application of χ^2 provides another solution to the problem of finding the probability of our observed result. A table with observed and expected frequencies, along with the appropriate calculations, follows. Notice that the one tied pair of observations has been omitted from consideration.

Positive differences	16 (9.5)
Negative differences	3 (9.5)

$$X_p^2 = \sum \frac{(\,|f_o - f_e|\, - 0.5)^2}{f_e}$$

$$X_p^2 = \frac{(6)^2}{9.5} + \frac{(6)^2}{9.5}$$

$$X_p^2 = 7.58$$

$$z = \sqrt{X_p^2} = \sqrt{7.58} = 2.75$$

Referring to the one-tail section of the table of the normal curve, we find that the probability is approximately 0.004. The data provide strong evidence against the null hypothesis.

The Wilcoxon matched-pair signed-rank test

For the sign test, only the direction of differences is considered. A test that involves not only the direction of differences but also the magnitude of differences in each direction is the Wilcoxon matched-pair signed-rank test. As a result, it is a more powerful test than the sign test. Even though our example has already provided a highly significant result using the sign test, we will use the same data to illustrate the Wilcoxon matched-pair signed-rank test.

The first step, just as in the sign test, is to find the differences between pairs of scores. If a directional test is intended, we next check to see whether the data are in the appropriate direction. If so, or if a nondirectional test is to be carried out, the absolute value of the differences is then ranked. In the case of ties, average ranks are assigned. Once all absolute differences have been ranked, signs are restored to the ranks. The sum of ranks is then obtained for positive or negative ranks, whichever is less frequent. The absolute value of the result is called T.

If the null hypothesis is true, there will be approximately an equal number of positive and negative ranks, and the magnitude of ranks within each set will also be approximately equal. If the null hypothesis is false, the majority of differences will be of one sign. For the example we have used, if jogging reduces pulse rate, subtracting C_i from B_i should provide a majority of positive differences. Any negative differences should be small, resulting in a low rank. The sum of the few negative ranks should be small. A low value for T is therefore evidence against the null hypothesis.

For paired samples of size 10 or more, the sampling distribution of T will be approximately normally distributed, with a mean and standard error as given in the following formulas:

$$\text{mean}_T = n(n + 1)/4$$

$$SE_T = \sqrt{n(n + 1)(2n + 1)/24}$$

A standard score can therefore be computed and evaluated by reference to a table of the normal curve.

Does the explanation just given sound familiar? It should. Once the differences are obtained, the calculations are exactly the same as those described for the Wilcoxon signed-rank median test. For that test, the differences that were ranked resulted from subtracting the hypothesized median from all of the scores in a one-sample experiment. For the present test, the differences result from subtracting one member of a pair of scores from the other member in a matched-pair two-sample test. In either case, if the null hypothesis is true, the result should be a set of differences in which approximately half are positive and half are negative. The absolute value of the sum of the ranks of negative and positive differences should also be about the same.

Table 14-5. Calculations for the Wilcoxon matched-pair signed-rank test

i	B	C	B − C	Rank	Signed rank	Fewer signed rank
1	54.3	51.3	3.0	2	2	
2	60	53.0	7.0	9	9	
3	82	86	−4.0	4	−4	−4
4	89.3	75.6	13.7	17	17	
5	76.3	69	7.3	10	10	
6	70	62	8.0	11	11	
7	85.6	82	3.6	3	3	
8	79	83.3	−4.3	5	−5	−5
9	88	71.3	16.7	18	18	
10	66	55.3	10.7	15	15	
11	84	79.6	4.4	6	6	
12	74.3	65	9.3	13	13	
13	78.6	72.6	6.0	8	8	
14	91	82	9.0	12	12	
15	73	67.3	5.7	7	7	
16	80.3	81	−.7	1	−1	−1
17	87	77.3	9.7	14	14	
18	78.6	61	17.6	19	19	
19	85	85	0	xx	xx	
20	70.3	59	11.3	16	16	
						$T = 10$

$$\text{Expected } T = \frac{n(n + 1)}{4} = \frac{(19)(20)}{4} = 95$$

$$SE_T = \sqrt{(n)(n + 1)(2n + 1)/24} = \sqrt{(19)(20)(39)/24} = 24.85$$

$$z = \frac{10 - 95}{24.85} = 3.42$$

Table 14-5 contains the scores and calculations for our example, using the scores of *B* and *C* from Table 14-2. Notice that for the case of a tie for pair number 19, the difference was zero; that result was dropped from the analysis, leaving an *n* of 19 rather than 20. Notice also that although the ranks being summed were associated with negative differences, the value of *T* was considered positive.

The sample sizes are large enough for *T* to be approximately normal. Therefore, we calculate a *z* score for the test statistic *T*.

Referring to the one-tail section of the table of the normal curve, we find that the *P* value of our test statistic is approximately 0.0003. As expected, the result using the Wilcoxon matched-pair signed-rank test is even more extreme than the result found using the sign test. The test can be used as a powerful nonparametric alternative to the matched-pair *t* test. For moderate-size samples, the probability for the observed value of *T* can be obtained using the normal curve, as demonstrated. For smaller samples, probabilities will have to be obtained from special tables provided in textboks for nonparametric statistics.

We have now provided nonparametric alternatives to the one-sample *t* test, the independent two-sample t test, and the matched-pair t test. The next lesson provides descriptions of nonparametric alternatives to the one-way ANOVA and the randomized-block ANOVA.

SUMMARY

This lesson began with a description of five conditions in which the usual data analysis involves tests of sample means. A table summarizing these tests and listing corresponding nonparametric tests was then provided. It was explained that these nonparametric tests are used when assumptions about the parameters of the populations to be compared are unwarranted. The first lesson described one-sample tests and two-sample tests for both paired and unpaired data. The following tests were described.

The one-sample median test is based on the expectation that approximately half of the scores in a sample will be above the population median and half below. How well observed scores fit this expectation can be evaluated using the binomial distribution for small samples and the χ^2 goodness of fit test for large samples. The test may be generalized to values other than the median and to situations in which the distribution is to be divided into more than two classes.

The Wilcoxon signed-rank median test is also a one-sample median test; however, it makes use of more of the data than the median test and is therefore more powerful. The hypothesized median is subtracted from each score in the sample. Both the sign and the magnitude of the differences are considered. A majority of differences, either positive or negative, constitutes evidence against a nondirectional null hypothesis. The absolute value of the differences are ranked, and the sum of ranks is then obtained for either positive or negative ranks, whichever is less frequent. This sum of ranks, designated T, is approximately normally distributed for samples of size 10 or more. A formula for SE_T was provided, making possible the use of the table of the standard normal curve.

The two-sample median test is used to test the null hypothesis that two populations have the same median. The test is based on the expectation that approximately half of the scores in each sample will be above the median of the populations. The median of the combined samples provides the best estimate of the common median. Pearson's chi-square statistic, X_p^2, provides a method of evaluating the discrepancies between observed and expected frequencies above and below the common median within each sample.

The Wilcoxon rank sum test and the Mann-Whitney U test, which were developed independently, are equivalent tests of the null hypothesis that two populations have the same distribution. The procedure of the Mann-Whitney U test was described. In this test, every score from one sample is compared with every score from the other sample. The number of comparisons in which a score from sample 1 exceeds a score from sample 2 is called U_1. Extreme values for U occur when the score distributions of the two samples fail to overlap. For moderate-size samples, U has a sampling distribution that is normal. The formula for SE_U was provided, making possible the use of the standard normal curve to evaluate the probability of a U as extreme as the one observed.

The sign test, which is used for two matched samples, is an alternative to the matched-pair t test. The test is based on the expectation that when one number of a pair is subtracted from the other (say, $A_i - B_i$), approximately half of the differences will be positive and half negative. Descrepancies between the observed and expected frequencies for positive and negative differences can be evaluated using the binomial distribution for small samples and the χ^2 distribution for large samples.

The Wilcoxon matched-pair signed-rank test is also an alternative to the matched-pair t test. It uses more of the data than the sign test and is therefore more powerful. Once again, differences between A_i and B_i scores are obtained. If the null hypothesis is true, both the frequency and the magnitude of positive and negative differences should be approximately equal. Once differences have been found, the absolute values of the differences are ranked. The calculations then become identical to those for the Wilcoxon signed-rank median test.

LESSON 1 EXERCISES

1. Consider only the first ten scores of variable D from Table 14-2, and use the one-sample median test to test the null hypothesis that they represent a random sample taken from a population in which the median is 75.
2. Consider the same ten scores, and use the Wilcoxon signed-rank median test to test the same hypothesis.
3. Consider the first ten scores of variables C and D; use the two-sample median test to test the null hypothesis that the two samples come from populations with the same median.
4. Use the Mann-Whitney U to test the same hypothesis.
5. Consider the first ten scores of variables C and D to be matched samples. Use the sign test to test the null hypothesis that the samples come from populations having the same distribution.
6. Use the Wilcoxon matched-pair signed-rank test to test the same hypothesis.

Lesson 2 □ Multiple-sample tests

Generally, when a one- or two-sample nonparametric procedure is used, it is because there is reason to doubt that populations are normally distributed. In the multiple-sample case, the reason may be either a lack of normality within populations or a lack of homogeneity of variances among populations. In either case, the assumptions of ANOVA would be violated. This lesson describes nonparametric procedures that may be used as alternative methods of analysis. The first section relates to independent samples, the second to samples that have been matched before the experiment is conducted.

TESTS APPLICABLE TO MULTIPLE INDEPENDENT SAMPLES

In the last lesson, you learned how to use the two-sample median test and the Mann-Whitney U test as substitutes for the independent-sample t test for differences between means. Alternatives that can be used when the number of independent samples is greater than two will now be described. These may be thought of as nonparametric alternatives to the one-way ANOVA.

The multiple-sample median test

This section begins with a further generalization of the median test. It should be apparent that nothing in the procedure described for the two-sample median test prevents its application to experiments involving three or more samples. Once more, if the null hypothesis is true, and all samples come from populations with equal medians, approximately half of the observations for each sample will be above the common median and half will be below it. The best estimate of the common median will be the median of all samples combined. The extent of discrepancies between observed and expected frequencies will again be measured by the magnitude of $X_p{}^2$. If samples are large enough, $X_p{}^2$ will have a sampling distribution approximately like the χ^2 distribution.

As an example, suppose that variables A, B, and C in Table 14-2 of the last lesson represent pulse rates for three separate groups of men. For group A, the measurements were made before the men began a jogging program. Members of groups B and C had been jogging 1 and 2 months respectively when the measurements were made. We wish to test the null hypothesis that median pulse rates are equal for the populations sampled.

For the three sets of scores, the overall median is found to be 78.8. The 2×3 table, containing observed and expected frequencies, looks like the one on the following page.

	A	B	C
Above 78.8	13 (10)	10 (10)	7 (10)
Below 78.8	7 (10)	10 (10)	13 (10)

The statistic X_p^2 is calculated as follows:

$$X_p^2 = \sum \frac{(f_o - f_e)^2}{f_e}$$

$$X_p^2 = \frac{3^2}{10} + \frac{0^2}{10} + \frac{(-3)^2}{10} + \frac{(-3)^2}{10} + \frac{0^2}{10} + \frac{3^2}{10} = 3.6 \text{ with } 2\,df$$

Referring to the table of χ^2, we find that discrepancies this great can be expected to happen somewhere between 10% and 25% of the time when the samples come from populations having the same median. We therefore conclude that there is insufficient evidence to reject the null hypothesis.

Even though the results are not statistically significant, the data of the table containing observed and expected values seem to support our belief about what should happen. It may be that the median test was not powerful enough to find the differences that exist. We therefore decide to look at the same data using a more powerful test.

The Kruskal-Wallis ANOVA by ranks

A multiple-sample, independent-sample test that uses more of the data than the median test is the Kruskal-Wallis ANOVA by ranks. It is an extension of the two-sample Wilcoxon rank sum test or the Mann-Whitney U test. As with these tests, the first step is to combine samples and obtain an overall ranking of all scores. The rankings are then identified with the original samples, and a sum of ranks is obtained for each sample. Dividing the sum of ranks for each sample by sample size provides an average of the ranks for that group. Dividing the *total* sum of ranks by the *total* number of observations provides an average of *all* ranks.

When samples are from populations with identical distributions, the average rank for each group should be approximately the same as the average of all ranks. The greater the discrepancies, the greater the evidence against the null hypothesis. If you have followed the reasoning this far, you will recognize that all that is needed now is an index that summarizes the discrepancies, along with knowledge of the sampling distribution for this index. The index used is arrived at by squaring the discrepancy for each group, weighting the result for sample size, and then summing for all samples. If n_j represents the size of a sample, R_j the average rank for a sample, and R_T the average rank for all observations, the index described is:

$$H' = \sum_{j=1}^{k} n_j(\bar{R}_j - \bar{R}_T)^2$$

where k is the number of samples.

When H' is multiplied by a factor that weights the index for total sample size, the result is designated H. For large samples, H is distributed in approximately the same way as χ^2 with $k-1$ df. The weighting factor is $12/n_T(n_T+1)$ where n_T is the total number of scores. The formula for H is given as Formula 14.2-1.

$$H = \frac{12}{n_T(n_T+1)} \sum_{j=1}^{k} n_j(\bar{R}_j - \bar{R}_T)^2 \qquad \textbf{14.2-1}$$

Most texts provide a computational formula that gives the same result. I fail to see any real advantage to the computational formula, in terms of the calculations required, and I see disadvantages, in terms of making the rationale for calculating H understandable. However, here is the computational formula, which you may use as a reference when you read other texts.

$$H = \left[\frac{12}{n_T(n_T+1)} \sum_{j=1}^{k} \frac{T_j^2}{n_j} \right] - 3(n_T+1) \qquad \textbf{14.2-2}$$

where T_j = sum of ranks for group j.

The calculations for obtaining H for variables A, B, and C from the last lesson appear in Table 14-6. For purposes of comparison, the solution has been obtained using both formulas. Notice that variables A, B, and C have been listed in order. This has simplified the process of obtaining the overall rankings. Since it is easy to make a clerical error, it is fortunate that a check is possible. The total of the ranks for each group has been obtained by adding each column of ranks. The sum of these totals is T, the total of all ranks. The total of all ranks can also be obtained by the formula $T = n_T(n_T+1)/2$. Agreement between these two ways of obtaining T provides a check for the overall rankings.

The average of all ranks, $\bar{R}_T$, is 30.5. The average of ranks for variable A is highest at 38.95; the average for variable B is next, at 30.95; and the average for variable C is lowest, at 21.6. The magnitude of the differences between these values and $\bar{R}_T$ provides a method for testing the null hypothesis that the three samples come from populations with identical distributions. The test statistic H expresses these discrepancies in an index that, for samples of sufficient size, has a χ^2 distribution. The test statistic has been calculated at the bottom of Table 14-6, using the two different formulas. The results are the same, except for rounding errors. With $k-1$, or 2, df, the index H is significant at the 0.01 level. We will conclude that the null hypothesis is false and that pulse rates are different for the populations from which the samples have been drawn.

Table 14-6. Calculations for the Kruskal-Wallis ANOVA by ranks

i	A	B	C	R_A	R_B	R_C
1	56	54.6	51.3	5	3	1
2	62.3	60.0	53.0	10	7	2
3	71	66	55.3	17	12	4
4	74	70	59	21	15	6
5	75.3	70.3	61	23	16	8
6	77.3	73	62	26.5	20	9
7	78	74.3	65	28	22	11
8	80.6	76.3	67.3	34	25	13
9	81	78.6	69	35.5	29.5	14
10	83	78.6	71.3	40	29.5	18
11	84	79.0	72.6	42.5	31	19
12	86	80.3	75.6	47.5	33	24
13	88	82	77.3	50.5	38	26.5
14	91	84	79.6	53.5	42.5	32
15	92	85	81	55	44.5	35.5
16	92.6	85.6	82	56	46	38
17	93.3	87	82	57	49	38
18	94	88	83.3	58	50.5	41
19	94.3	89.3	85	59	52	44.5
20	97	91	86	60	53.5	47.5
				$T_A = 779$	$T_B = 619$	$T_C = 432$
				$\overline{R}_A = 38.95$	$\overline{R}_B = 30.95$	$\overline{R}_C = 21.6$

Total: all samples $= n_T = 60$

Total: all ranks $= T = \dfrac{n_T(n_T + 1)}{2} = \dfrac{(60)(61)}{2} = 1830$

Average: all ranks $= \dfrac{T}{n_T} = \dfrac{1830}{60} = 30.5$

Weighting factor $= \dfrac{12}{n_T(n_T + 1)} = \dfrac{12}{3660} = 0.00328$

Using Formula 14.2-1:

$$H = \frac{12}{n_T(n_T + 1)} \sum_{j=1}^{k} n_j(R_j = \overline{R}_T)^2$$

$H = 0.00328[20(38.95 - 30.5)^2 + 20(30.95 - 30.5)^2 + 20(21.6 - 30.5)^2] =$
$0.00328[1428.05 + 4.05 + 1584.2] = 0.00328(3016.3) = 9.89$ with 2 df

Using Formula 14.2-2:

$$H = \left[\frac{12}{n_T(n_T + 1)} \sum_{j=1}^{k} \frac{T_j^2}{n_j}\right] - 3(n_T + 1)$$

$H = [0.00328]\left[\dfrac{779^2}{20} + \dfrac{619^2}{20} + \dfrac{432^2}{20}\right] - 3(61) = (0.00328)(58831.3) - 183 =$

$192.97 - 183 = 9.97$ with 2 df

The effect of ties. In this example of the use of the Kruskal-Wallis ANOVA by ranks, there were seven instances in which two scores were tied. There was one case in which three scores were tied. The procedure described is based on the assumption of no ties. A modification in the formula to adjust for ties is possible. It involves dividing H by a correction factor. The correction factor is a decimal value that approaches one as the number of ties approaches zero.

The effect of dividing H by a decimal value is to increase H, thereby increasing the probability of finding a significant value. Unless sample sizes are small and the number of ties is numerous, the effect of the modification is of little consequence. For our example, the value of the correction factor is 0.9997, which will have virtually no effect on H. The modification of H to adjust for ties can be ignored for problems involving a total sample size of 20 or more. For smaller samples, be warned that the use of χ^2 to evaluate H will not be appropriate anyway. It will be necessary to refer to special tables in a textbook of nonparametric statistics. The formula that incorporates the correction factor can be obtained from the same source.

Individual comparisons. You will remember that when an F value was found to be significant in á one-way ANOVA, the analysis did not stop. The Scheffé procedure was used to find where the differences occur. A similar procedure of setting up confidence intervals for differences between average ranks is applicable when H is found to be significant. The procedure in such a case is similar to that described in conjunction with ANCOVA.

Again, the comparison is considered to be statistically significant when the confidence interval does not include zero. The procedure allows the experimenter to make all possible pairwise comparisons in a post hoc setting, while holding the overall alpha level at a designated value. In this last sentence the term *overall alpha level* refers to the probability that at least one of the possible pairwise comparisons will result in a Type I error.

To illustrate the procedure, let us compare the three values of $\overline{R}_J$ from Table 14-6. We decide to make all possible comparisons, while holding the overall alpha level to 0.05 or less. As you learned in Chapter 13, the total number of possible pairwise comparisons can be found by using Formula 13.1-4.

$$m = \frac{k(k-1)}{2}$$

where k is the number of groups.

For our example, $m = 3$.

Now that we have decided to calculate three confidence intervals, what level of confidence shall we use for each in order to hold the overall alpha level to 0.05? Individual confidence intervals cannot be set at 0.95. Instead, we will need to use some higher level. To find the appropriate individual confidence levels, we will use Formula 13.1-5.

$$\text{Individual confidence levels} = 1 - \frac{\text{overall alpha level}}{m}$$

where m is the number of comparisons made.

For the three comparisons we plan to make while holding the overall alpha level to 0.05, we get 0.9833 from Formula 13.1-5. For the first set of parentheses of the formula for confidence intervals, we will look for a tabled value that leaves $1 - 0.9833$, or 0.0167, in the tails of the distribution. When samples are of moderate size, the difference between any two sample values of $\bar{R}_j$ will be normally distributed. From Table E-6, in Appendix E, we find the value of 2.39.

The standard error of the difference between any two sample values of $\bar{R}_j$ can be calculated by Formula 14.2-3.

$$SE_{\bar{R}_1 - \bar{R}_2} = \sqrt{\frac{n_T(n_T + 1)}{12}\left(\frac{1}{n_1} + \frac{1}{n_2}\right)} \qquad \textbf{14.2-3}$$

where n_T is the total sample size.

For our example, $n_T = 60$, and each individual sample contains 20 observations.

$$SE_{\bar{R}_1 - \bar{R}_2} = \sqrt{\frac{60(61)}{12}\left(\frac{1}{20} + \frac{1}{20}\right)} = 5.52$$

For our example, in which samples are all of equal size, the confidence interval for each comparison is therefore calculated as follows:

Confidence interval = observed difference $\pm$ (z for a two-tail probability of 0.0167)($SE_{\bar{R}_1 - \bar{R}_2}$) = $(\bar{R}_1 - \bar{R}_2) \pm (2.39)(5.52) = \bar{R}_1 - \bar{R}_2 \pm 13.19$

The three confidence intervals are as follows:

$$\bar{R}_A - \bar{R}_B = (38.95 - 30.95) \pm 13.19 = -5.19 \text{ to } 21.19$$

$$\bar{R}_A - \bar{R}_C = (38.95 - 21.6) \pm 13.19 = 4.16 \text{ to } 30.54$$

$$\bar{R}_B - \bar{R}_C = (30.95 - 21.6) \pm 13.19 = -3.84 \text{ to } 22.54$$

Only the second confidence interval fails to include zero. With an overall alpha level of 0.05, we could only conclude that the pulse rates of men who have been jogging for 2 months are different from those of men who are about to begin the program. It should be mentioned that the procedure used was designed to hold the overall alpha level to 0.05 when all possible *nondirectional* comparisons are to be made. In our example, in which the comparisons would logically be directional, the procedure is overly conservative. To apply the procedure to our situation, it would be appropriate to use a z value corresponding to a *one-tail* probability of 0.0167. In our example, changing the z value in the relevant parentheses to 2.13 would not change our result, but the point needs to be considered in other applications.

You will find that often when the procedure just described is used, the overall alpha level is set at a value less extreme than that which is set in making the original evaluation of H. This is because guarding against any Type I errors for *many* comparisons amounts to setting up a more stringent criterion for finding statistical significance. An overall H, found to be significant at the 0.05 level, will often result in *no* significant individual comparisons being found if the procedure just described is used with an overall alpha level set at 0.05. Therefore, it is common to set the overall alpha level at some higher level, such as 0.15 or 0.20.

COMPARISON OF SEVERAL MATCHED SAMPLES

Suppose that the variables A, B, and C represent the pulse rates for a single group of 20 men. The pulse rates represented by variable A are measured before the men begin the jogging program; the pulse rates represented by variable B are measured after 1 month of participation; and the pulse rates represented by variable C are measured after 2 months of participation. The measurements are those originally reported in Table 14-2 of the last lesson. There is a high correlation between sets of scores.

The Friedman test for multiple matched groups provides a test of the null hypothesis that sets of correlated measurements are from identical distributions. It is an extension of the Wilcoxon matched-pair signed-rank test, and it corresponds to a randomized-block ANOVA. It is also directly related to the coefficient of concordance discussed previously as a measure of relationship.

In the Friedman test, the several scores for each individual are ranked. An average of ranks $(\bar{R}_j)$ is then obtained for each of the ranked measurements, in our case measurements A, B, and C. The overall average of ranks, $\bar{R}_T$, will be $(k + 1)/2$ (for our example, 2.0). If the null hypothesis is true, the values for $\bar{R}_j$ should be approximately the same as $\bar{R}_T$. The greater the discrepancies between observed values of $\bar{R}_j$ and $\bar{R}_T$, the greater the evidence against the null hypothesis. This time, the test statistic is given the symbol S. For large samples, S is distributed as χ^2 with $k - 1$ df. The formula used for the Friedman test is similar to the formula for H in the Kruskal-Wallis test.

$$S = \frac{12n}{k(k + 1)} \sum_{j=1}^{k} (\bar{R}_j - \bar{R}_T)^2 \qquad \text{14.2-4}$$

where n is the number of scores in a group,
   k is the number of groups,
   $\bar{R}_j$ is the average rank for a group, and
   $\bar{R}_T$ is the overall average rank.

There is also a computational formula.

$$S = \left[\frac{12}{nk(k + 1)} \sum_{j=1}^{k} T_j^2\right] - 3n(k + 1)$$

where T_j is the sum of ranks for a group.

Table 14-7. Calculations for the Friedman test for multiple, matched groups

i	A	B	C	R_A	R_B	R_C
1	56.3	54.3	51.3	3	2	1
2	62.3	60	53	3	2	1
3	97	82	86	3	1	2
4	86	89.3	75.6	2	3	1
5	81	76.3	69	3	2	1
6	77.3	70	62	3	2	1
7	92.6	85.6	82	3	2	1
8	94.3	79	83.3	3	1	2
9	83	88	71.3	2	3	1
10	71	66	55.3	3	2	1
11	91	84	79.6	3	2	1
12	78	74.3	65	3	2	1
13	84	78.6	72.6	3	2	1
14	93.3	91	82	3	2	1
15	80.6	73	67.3	3	2	1
16	92	80.3	81	3	1	2
17	88	87	77.3	3	2	1
18	75.3	78.6	61	2	3	1
19	94	85	85	3	1.5	1.5
20	74	70.3	59	3	2	1
				$T_A = 57$	$T_B = 39.5$	$T_C = 23.5$
				$\overline{R}_A = 2.85$	$\overline{R}_B = 1.975$	$\overline{R}_C = 1.175$

Number of groups = 3
Number in each group = n = 20

Total: all ranks = $T = \dfrac{nk(k+1)}{2} = 120$

Average: all ranks = $\overline{R}_T = \dfrac{k+1}{2} = 2$

Weighting factor = $\dfrac{12n}{k(k+1)} = 20$

$S = \dfrac{12n}{k(k+1)} \sum\limits_{j=1}^{k} (\overline{R}_j - \overline{R}_T)^2 = \dfrac{12(20)}{(3)(4)}\Sigma[(2.85-2)^2 + (1.975-2)^2 + (1.175-2)^2] = 28.08$ with 2 *df*

Table 14-7 contains the scores and calculations for Friedman's test for variables *A*, *B*, and *C*. Notice that the calculations for ranks can be checked by obtaining *T*, the total of all ranks, in two different ways. The sum of the totals at the bottom of columns R_A, R_B, and R_C is 120. The same value is obtained by the formula $nk(k+1)/2$. A discrepancy would have meant an error somewhere in the calculations.

The average of all ranks is $(k+1)/2$, or 2.0. The average for ranks for the three groups *A*, *B*, and *C* is in the direction anticipated: 2.85, 1.975, and 1.175 respectively. The magnitude of the discrepancies between these group averages and the overall average provides a method of testing the null hypothesis that the three samples come from populations with identical distributions. For large samples (say, ten or more), the test statistic S is distributed in approximately the same way as χ^2 with $k - 1$ df. The value 28.08 with 2 df is convincing evidence against the null hypothesis (probability less than 0.001). We will undoubtedly conclude that the samples do not come from populations in which the distributions of pulse rates are identical.

Notice that when the same set of data was analyzed using the Kruskal-Wallis test, the value to be evaluated as χ^2 was 9.9, compared with the value of 28.08 obtained using the Friedman test. The difference is that when the Kruskal-Wallis test was used, the correlation between variables A, B, and C was ignored. This is similar to what we do if we analyze data from a randomized-block ANOVA with a simple one-way ANOVA.

We can use a process similar to that used in the Kruskal-Wallis test to find which of the possible pairwise comparisons are significant. Before the calculations are shown, it should be noted that for individual 19, there was a tie for variables B and C. An average rank of 1.5 was assigned in each case. As with the Kruskal-Wallis test, in the case of ties an adjustment that increases the value of S is appropriate. The adjustment has little effect and can be ignored for moderate-size samples with few ties. For small samples, the χ^2 evaluation of S will not be appropriate, and the reader will have to refer to tables in a nonparametric statistics book.

To test which comparisons are significant, we will once more set a confidence interval for the difference between average ranks. For moderate-size samples, the difference will be normally distributed, with a standard error calculated using Formula 14.2-5.

$$SE_{\bar{R}_1 - \bar{R}_2} = \sqrt{\frac{k(k + 1)}{6n}} \qquad \textbf{14.2-5}$$

For our data:

$$SE_{\bar{R}_1 - \bar{R}_2} = \sqrt{\frac{(3)(4)}{(6)20}} = 0.316$$

We decide to make all possible comparisons, while holding the overall alpha level at 0.05. Using Formula 13.1-4, we find that the number of possible comparisons is again three. Each individual confidence interval will therefore be set at 0.9833 (see the explanation given with the Kruskal-Wallis test). The confidence intervals will each be based on the formula:

Observed difference $\pm$ (z for two-tail probability of 0.0167)($SE_{\bar{R}_1 - \bar{R}_2}$)

The confidence intervals are:

$\bar{R}_A - \bar{R}_B = (2.85 - 1.975) \pm (2.39)(0.316) = 0.875 \pm 0.755 = 0.12$ to 1.63

$\bar{R}_A - \bar{R}_C = (2.85 - 1.175) \pm (2.39)(0.316) = 1.675 \pm 0.755 = 0.92$ to 2.43

$\bar{R}_B - \bar{R}_C = (1.975 - 1.175) \pm (2.39)(0.316) = 0.80 \pm 0.755 = 0.045$ to 1.555

This time none of the confidence intervals includes zero. We would therefore conclude that pulse rates are reduced after 1 month in the program and that they are reduced even further after 2 months in the program. Once again, if directional comparisons were planned in advance, a value of z from the one-tail section of the normal curve would be appropriate.

THE USE OF NONPARAMETRIC TESTS

In this and the previous lesson, you have learned about a number of nonparametric tests that can be used as alternatives to corresponding parametric tests. Since it is usually suggested that they be used when the assumptions for the corresponding parametric test are subject to question, many students conclude that nonparametric tests are somehow not as "good" as parametric tests. They are often thought to be less scientific or less rigorous. Although a researcher may do a sign test as a quick method of comparing matched samples, he or she will often plan to do a matched-pair t test later for the publication of the study. The researcher then follows through with this plan, even though the sign test is significant at the 0.001 level.

It is wrong to think that nonparametric tests are somehow inferior to parametric tests. Often a nonparametric test, such as the sign test, provides an answer to the question of interest at a level of significance satisfactory to the investigator. The ease with which the calculations can be carried out and reported is a positive feature of the test, not a reason for feeling that something more should be done.

It is usually true that nonparametric tests are less powerful than the corresponding parametric tests. But even this distinction becomes less clear for tests such as the Mann-Whitney U test, the Kruskal-Wallis test, and the Friedman test. In any case, if the result using a nonparametric test turns out to be significant evidence against the null hypothesis, what is the advantage of using a parametric test solely because it is more powerful? If the more powerful parametric test is needed to identify a difference that exists, or if the parametric test provides a more specific test of the hypothesis, then of course it should be used. Otherwise, the nonparametric test is not only legitimate, it may in fact be preferable, since one need not worry about the characteristics of the populations being sampled.

SUMMARY

The second lesson concerning nonparameric statistics described several tests that are designed for experiments involving multiple groups. The techniques represent nonparametric alternatives to ANOVA procedures. The first section of the lesson described tests used for independent samples; the second section described a single test used for multiple matched samples. A procedure for making individual comparisons was also presented. The procedure is designed to hold the overall alpha level to a designated value while we make all possible pairwise comparisons. The procedure is similar to the method described in the chapter on ANCOVA. It is a conservative approach when used for planned comparisons.

The following tests were presented in the lesson.

The multiple-sample median test is a direct extension of the two-sample median test and uses the same procedure.

The Kruskal-Wallis ANOVA *by ranks test,* like the median test, is appropriate when we are comparing several populations in a situation in which samples have been drawn independently from each population. It is an extension of the two-sample Mann-Whitney *U* procedure or the Wilcoxon rank sum test. Samples are combined, and an overall ranking is obtained. The rankings are then identified with their original samples, and an average rank is calculated for each sample. An overall average rank is also calculated. The statistic *H* is used as an index of the magnitude of the discrepancies between the average ranks for samples and the overall average rank. It is evaluated using the table of χ^2.

The Friedman test for multiple matched groups is used when there are several groups of correlated scores. It is an extension of the Wilcoxon matched-pair signed-rank test and corresponds to a randomized-block ANOVA. The rationale is similar to that of the Kruskal-Wallis test, except that scores are ranked for each matched set of measurements. Once more, the averages of ranks for each variable are compared with the overall average rank. The index is called *S,* and it is evaluated using the table of χ^2.

LESSON 2 EXERCISES

1. Use the first 10 scores of variables *A* and *B* from Table 14-2 and all 20 scores of variable *C* to perform a three-sample median test.
2. Use all 20 scores of variable *C* and the first 10 scores of variables *D* and *E* to perform a Kruskal-Wallis ANOVA by ranks test. If the result is significant, make all possible comparisons while holding the overall alpha level to 0.20.
3. Use the first 10 scores of variables *A, B, C, D,* and *E* to perform a Friedman test for multiple matched groups. If the result is significant, make all possible comparisons that involve variable *A* while holding the overall alpha level to 0.20.

CHAPTER 15

Study design and interpretation of results

Lesson 1

You have now spent a great amount of time and energy learning how to compare populations by examining statistics computed for random samples chosen from those populations. The understanding of these principles of inferential statistics is a noteworthy achievement. It is important, however, to recognize the limitations of the knowledge you have gained. The limitations referred to do not relate to the extent of *your* introduction to statistics. You will find that this introduction is adequate for you to understand the statistical analyses of a large number of the studies reported in scientific literature.

Rather you should be cautioned about the meaning of the inference made in a statistical test and the common misinterpretations of this inference. There are two major problems. One stems from a misunderstanding of the nature of the inference. The second arises from a misunderstanding of the real nature of the treatment that distinguishes the test population. These statements contain many implications that require explanation. An example will clarify the problems.

Suppose that you wish to learn about the effectiveness of a remedial reading program. You begin by identifying all eighth-graders in Chicago public schools who are reading more than 1 year below their grade level. You then select a random sample of 100 of these students, who will take part in a special reading class using computerized self-instructional materials. After 6 months of participation in the program, you again evaluate the students' reading skills. You might set a confidence interval for the mean gain, or you might test the null hypothesis that the mean gain score is zero. In either case, you will be making an inference about the mean gain score resulting from the computerized instruction. Suppose that following a hypothesis test with an alpha level of 0.01, you reject the null hypothesis and conclude that the average change in reading level is greater than zero. You therefore recommend the computerized instruction for students who have difficulty with reading. Let us examine some potential problems with the validity of this recommendation.

366

MEANING OF THE INFERENCE MADE IN A STATISTICAL TEST

It is necessary to distinguish between two types of inferences in any analysis of the data from a research study—the statistical inference and the logical inference.

The statistical inference

In hypothesis testing, you decide whether the parameter value of the *test* population is different from the parameter value of the *standard* population. The test population is that population from which the random sample is drawn. This is a very restrictive statement. In the study just described, eighth-graders were chosen as the group to be used in testing the effectiveness of the program. Not all eighth-graders were included. The test population was made up of eighth-graders who: (1) attended Chicago public schools during the year of the study, and (2) were identified as reading at least 1 year below grade level. Your rejection of the null hypothesis says that you are convinced that if you had used the remedial program with all of *these* students, the mean gain score would have been greater than zero. The statistical inference does not extend to any other group.

The logical inference

If scientific studies were to end with the statistical inference, most results would be of interest only to historians. It is of little interest to know that the remedial program would have worked with the entire population of slow-reading eighth-graders in Chicago schools during the past year. Ideally, we want to know whether it will work during the coming year, for slow readers at various grade levels, for school systems throughout the United States. Such a generalization of the results is not a statistical inference from a sample statistic to a population parameter. It is a logical inference from the parameter of one population to the parameter of another. If you wish to make a statistical inference to slow readers in grades 5 through 8, you will have to select your random sample from this population.

The basis of logical inferences. You have spent a lot of time learning how to make a statistical inference only to be told that the statistical inference *by itself* may merely be of historical interest. The key words are *by itself*. No one expects the research study to stop with the statistical inference. You, as the investigator, and all of your readers are expected to generalize the results. The importance of making the correct statistical inference is in the fact that it will serve as the basis for all of the logical inferences that follow. You and many others will be saying, ''Since the treatment had the desired effect with this test population, it can also be expected to have the desired effect with these other populations.'' Of course, if you have made a Type I error and the treatment does not have the stated effect even in your test population, all of the logical inferences will be without basis.

Assuming that you have made the correct statistical inference, how do you decide to what other populations the results can be generalized? There is no exact answer, only a general principle. The general principle is that generalization from the test population to other populations is valid to the extent that other populations resemble the test population in all variables related to the outcome of the experiment. The application of this principle has two requirements: identification of those variables that relate to the outcome of the experiment, and knowledge of the distribution of these variables, both within the test population and within the additional populations to which you wish to generalize.

To decide whether to generalize our results from eighth-graders in the Chicago public schools to slow readers in the school system of a middle-class suburb, we must decide first which variables are likely to affect the outcome of the experiment. We will then compare the distributions of these variables within Chicago public schools with the distributions within the middle-class suburb. As an example, we probably will conclude that academic motivation is a factor affecting the outcome of this experiment. Therefore, if the distributions of the two populations differ greatly in the variable "academic motivation," the generalization will be risky. On the other hand, we will probably conclude that swimming skill has little to do with the outcome, and we will not be concerned if the two populations differ greatly in the distribution of swimming skills.

It is important to recognize that the process of making logical inferences is a joint venture. You and your readers will both participate in the process. Since someone else may not agree with you concerning all the variables that are relevant to the outcome, it becomes important for you to provide as much information as you can about the test population. Others will then be able to compare your test population with another population in which they are interested, even for variables that you might not recognize as important. This is the reason that research reports almost always include a paragraph headed "Subjects," which describes the persons included in the study. The more exact and complete your description of the test population, the more information you provide for a reader to use in making logical inferences.

NATURE OF THE TREATMENT

In an ideal one-sample scientific study, the test population is exactly like the standard population in all variables except for the treatment, and all of the details of the treatment are thoroughly explained. In two-sample experiments, the two populations being compared are alike in all variables except for the treatment. In either case, when a difference is found to be statistically significant, you will then be able to state that the difference observed is related to the treatment. Since the conclusions reached from a research report are based on the assumption that the treatment is the only variable that distinguishes the test population, problems of misinterpretation will occur either when the treatment is inadequately or inaccurately described or when the test population differs in ways other than the treatment. The reader will conclude that your treatment, *as you have described it*, is the explanation for any differences that were found to be significant.

DESCRIPTION OF THE TREATMENT

Failure to adequately and accurately describe the treatment is a major source of error in the interpretation of research results. Care in the early stages of planning the study can do much to eliminate this problem.

Every study should start by defining just what question the study is intended to answer. The choice of the question is made by the individual investigator. Some questions are worth investigating because the answer will affect a large number of people (a cure for the common cold). Other questions are important even though only a few persons are affected, because the effect of the answer is great for those few who are affected (a cure for a rare form of cancer).

Virtually every study involves the search for a relationship between two or more variables. In most studies, it is possible to identify one variable as the dependent variable. This is the one that the researcher hopes to learn more about or how to modify in a predictable fashion; for example, the study of the Salk vaccine was an attempt to reduce the incidence of polio. The incidence of polio was the dependent variable. There may also be one or more independent variables. These are variables that the researcher believes to be related in some way to changes in the dependent variable. A clear statement of the research question will indicate which is the dependent variable and which independent variables are being studied in hopes of finding a relationship.

One of the hardest tasks in the design of the study will be to define precisely all of the terms in the question in such a way as to prevent any possible misunderstanding about their meaning. The goal should be to describe each variable in terms so precise that someone else can merely read the study and be able to describe how he would make an exact replication of it. For example, in the study described earlier, the initial question might be, "What is the relationship between student participation in a remedial reading program and gains in reading scores?" The dependent variable is the reading score and the independent variable is participation in the remedial reading program. If the question were left in this form, each investigator setting up a study to examine the question would design a somewhat different study. The goal is to define all of the terms in the question so exactly that this would not happen. For example, the question might be made more specific in the following manner: "If eighth-grade children in the Chicago public schools who read more than 1 year below their grade level participate in the CRI (computerized reading instruction) remedial reading program for a 6-month period, will there be an average gain of more than 6 months in their reading levels?" It would be necessary to identify the test to be used in assessing reading level and to provide full descriptions of the CRI. Once this had been done, there would be good agreement as to how the study is to be carried out.

Often the question is stated in the form of a research hypothesis, which ends up as the alternate hypothesis in the statistical analyses. The research hypothesis is stated in the "If . . . , then . . ." form. The research hypothesis for our example, stated in this form, can be obtained by simply replacing the words *will there be* in the statement of the question with the words *then there will be*.

Even if the research question and hypothesis have been clearly stated and all terms have been clearly defined, there is still the possibility that the conclusion from the study will be in error, because some other variable becomes confounded with the treatment in the study design.

Variables confounding the treatment

As just explained, the true nature of the test population will be misrepresented if the treatment is inadequately or inaccurately described. It will also be misrepresented when the populations being compared differ in ways other than the treatment. In the case of the reading improvement program, all of the eighth-grade students are involved in many activities that might improve their reading skills. They are attending other classes, reading school materials and a wide range of other printed material, and engaging in other activities. They are even taking other tests that might improve their test-taking ability. All these activities are possible explanations for any gain in scores that is identified. In other words, at the end of 6 months, the students differ in many ways besides the treatment from the students who were identified at the start of the school year as slow readers. It is possible to identify several types of variables that can interfere with a pure measure of the effect of treatment. There are several types of study designs intended to eliminate these extraneous variables.

Types of study designs

It is not an easy task to plan a study so as to isolate the relationship that exists between the variables in which you are interested. Other forces get in the way when an effort is made to examine the relationship. As a very simple example, consider the following design, which is called an "after only with no control design."

Test refers to any measurement or observation made in an attempt to record the effect of the treatment. In this study design, a group of people is chosen and given a treatment; following this, group members are tested. For example, a group of volunteers who have high blood pressure is given a medication designed to reduce blood pressure. Even if testing following the treatment shows that the average blood pressure is only slightly above normal, there is no real evidence that the treatment has had an effect. It is possible that the group mean was only slightly above average before treatment. An "after only with no controls design" cannot provide meaningful results unless there is well-established prior data about how the subjects could have been expected to test without the treatment.

A "before and after with no controls" design may be graphically described as follows.

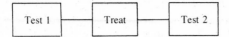

Here, a test is made before treatment and compared with a test after treatment to assess the effect of the treatment. The difficulty is that other events occurring at the same time as the treatment may contribute to any differences that exist between the results of test 1 and those of test 2. For example, having voluntered to participate in the study, the subjects may get interested in the causes of high blood pressure and may modify their eating and living habits during the course of treatment. The longer the period of treatment, the greater the possibility that extraneous factors will confound the effects of treatment using a "before and after with no controls design."

The "after only with one control" design avoids many of the difficulties of the two designs previously described.

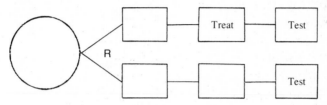

In this design, a population of potential subjects is identified. Those chosen for the study are randomly assigned to the control or the experimental group. The comparison is then between the test results of the experimental group and those of the control group. Notice that the groups can be expected to be alike initially and that any contemporary events might be expected to affect both groups similarly and, therefore, not affect the differences between test results for the two groups.

More complex designs are available that provide a measure for any possible effects of contemporary events.

EXPERIMENTAL VERSUS CORRELATIONAL STUDIES

A chapter such as this cannot be concluded without a few words about the difference between experimental and correlational studies. Independent variables may be treated in one of several ways. Ideally, independent variables are under the *direct manipulative control* of the researcher. This means that the researcher chooses a population of subjects and then decides which are to be members of the treatment group and which are to be members of the control group. This ability to directly manipulate the independent variable is required for a true *experimental* study. Since direct manipulation is not generally possible when working with human beings, the researcher often has to satisfy himself with *observation* of the independent variable. The result is a *correlational* study.

A great deal of caution is needed in interpreting the result of any correlational study.

For example, suppose that a study is to be made to determine the relationship between the risk of heart attacks and smoking. The dependent variable is the incidence of heart attacks and the independent variable is the past smoking history. Obviously, you cannot choose a population of people, randomly assign them to one of the two groups (smokers and nonsmokers), and then observe them for 30 years to compare the incidence of heart attacks among smokers and nonsmokers. Instead, you must be satisfied to do a correlational study and place persons in one of the two groups on the basis of their own reports of smoking history. The difficulty is that the groups will vary in a great many other characteristics as well as smoking history. Although you can sometime match for variables that are related to the experiment, you will never be able to match all of the variables that someone might believe to be related. The result is that correlational studies never give unequivocal evidence of a cause-and-effect relationship.

Notice the difference when a true experimental study is conducted. When the subjects are randomly assigned to the different groups, they will be alike, except for sampling variations in *all* variables. Since the statistical procedure will allow for such chance variations, initial differences between the groups will not be an alternative explanation for observed differences at the end. The experimental study that is carefully designed and carried out can therefore provide evidence of a causal relationship between the independent and dependent variables. The key words in the preceding statement are *can* and *carefully planned and carried out*.

In fields of public health, studies that are carried out as experimental studies are generally referred to as clinical trials. Their most common use is for testing the effects of newly developed drugs. Correlational studies are often referred to as epidemiological studies. A whole field of specialization has developed dealing with epidemiological investigations. Epidemiological studies may be either prospective or retrospective. In either case, their distinguishing feature is that the independent variable results from measurement rather than from direct experimental manipulation.

SUMMARY

This chapter emphasized the limitations of inferential statistics. It pointed out that the statistical inference is only to the population from which the sample has been randomly chosen. Logical inferences from this population to others that are similar are made both by the investigator and the readers of the report. It was pointed out that misinterpretation of study results is likely when the treatment is not clearly defined or when there is a misconception about the test population. The test population is supposed to differ from the standard population only with respect to treatment. If the treatment is not clearly and accurately described, or if the test population differs in other respects, misinterpretation will result. Avoiding these problems requires careful statement of the research question (including all needed definitions) and great care in the selection and implementation of a study design. The difficulty of determining causal relationships on the basis of correlational studies and the advantage of true experimental studies for this purpose were also pointed out.

Appendixes

Appendixes

APPENDIX A

Public health data and analyses

Lesson 1 □ Sources of public health data

Persons working in the fields of health benefit by having available to them data relating to a wide variety of problems and topics. These data are based on measurements that have been made in a standard fashion on large segments of the U.S. population. Just to illustrate, the following questions can all be answered by referring to data on tapes made available from the National Center for Health Statistics.

1. What was the average length of hospitalization for patients undergoing appendectomies in short-stay hospitals of the Middle Atlantic region for the year 1974?
2. What was the distribution of nursing homes in the United States in 1973-1974?
3. What percentage of babies born in the United States during 1972 were born to mothers who had made fewer than two prenatal visits to a doctor?
4. What percentage of the marriages taking place in Minnesota in the year 1971 involved partners more than 10 years apart in age?
5. What proportion of married women less than 45 years old were using oral contraceptives in the United States during 1973?
6. What were the types of anesthetics used during childbirth in the United States during 1972, what percentage of each was used, and what was the average length of labor associated with each?
7. Was there a difference between smokers and nonsmokers in the percentage of miscarriages among pregnant women in the United States in 1967?
8. What percentage of the population of each of the regions of the United States was covered by hospital insurance during 1974?
9. What was the average income, by occupation, for residents of the United States during 1960?
10. What was the normal growth patterns of children in the United States?
11. What are the dietary habits of diabetic persons in the United States?
12. What is the age distribution in each of the regions of the United States, and how did the distribution change between 1965 and 1975?

From this sampling, it can be seen that the data available are truly extensive. How

375

does such a data-gathering system operate, how good are the data, and where can the data be found? There are three basic systems for gathering data. One is a decennial census, the second consists of periodic national surveys, and the third is the system of registries maintained by the states.

DECENNIAL CENSUS

For centuries, governments have been attempting to maintain data on their populations through the use of national censuses. The first decennial census in the United States was conducted in 1790. The goal of the census is to enumerate every member of the population through contact with every household. The logistics involved in such an undertaking are staggering in a nation of over 200 million people scattered over more than 3½ million square miles. Add the requirement that the data be gathered at a single time, and the task becomes almost impossible. It is no wonder that, starting with the 1960 census, the switch from personal contacts to mailed questionnaires was initiated. The result is that our decennial census is now based on "self-enumeration" rather than direct interview. Our census also uses the "de jure" method as opposed to a "de facto" method of enumeration. Using the de jure method, the census counts people on the basis of where they normally live; under the de facto method, people are counted on the basis of where they are at the time the census takes place.

In addition to simply enumerating our population, the decennial census has become an occasion for gathering a large amount of demographic data on the populace of the United States. Certain pieces of information are required from every head of household (the short form), whereas other data are required only from carefully chosen samples of the total populace (the long form). At a minimum, each respondent is required to provide information about residence and to answer, for each member of the household, questions about relationship to the head of the household, age, sex, race, marital status, education, place of birth, native language, citizenship, occupation, employment, and military service status. Data on the long form change periodically and have occasionally provoked the public's anger over invasion of privacy. All persons receiving census questionnaires are expected to cooperate, and failing to do so is regarded as a serious matter.

The advent of high-speed computers has greatly enhanced the analysis of the data from the national census. It has also provided ready access to the data by interested persons. The U. S. Bureau of the Census publishes a large amount of summary census data in volume 1 of *Characteristics of the Population*. In addition, data tapes that contain much data not tabled in this publication can be purchased by legitimate users.

As with all measurements, one needs to be aware of the possibility of errors in the data reported for each census. The errors fall into three groups: (1) errors of coverage, (2) errors in content, and (3) sampling errors. Errors of coverage occur when a person is not counted or is counted twice. Certain groups of persons are more likely to go uncounted than others. In general, these groups include persons who move about frequently. The estimate of the number of persons missed in the 1970 census was less than 2% of the population.

Errors of content occur when the person contacted either unintentionally or deliberately provides incorrect responses to the questions. This can occur if questions are poorly designed or if they call for information that the respondent considers to be sensitive. The design of the questionnaires is done with great care to avoid misunderstanding. In fact, the questions that deal with general topics, such as age and occupation, are good models for persons constructing questionnaires that include demographic items. To seek to reassure persons that sensitive items on the questionnaire can be answered without any adverse effects, the Bureau of the Census acts as an independent agency and does not supply data on individuals to other agencies of the government.

In spite of all the efforts taken, there are errors of content. For example, the total age distribution of the population derived from census figures shows dips and bulges that can be associated with certain ages. Persons tend to report their ages near numbers like 25, 30, and 35, rather than their actual ages. Bulges also appear near specific retirement ages.

The third type of error relates to the samples taken for data not to be gathered from all members of the population. As we have learned, there is always some element of error involved when samples are chosen to represent a large population. The sampling errors in the national census are generally small because of the care with which the census is done and the large numbers involved even in the samples.

SURVEYS

In addition to the regular decennial census, the government also conducts extensive, periodic surveys of the population. Generally, these are sample surveys rather than contacts with the entire population. Each survey is for a specific area. Some of the surveys are:

The Current Population Survey
The Health Interview Survey
The Health Examination Survey
The Vital Records Follow-Back Surveys
Institutional Surveys
The Hospital Discharge Survey
The National Ambulatory Medical Care Survey
The Family Growth Survey
Special Epidemiological Surveys

As with the data from the census, an extensive amount of data is gathered, and only a limited amount is published in books, such as the Rainbow series of the National Center for Health Statistics. The raw data are stored on tapes and are available for purchase by legitimate users.

REGISTRIES

Unlike the census and the national surveys, which are static, the several registries are dynamic systems, which receive data on a continuous basis. The four vital registries are natality, mortality, marriage, and divorce. Standards for completeness of data must be met before data from a state can be included in the national registries.

Natality. Data on natality include all births covered by the birth registration system. All 50 states are included. Data for each birth are available on data tapes. The data are extensive, including information about the parents, the pregnancy, the delivery, and the child. Regular publications summarize the data by state.

Mortality. Data on mortality include all deaths recorded in the death registration system. All 50 states are included. Demographic and medical information is included for each death. Cause of death is classified in accordance with the *International Classification of Diseases, Adapted for Use in the United States* (Public Health Service Publication No. 1693). Death certificate numbers are not on the tapes.

Marriage. The marriage registry is incomplete, with only 41 states having met the requirement for inclusion by 1971. Demographic data such as state, age, race, sex, and previous marital status for each partner are included. Summaries of the data are published annually. Detailed data are available on tapes.

Divorce. The divorce registry is incomplete, with only 29 states having met the requirements for inclusion by 1971. Demographic data such as state, age, race, sex, and previous marital status for each partner are included. Summaries of the data are published annually. Detailed data are available on tapes.

INTERNATIONAL DATA

Data from censuses, surveys, and registries vary in completeness and accuracy from nation to nation. In general, data are less complete in developing countries. The United Nations and the World Health Organization work to standardize, evaluate, and report data.

PUBLICATIONS

In addition to the tapes that are available from government agencies, there are many printed sources of data. Primary sources are the publications of those agencies that themselves have the responsibility for gathering data. Secondary sources are publications from agencies that analyze and present data gathered by others.

LESSON 1 EXERCISE

Purpose: To become acquainted with sources of population and health statistics.

Choose a state of the United States having a first initial that is the same as the third letter of your last name. If none is available, go to the fouth letter, fifth letter, and so forth, of your last name or to the third letter of your first name. Give brief answers to the following questions concerning the chosen state, listing the source of your answers.

1. What is the size of the population as of 1970?
2. What is the rate of change in the size of the population?
3. What percentage of the population is 65 years old or older?
4. What percentage of the population is under 15 years old?
5. Compare the above percentages with the same percentages for the United States as a whole.
6. What is the infant mortality for the state? How does it compare with the infant mortality for the United States as a whole?
7. How many physicians per 1000 persons are there in the state?
8. What is the median family income for the state? Is this below the median family income for the United States as a whole?
9. What is the birth rate for the state? Is it increasing or decreasing? How does it compare with the birth rate for the United States as a whole?
10. Is the state in the divorce registration area? If so, what is the divorce rate per 1000 persons for the state, and what is its rate of change? Compare this with the equivalent statistics for the United States as a whole.

Lesson 2 □ Rates and their direct adjustment

In the remaining lessons of Appendix A, several techniques that have been developed for use in the health field will be discussed. This lesson will introduce the idea of rates and explain how rates can be adjusted (using the direct method) to make them more comparable from one population to another. The next lesson will present a method of rate adjustment (the indirect method) that is applicable for certain situations in which the data are incomplete. The following lesson will introduce the construction of ordinary life tables. The final lesson will describe the method for constructing follow-up life tables.

RATES

Suppose that you were charged with the responsibility of designing and implementing programs for improving the health of a specified population of people. Where would you begin? Probably the first logical step would be to assess the current state of health to identify problems that needed solution. You would want to know how the health of your population compares with similar groups and whether there are unusual health problems that require attention. For example, an unusually large percentage of persons in your population might suffer from lung diseases. How could you expect to discover this fact? You would need some means of comparing the incidence of lung disease in your population with the incidence of the disease in other, similar populations. To make this comparison, a common method of identifying lung diseases, a reliable reporting mechanism, and comparable methods of organizing the data are required. Since this is a book dealing with statistics, we will concentrate on the requirement of comparable data-organizing methods.

Demographers and epidemiologists make use of rates as a means of summarizing the experience of a population with regard to designated phenomena such as births, deaths, marriages, accidents, diseases, and cures. A rate such as miles per hour tells how many miles are completed during an hour of travel at the designated speed. In a similar fashion, a death rate reports the number of deaths occurring during a designated period of time. Obviously, for rates to have meaning from one population to another, the size of the population needs to be included when rates are reported. For example, a death rate is expressed in terms of deaths per thousand per year. The choice of the time period and the size of the population base (100, 1000, 100,000, etc) are arbitrary. However, in practice, a year is usually used as the length of the time period, and the number for the population base is generally chosen to be large enough so that the rate contains at least one whole number. We will look at some traditional rates shortly. First, here is a formula that represents a rate:

$$\text{Rate} = (D/P)(k) \qquad\qquad \textbf{A.2-1}$$

where   rate is the frequency of occurrence of the event of interest per k individuals in a defined population during a specified period of time,

D is the estimated total number of events occurring in the population during a specified period of time,

P is the estimated average size of the population at risk during the period (usually the estimated size of the population at midyear), and

k is the number of individuals to which the rate applies, that is, the population base (100, 1000, 100,000, etc.).

The data that are used for D come from registries that are continuously being updated. The data for P come from a census and may require interpolation, since the census data are only available for 10-year intervals. The choice of k, as stated earlier, is arbitrary.

When the figure for P in the formula represents the total population, the rate is referred to as a *crude* rate. If P is based on some segment of the population (such as persons between the ages of 20 and 29), the rate is called a *specific* rate. It is common to see rates that are age-specific, sex-specific, race-specific, or some combination, such as "deaths per thousand for 50- to 54-year-old men."

A distinction is often made between *incidence* and *prevalence* rates of diseases. Incidence reports the number of new cases over a period of time. In contrast, a prevalence rate reports the total number of cases, both old and new. Prevalance rates may be either point prevalence rates or period prevalence rates. Point prevalence rates indicate the number of cases existing at a point in time, whereas period prevalence rates indicate the total number of cases that existed during a period of time.

Here are some examples of common rates:

1. Maternal death rate

$$\frac{\text{Number of deaths from puerperal cases}}{\text{Number of live births during the year}} \times 1000$$

2. Neonatal mortality

$$\frac{\text{Number of deaths of children} < 28 \text{ days old}}{\text{Number of live births during the year}} \times 1000$$

3. Total infant mortality

$$\frac{\text{Number of deaths of children} < 1 \text{ year old}}{\text{Number of live births during the year}} \times 1000$$

4. Disease-specific mortality

$$\frac{\text{Number of deaths due to cancer}}{\text{Total population as of July 1}} \times 100,000$$

5. Crude birth rate

$$\frac{\text{Number of live births in a year}}{\text{Total population as of July 1}} \times 1000$$

6. Age-sex–specific birth rate

$$\frac{\text{Number of live births in a year}}{\text{Number of 15- to 45-year-old women as of July 1}} \times 1000$$

ADJUSTMENT OF RATES

A comparison of the crude rates in two population can be misleading when there is a variable that is related to the event in which you are interested and the distribution of this variable differs for the two populations. The simplest example is the crude death rate. Obviously, age is a variable that is related to mortality. If two populations have very different age distributions, one would expect their crude death rates to differ. The population with a larger proportion of older people often will have a higher crude death rate. But suppose you wanted to know whether there were other factors that were causing a difference in mortality in the two populations. You would want to answer the question, ''If the two populations had the same age distributions, what then would be the comparison in crude death rates?'' You would need a method of removing the effect of age differences in the two populations. The technique for accomplishing this is called adjustment of rates.

There are two methods of adjusting rates: the direct method and the indirect method. The direct method is, as the name implies, straightforward, and it will be described in the remainder of this lesson. The indirect method is a bit more difficult to understand. It will be presented in the next lesson. Both methods can be used in a wide variety of situations. The event being measured is not limited to deaths, but can be almost anything you wish to study, for example, births, marriages, divorces, or pregnancies. The related variable can also be almost any characteristic of the population to be studied, for example, age, sex, race, occupation, or years of schooling.

Direct adjustment

Suppose you want to compare two populations as to the rate of some phenomenon, such as death, pregnancy, or coronary attack. You know, however, that there is another variable that is related to the one you wish to study (for example, age or sex) and that the populations you are studying differ in this variable. Let us call one the study variable and the other the related variable.

Both populations are divided into categories or classes for the related variable (for example, age groups or sex classifications). The ideal situation is one in which both the number of persons in each category and the number of persons for which the phenomenon has occurred in each category are known. This makes it possible to use the direct method of adjustment. The data might appear as in Table A-1.

In the example shown in Table A-1, the phenomenon being investigated is death and the related variable is age. It is apparent that the two cities have different age distributions. As a result, comparing crude rates for the two cities (3.33 per 1000 vs. 3.70 per 1000)

Table A-1. Comparison of crude death rates in two cities

	City A		City B	
Age	Population number	Number of deaths	Population number	Number of deaths
20-29	45,000	90	30,000	45
30-39	40,000	120	30,000	105
40-49	35,000	140	40,000	180
50-59	30,000	150	50,000	225
TOTAL	150,000	500	150,000	555

Crude rate for city A = 500/150,000 × 1000 = 3.33 per 1000
Crude rate for city B = 555/150,000 × 1000 = 3.70 per 1000

might be deceptive. The question is how the crude rates would have compared if the age distributions of the two cities (study populations) had been the same.

Since this is the ideal situation, all the necessary data are available for adjusting the rates by the direct method. The general rationale of the direct method of adjustment is to choose some standard population and then answer the question, "If both populations I am comparing had the same distributions of the related variable as that of the standard population, what would their crude rates be?"

First, let us consider the choice of the standard population and then see how the question would be answered. The standard population can be any population you choose, so long as the distribution of the related variable is known. Usually, it is some larger population, of which the two populations to be compared are subpopulations. You might use the population of the United States as a standard when adjusting crude rates for the population of two states. Or you might use the population of a state as the standard when comparing two counties within a state. Sometimes, one of the populations of interest is used as the standard for the other population. In other instances, a hypothetical standard population is formed by combining the two populations being studied. For simplicity, this is the standard population that will be used for our example.

Once a standard population is chosen, how can we determine the crude rates that would have occurred if the distributions in our study populations had been similar to those of the standard population? The process involved is one in which the specific rate for each class of the frequency distribution is determined for each of the study populations and these specific rates are then applied to the corresponding classes of the standard population to get two sets of the expected number of events in the standard population. It sounds complicated when described in words but is not difficult when shown in an example, with each step described and illustrated.

For the example presented in Table A-1, the five steps in the direct adjustment of rates are as follows.

STEP 1   Choose a population with a known distribution of the related variable, age. (A standard population may be formed by combining the two test populations.)

STEP 2   Compute the specific rate for each category for each test population.

STEP 3   Apply the category-specific rates for each of the test populations to the standard popultion to get the number of expected deaths within each category for each test population.

STEP 4   For each test population, add the number of expected deaths for each category to get the total number of expected deaths.

STEP 5   Divide each total number of expected deaths by the total number of persons in the standard population to get the adjusted crude rates. These adjusted crude rates are the rates that would be expected in each of the test populations if they had had the same age distribution as the standard population. Since both rates are based on the same standard age distribution, they can be compared directly.

When a standard population is formed by combining the two study populations, the computations for the data in Table A-1 would be as shown in Table A-2.

The only computation not shown in Table A-2 is the method for arriving at expected deaths. For each class within each test population, the expected deaths are obtained by dividing the number in the standard population by the value of k used to find specific rates and then multiplying by the specific rate in the study population. In the example $k = 1000$. To illustrate the computations, look at the last age group. For the standard population, there are 80,000 people in the age group 50 to 59. Dividing by 1000, you get 80. Multiplying 80 by 5, you get 400 expected deaths in city A. Multiplying 80 by 4.5, you get 360 expected deaths in city B.

Notice that the crude rates for cities A and B were 3.33 per 1000 and 3.73 per 1000 respectively, whereas the adjusted crude rates are nearly identical. The relatively fewer numbers in the high-risk classes in city A resulted in a low *unadjusted* crude rate. As a generalization, if a study population has relatively fewer members in high-risk categories than does the standard population, the adjusted crude rate will be higher than the unad-

Table A-2. Comparison of adjusted crude death rates in two cities

Age	Number in standard population	City A Specific rate per 1000	City A Expected deaths	City B Specific rate per 1000	City B Expected deaths
20-29	75,000	$(90/45,000 \times 1000) = 2$	150	$(45/30,000 \times 1000) = 1.5$	112.5
30-39	70,000	$(120/40,000 \times 1000) = 3$	210	$(105/30,000 \times 1000) = 3.5$	245
40-49	75,000	$(140/35,000 \times 1000) = 4$	300	$(180/40,000 \times 1000) = 4.5$	337.5
50-59	80,000	$(150/30,000 \times 1000) = 5$	400	$(225/50,000 \times 1000) = 4.5$	360
TOTAL	300,000		1060		1055

Adjusted rate for city $A = 1060/300,000 \times 1000 = 3.53$ per 1000
Adjusted rate for city $B = 1055/300,000 \times 1000 = 3.52$ per 1000

justed crude rate. If there are relatively more members in the high-risk classes of the study population than in the same classes of the standard population, the adjusted crude rate will be lower.

The following exercise will give you practice in the direct adjustment of rates. In the next lesson, you will learn about the indirect method, which is used when specific rates are not obtainable for the study population.

LESSON 2 EXERCISE

The following data are available regarding pregnancies among high-school girls in two cities:

Age group	City A		City B	
	Number of female high-school students	Number of pregnancies	Number of female high-school students	Number of pregnancies
15	2500	55	4000	72
16	2300	92	3000	120
17	2200	121	2000	120
18	2100	210	1000	90
TOTAL	9100	478	10,000	402

1. What are the crude pregnancy rates among high-school girls in the two cities?
2. For this data set, what is the event being measured, what is the related variable, and which city has a larger proportion in high-risk categories? What would you expect to happen if you were to compute adjusted crude rates?
3. Using the standard population that results from combining the populations of cities A and B, use the direct method to arrive at adjusted rates for each city.
4. What do you conclude about the rates of pregnancies among high-school girls in cities A and B?

Lesson 3 □ Indirect adjustment of rates

To produce standardized or adjusted rates by the direct method for each of several populations being compared, two things are necessary: a standard population and category-specific rates for each of the populations being compared. Often, however, category-specific rates are not available for each of the populations, and the direct method cannot be used. In other instances, the population of interest is so small that when it is divided into classes of the related variable, the number of events in each class is extremely small. In this event, random fluctuations cause specific rates to be unreliable. This situation also calls for the indirect method of adjusting rates. Notice that there is no difference in purpose between direct and indirect adjusting of rates. The goal is still to allow the comparison of the rates of some phenomenon in two populations that have different distributions of a related variable.

When the direct method of adjustment is used in a study of mortality, for example, specific rates from the study population are applied to the standard population to get the number of expected deaths. When the indirect method is used, the process is reversed. Specific rates from the standard population are applied to each category of the study populations to get the expected number of deaths in each study population. A comparison between the total number of expected deaths and the actual number of deaths in each study population will then give an indication of the *forces of mortality* in each study population, compared with those forces in the standard population. This comparison is made by forming a ratio, with the actual number of deaths in the numerator and the expected number in the denominator. The value of this ratio is called the *standard mortality ratio (SMR)* and is often the desired end product of the indirect method. If the *SMR* is greater than one, it indicates that forces of mortality in the study population exceed those in the standard population (with the effects of the related variable removed). If the *SMR* is less than one, forces of mortality are less in the study population than in the standard population.

It is also possible to continue the computations to arrive at an adjusted crude rate, as was done when we used the direct method. There is a long way and a short way of doing this. The short way is to multiply the *crude* rate in the *standard* population by the *SMR* for the study population to get the adjusted crude rate for the study population. The long way produces two interim results that are themselves meaningful: the *index rate* and the *handicap factor*. In the worked-out example, the long way will be illustrated in the steps that are outlined.

Before we go into an example of the indirect method, it should be noted that often the indirect method is used when a small subpopulation is to be compared with some larger population. For example, a county or township may be compared with an entire state. In this instance, the large population is used as the standard population, and there is only a

Table A-3. Examples for indirect adjustment of rates

Smoking history	Coffee drinkers			Nondrinkers of coffee		
	n	Number of deaths	Specific rate	n	Observed number of deaths	Expected number of deaths
<Pack per week	10,000	10	1 per 1000	4000	?	4
<Pack per day	30,000	60	2 per 1000	3000	?	6
<Two packs per day	40,000	120	3 per 1000	2000	?	6
>Two packs per day	10,000	40	4 per 1000	1000	?	4
TOTAL	90,000	230		10,000	18	20

single study population. We will use this as the situation for our example of the indirect method.

To repeat, the requirements for this indirect method are: (1) a standard population with a known distribution of the related variable and a known number of events within each category, and (2) a study population with a known population distribution for the related variable and a known total number of events experienced by the study population. Again, this is seen more easily by examining the procedure step by step and following a worked-out example. The steps are as follows.

STEP 1 Compute the crude rates for the standard population and for each of the study populations.

STEP 2 Compute the specific rates for the standard population.

STEP 3 Apply the specific rates from the standard population to each category of the study populations to get the expected deaths for each category.

STEP 4 For each study population, add the expected deaths in each category to get the total number of expected deaths.

STEP 5 Divide the total observed deaths in each study population by each total of expected deaths to get the *SMR*'s.

STEP 6 Divide the total expected deaths in each study population by the number in each study population, and multiply by an appropriate constant *(k)*. This is called the *index rate*.

STEP 7 Divide the crude rate in the standard population by each index rate to get the *handicap factor* for each study population.

STEP 8 Multiply the crude rate in each study population by the handicap factor to get the adjusted crude rates.
Note: A shorter method is to simply multiply the crude rate in the standard population by each *SMR* to get adjusted crude rates for the study populations.

The data and computations might appear as shown in Table A-3. The two populations are coffee drinkers and nondrinkers. The related variable is smoking history. In this example, as is often true, a single study population (nondrinkers of coffee) is compared with a standard population (coffee drinkers) that is much larger. The study population is so small that specific rates computed from observed deaths within each category would be unreliable. It is known, however, that the total number of deaths among nondrinkers of coffee is 18. The steps taken in calculating the rates adjusted by the indirect method are as follows.

STEP 1   Crude rate (standard population) $= \dfrac{\text{total deaths}}{\text{total population}} \times k = \dfrac{230}{90,000} \times 1000 = 2.55$

Crude rate (study population) $= \dfrac{\text{total deaths}}{\text{total population}} \times k = \dfrac{18}{10,000} \times 1000 = 1.80$

STEP 2   Specific rates (standard population) $= \dfrac{\text{number of deaths}}{\text{population}} \times k$

$= \dfrac{10}{10,000} \times 1000 = 1 \text{ per } 1000$

$= \dfrac{60}{30,000} \times 1000 = 2 \text{ per } 1000$

$= \dfrac{120}{40,000} \times 1000 = 3 \text{ per } 1000$

$= \dfrac{40}{10,000} \times 1000 = 4 \text{ per } 1000$

STEP 3   Expected deaths (study population) $= (\text{rate})(\text{number})/k$
$= (1)(4000)/1000 = 4$
$= (2)(3000)/1000 = 6$
$= (3)(2000)/1000 = 6$
$= (4)(1000)/1000 = 4$

STEP 4   Total expected deaths $= 4 + 6 + 6 + 4 = 20$

STEP 5 $SMR = \dfrac{\text{observed deaths in study population}}{\text{expected deaths in study population}} = \dfrac{18}{20} = 0.9$

STEP 6   Index rate $= \dfrac{\text{expected deaths in study population}}{\text{number study population}} \times k = \dfrac{20}{10,000} \times 1000 = 2.0$

STEP 7   Handicap factor $= \dfrac{\text{crude rate in standard population}}{\text{index rate}} = \dfrac{2.55}{2} = 1.275$

STEP 8   Adjusted rate $=$ (crude rate in study population)(handicap factor) $= (1.8)(1.275) = 2.295$ or, alternatively: Adjusted rate $=$ (crude rate in standard population)$(SMR) = (2.55)(0.9) = 2.295$

Since this process of the indirect adjustment of rates is rather involved, it is worthwhile to summarize the meaning of the *SMR*, the index rate, the handicap factor, and the adjusted crude rate. Think about each of the following paragraphs and see if you can understand why each of the statements makes sense.

The *SMR* indicates how forces of mortality compare between each *study* population and the *standard* population. If the forces are the same, the *SMR* = 1.00. If the forces of mortality are greater in the study population, the *SMR* is greater than 1.00. If the forces of mortality are less in the study population, the *SMR* is less than 1.00.

The index rate indicates what the crude rate would be in the study population if the forces of mortality were the same in the study population as in the standard population. Also, the index rate would equal the crude rate of the standard population if the standard population and study population have the same distributions. However, an adjustment would not be done in such a case. Why?

The handicap factor indicates how the distribution of the study population compares with the distribution of the standard population. If the two distributions are identical, the handicap factor = 1.00. If a larger proportion in the study population is in high-risk categories, the handicap factor will be less than 1.00. If a larger proportion is in low-risk categories, the handicap factor will be greater than 1.00.

The adjusted crude rate indicates what the crude rate would be in the study population if the distribution of the related variable were the same in the study population as in the standard population.

You will now have an exercise similar to that of the preceding lesson, except that this time an indirect adjustment is called for. Then, in the next lesson, you will learn about life tables and their construction.

LESSON 3 EXERCISE

The following data are available for city *C*.

	Population and pregnancy data	
Age group	Female student population	Number of pregnancies
15	1000	↑
16	1000	Unknown
17	1000	
18	2000	↓
TOTAL	5000	300

1. What is the crude pregnancy rate for female students in city *C?*
2. Use City *A* from the exercise in the last lesson as the standard population. Which city has proportionately more students in high-risk status? Do you expect the handicap factor to be greater or less than 1.00?
3. Use the indirect method to get an adjusted crude rate for city *C*. Do this both ways and indicate the meaning of the *SMR*, the index rate, and the handicap factor.
4. What do you conclude about the rate of pregnancy in city *C* in comparison to the rate of pregnancy in the standard population (city *A*)?

Lesson 4 □ Ordinary life tables

A life table is a construct used primarily for measuring various functions of mortality, such as expected length of life remaining at various ages and probability of survival during a specified length of time. In addition, it has proved to be a useful device to actuaries, statisticians, demographers, and health scientists in studying phenomena other than mortality, such as fertility, migration, population growth, working life, and many others. A detailed study of life table uses and methods is included in books dealing with demography. We will merely skim the surface in this book.

A life table for the total population of the United States appears as Table A-4. As we will see shortly, there are two methods of constructing a life table: current versus cohort. The one provided as Table A-4 is a current life table and is based on information from the 1960 census and from death registries for the corresponding period. In Table A-4, column (6), with the symbol $\overset{o}{e}_x$ at the head, is the one of interest. This column gives the average life expectancy for persons who are alive at each specified age. As examples, a person reaching his or her tenth birthday is said to have a life expectancy of 62.19 more years; on his or her sixtieth birthday, a person's life expectancy is 17.71 more years. The table shown is for the total population. Specific life tables are also available by race, sex, and other categories. A complete life table gives the data for each year. There are also abridged life tables, which use periods of 5 years or more as the basis for classes.

Before going into the methods of construction of life tables, let us pause to examine each of the other columns in Table A-4. The columns have both numbers and symbols. The symbols are the ones traditionally used, so we will also make use of them in our examples. The first column gives the classes for the frequency distribution. Following our suggestion in the lesson on frequency distributions, we would eliminate the second number describing each class. Do you remember why?

Column (1) has the symbol $_nq_x$. It lists the probability of a person's dying during period x of length n. Column (2) is labeled l_x. It corresponds to the frequency column in a frequency distribution and reports the number of persons alive at the start of period x. Column (3) has the symbol $_nd_x$. It tells the number dying during period x of length n. Column (4) has the heading $_nL_x$. It is a bit harder to understand. It is intended to represent the total number of years of life experienced during the period by all those who survive to start the period. Column (5), labeled T_x, represents the total number of years experienced both within period x and in all subsequent periods. We have already discussed the last column. The meaning of each of the columns will become clearer as you learn how the entries are computed.

Table A-4. Complete life table for the total population of the United States: 1959-61

Age interval	Proportion dying	Of 100,000 born alive		Stationary population		Average remaining lifetime
Period of life between two exact ages stated in years (x to x + n)	Proportion of persons alive at beginning of age interval dying during interval ($_nq_x$)	Number living at beginning of age interval (l_x)	Number dying during age interval ($_nd_x$)	In the age interval ($_nL_x$)	In this and all subsequent age intervals (T_x)	Average number of years of life remaining at beginning of age interval ($\overset{o}{e}_x$)
	(1)	(2)	(3)	(4)	(5)	(6)
0-1	.02593	100,000	2,593	97,815	6,989,030	69.89
1-2	.00170	97,407	165	97,324	6,891,215	70.75
2-3	.00104	97,242	101	97,192	6,793,891	69.87
3-4	.00080	97,141	78	97,102	6,696,699	68.94
4-5	.00067	97,063	65	97,031	6,599,597	67.99
5-6	.00059	96,998	57	96,969	6,502,566	67.04
6-7	.00052	96,941	50	96,916	6,405,597	66.08
7-8	.00047	96,891	46	96,868	6,308,681	65.11
8-9	.00043	96,845	42	96,824	6,211,813	64.14
9-10	.00039	96,803	38	96,784	6,114,989	63.17
10-11	.00037	96,765	36	96,747	6,018,205	62.19
11-12	.00037	96,729	36	96,711	5,921,458	61.22
12-13	.00040	96,693	39	96,674	5,824,747	60.24
13-14	.00048	96,654	46	96,630	5,728,073	59.26
14-15	.00059	96,608	57	96,580	5,631,443	58.29
15-16	.00071	96,551	68	96,517	5,534,863	57.33
16-17	.00082	96,483	80	96,443	5,438,346	56.37
17-18	.00093	96,403	89	96,358	5,341,903	55.41
18-19	.00102	96,314	98	96,265	5,245,545	54.46
19-20	.00108	96,216	105	96,163	5,149,280	53.52
20-21	.00115	96,111	110	96,056	5,053,117	52.58
21-22	.00122	96,001	118	95,942	4,957,061	51.64
22-23	.00127	95,883	122	95,822	4,861,119	50.70
23-24	.00128	95,761	123	95,700	4,765,297	49.76
24-25	.00127	95,638	121	95,578	4,669,597	48.83
25-26	.00126	95,517	120	95,456	4,574,019	47.89
26-27	.00125	95,397	120	95,337	4,478,563	46.95
27-28	.00126	95,277	120	95,217	4,383,226	46.00
28-29	.00130	95,157	123	95,095	4,288,009	45.06
29-30	.00136	95,034	129	94,970	4,192,914	44.12
30-31	.00143	94,905	136	94,836	4,097,944	43.18
31-32	.00151	94,769	143	94,698	4,003,108	42.24
32-33	.00160	94,626	151	94,551	3,908,410	41.30

Source: U.S. National Center for Health Statistics. *Life Tables. 1959-61*, vol. 1, no. 1, "United States Life Tables: 1959-61," December 1964, pp. 8-9.

Continued.

Table A-4. Complete life table for the total population of the United States: 1959-61—cont'd

Age interval	Proportion dying	Of 100,000 born alive		Stationary population		Average remaining lifetime
Period of life between two exact ages stated in years (x to x + n)	Proportion of persons alive at beginning of age interval dying during interval ($_nq_x$)	Number living at beginning of age interval (l_x)	Number dying during age interval ($_nd_x$)	In the age interval ($_nL_x$)	In this and all subsequent age intervals (T_x)	Average number of years of life remaining at beginning of age interval ($\overset{o}{e}_x$)
	(1)	(2)	(3)	(4)	(5)	(6)
33-34	.00170	94,475	160	94,395	3,813,859	40.37
34-35	.00181	94,315	171	94,229	3,719,464	39.44
35-36	.00194	94,144	183	94,053	3,625,235	38.51
36-37	.00209	93,961	196	93,863	3,531,182	37.58
37-38	.00228	93,765	214	93,658	3,437,319	36.66
38-39	.00249	93,551	232	93,435	3,343,661	35.74
39-40	.00273	93,319	255	93,191	3,250,226	34.83
40-41	.00300	93,064	279	92,925	3,157,035	33.92
41-42	.00330	92,785	306	92,632	3,064,110	33.02
42-43	.00362	92,479	335	92,311	2,971,478	32.13
43-44	.00397	92,144	366	91,961	2,879,167	31.25
44-45	.00435	91,778	400	91,578	2,787,206	30.37
45-46	.00476	91,378	435	91,161	2,695,628	29.50
46-47	.00521	90,943	473	90,707	2,604,467	28.64
47-48	.00573	90,470	519	90,210	2,513,760	27.79
48-49	.00633	89,951	569	89,667	2,423,550	26.94
49-50	.00700	89,382	626	89,069	2,333,883	26.11
50-51	.00774	88,756	687	88,412	2,244,814	25.29
51-52	.00852	88,069	751	87,693	2,156,402	24.49
52-53	.00929	87,318	811	86,913	2,068,709	23.69
53-54	.01005	86,507	870	86,072	1,981,796	22.91
54-55	.01082	85,637	926	85,174	1,895,724	22.14
55-56	.01161	84,711	983	84,220	1,810,550	21.37
56-57	.01249	83,728	1,047	83,204	1,726,330	20.62
57-58	.01352	82,681	1,117	82,123	1,643,126	19.87
58-59	.01473	81,564	1,202	80,962	1,561,003	19.14
59-60	.01611	80,362	1,295	79,715	1,480,041	18.42
60-61	.01761	79,067	1,392	78,371	1,400,326	17.71
61-62	.01917	77,675	1,489	76,930	1,321,955	17.02
62-63	.02082	76,186	1,586	75,393	1,245,025	16.34
63-64	.02252	74,600	1,680	73,760	1,169,632	15.68
64-65	.02431	72,920	1,773	72,033	1,095,872	15.03
65-66	.02622	71,147	1,866	70,214	1,023,839	14.39
66-67	.02828	69,281	1,959	68,302	953,625	13.76
67-68	.03053	67,322	2,055	66,295	885,323	13.15
68-69	.03301	65,267	2,155	64,189	819,028	12.55
69-70	.03573	63,112	2,255	61,985	754,839	11.96
70-71	.03866	60,857	2,352	59,681	692,854	11.38
71-72	.04182	58,505	2,447	57,282	633,173	10.82

Table A-4. Complete life table for the total population of the United States: 1959-61—cont'd

Age interval	Proportion dying	Of 100,000 born alive		Stationary population		Average remaining lifetime
Period of life between two exact ages stated in years (x to x + n)	Proportion of persons alive at beginning of age interval dying during interval ($_nq_x$)	Number living at beginning of age interval (l_x)	Number dying during age interval ($_nd_x$)	In the age interval ($_nL_x$)	In this and all subsequent age intervals (T_x)	Average number of years of life remaining at beginning of age interval ($\overset{\circ}{e}_x$)
	(1)	(2)	(3)	(4)	(5)	(6)
72-73	.04530	56,058	2,539	54,788	575,891	10.27
73-74	.04915	53,519	2,631	52,204	521,103	9.74
74-75	.05342	50,888	2,718	49,529	468,899	9.21
75-76	.05799	48,170	2,794	46,773	419,370	8.71
76-77	.06296	45,376	2,857	43,948	372,597	8.21
77-78	.06867	42,519	2,920	41.059	328,649	7.73
78-79	.07535	39,599	2,983	38,108	287,590	7.26
79-80	.08302	36,616	3,040	35,096	249,482	6.81
80-81	.09208	33,576	3,092	32,030	214,386	6.39
81-82	.10219	30,484	3,115	28,926	182,356	5.98
82-83	.11244	27,369	3,078	25,830	153,430	5.61
83-84	.12195	24,291	2,962	22,811	127,600	5.25
84-85	.13067	21,329	2,787	19,935	104,789	4.91
85-86	.14380	18,542	2,666	17,209	84,854	4.58
86-87	.15816	15,876	2,511	14,620	67,645	4.26
87-88	.17355	13,365	2,320	12,205	53,025	3.97
88-89	.19032	11,045	2,102	9,995	40,820	3.70
89-90	.20835	8,943	1,863	8,011	30,825	3.45
90-91	.22709	7,080	1,608	6,276	22,814	3.22
91-92	.24598	5,472	1,346	4,799	16,538	3.02
92-93	.26477	4,126	1,092	3,580	11,739	2.85
93-94	.28284	3,034	858	2,605	8,159	2.69
94-95	.29952	2,176	652	1,849	5,554	2.55
95-96	.31416	1,524	479	1,285	3,705	2.43
96-97	.32915	1,045	344	873	2,420	2.32
97-98	.34450	701	241	580	1,547	2.21
98-99	.36018	460	166	377	967	2.10
99-100	.37616	294	111	239	590	2.01
100-101	.39242	183	72	147	351	1.91
101-102	.40891	111	45	89	204	1.83
102-103	.42562	66	28	52	115	1.75
103-104	.44250	38	17	29	63	1.67
104-105	.45951	21	10	17	34	1.60
105-106	.47662	11	5	8	17	1.53
106-107	.49378	6	3	5	9	1.46
107-108	.51095	3	2	2	4	1.40
108-109	.52810	1	0	1	2	1.35
109-110	.54519	1	1	1	1	1.29

CONSTRUCTION OF LIFE TABLES

As mentioned earlier, there are two methods of constructing life tables: cohort and current. With the *cohort* method, you gather data on a single group of persons throughout their entire life span. Using the *current* method, you gather data over a 1-year period on groups that represent the entire life span. Since the cohort method is the easiest to understand, let us start with it.

Cohort life tables

Suppose that we have a population of 100 newborn guinea pigs and that we observe them until they are all dead. Suppose that 20 animals die during the first year (before reaching their first birthday). Of the 80 remaining guinea pigs, suppose that 40 die during the second year of observation. Of the remaining 40 guinea pigs, suppose that 30 die during the third year of observation. Finally, suppose that of the 10 remaining guinea pigs, all die during the fourth year of observation. The mortality experience of these 100 guinea pigs is summarized in Table A-5.

We note that column l_x is obtained by counting the number of guinea pigs that are alive at the start of each period x. For example 80 guinea pigs remain alive 1 year after birth, so $l_1 = 80$. Column d_x is obtained by counting the number of guinea pigs that die during the interval x. For example, at $x = 1$, we note that 40 guinea pigs die between 1 and 2 years after birth; so $d_1 = 40$. The entries in column q_x are obtained by dividing d_x by the entry in the l_x column. At $x = 1$, we note that $d_1 = 40$ and $l_1 = 80$; hence $q_1 = 40/80 = 0.50$. The column L_x represents the amount of time spent in the interval x by those members of the cohort surviving to age x. To compute this, we must make some assumption about the amount of time spent in the interval by those animals that die during the interval. For these data, we will assume that the deaths are uniformly distributed throughout the interval, which is arithmetically equivalent to assuming that they all die at the midpoint of the interval. Each animal is therefore counted as contributing ½ year to the total. For example, to compute L_x for $x = 2$, our reasoning would be as follows.

Of the 40 guinea pigs alive at the beginning of the interval, 10 survive the interval and the other 30 die during it. The 10 who survive the interval each spend 1 year in the

Table A-5. Illustrative cohort life table for 100 guinea pigs

Age in years = x	l_x	d_x	$q_x = d_x/l_x$	L_x	T_x	$\overset{\circ}{e}_x = T_x/l_x$
0	100	20	0.20	$80 + \dfrac{20}{2} = 90$	180	$180/100 = 1.8$
1	80	40	0.50	$40 + \dfrac{40}{2} = 60$	90	$90/80 = 1.125$
2	40	30	0.75	$10 + \dfrac{30}{2} = 25$	30	$30/40 = 0.750$
3	10	10	1.00	$10/2 = 5$	5	$5/10 = 0.5$

interval, so they contribute $10 \times 1 = 10$ "guinea pig–years" to it. According to our assumptions, the 30 that die during the interval each spend $1/2$ year in it, so they contribute $30 \times 1/2 = 15$ guinea pig–years to the interval. Thus, $L_2 = 10 + 15 = 25$ guinea pig–years. Therefore, the 40 guinea pigs that survived to age 2 experienced a total of 25 guinea pig–years during the interval of 2 to 3 years of age.

You will remember that the T_x column reports the total years of life experienced both during the period x and in all subsequent periods by those who survive to start period x. The easiest way to calculate the entries for column T_x is to start with the last period. Since there are no further periods, the entry for T_x for the last period is the same as for L_x. For the next-to-last period, T_x is obtained by adding L_x for the next-to-last period to T_x for the last period. This process is continued all the way up to the first period; each time, the L_x for the designated period is added to the T_x just computed for the next period.

Since the $\overset{\circ}{e}_x$ column represents the average years of life remaining for a person who survives to start the period in question, it is reasonable to obtain this figure by dividing the entry in the T_x column by the entry in the l_x column.

Although the construction of a cohort life table is fairly straightforward, there is one rather important problem. Since it takes so long to gather the data, the table may be irrelevant by the time it is completed. Conditions affecting mortality may have changed drastically. The *current* life table is designed to solve this problem.

Before turning our attention to the construction of a current life table, let us pause briefly to examine the assumption that deaths during a period are evenly spread throughout the period. This assumption is probably valid during the middle years of the life span; however, it is far from valid during the first year. The majority of infants who die before reaching their first birthday do so during the first month of life. If you examine Table A-4, you will see that an adjustment has been made for this. Those who die (2593) during the first year are credited with less than one-half year of life, as can be seen by comparing L_x for the first period (97,815) with l_x for the second period (97,407). The difference is only 408, rather than half the number dying during the first period (2593). Later in the table, the difference is one-half the number dying during each preceding period.

Current life tables

The purpose of a current life table is the same as that of a cohort life table, and the columns are also the same. The difference is in the length of time required for data gathering and in the sources for the entries in the initial columns. If you think of the process of constructing a current life table as a two-stage process, you will avoid an error that is very common. The first stage is data gathering on a *real* population of persons chosen to represent the entire span of human life. For each age period, you record the number of persons observed and the number dying during the period. From these two figures, you compute $_nq_x$ for each age period. This $_nq_x$ column is the goal of the data-gathering stage.

The second stage is the construction of the life table using the $_nq_x$ values obtained

Table A-6. First-stage data for construction of a current life table for guinea pigs

Age in years	Number at beginning of year	Number dying during year	Proportion dying during year
0	100	20	0.20
1	78	39	0.50
2	80	60	0.75
3	45	45	1.00

from the first stage and a *hypothetical* population of arbitrarily chosen size. Starting with the number in the hypothetical group as the initial l_x, you apply the $_nq_x$ value that you computed for the first period to get the number of deaths expected during the first period. Then, subtracting this number from the first l_x, you obtain the hypothetical number that would survive to start the second period. Repeating this process results in a complete set of hypothetical figures for the l_x and $_nd_x$ columns. Once these columns are complete, the construction of the remainder of the table is the same as for a cohort table. This sounds very complicated. Let us show a worked example using guinea pigs.

In the previous example, we constructed a life table representing the longevity experience of a cohort of 100 guinea pigs, all born at the same time. Their numbers were gradually depleted by deaths until the last animal died sometime during the fourth year of life. It would have taken, therefore, between 3 and 4 years of observation to construct this table.

Suppose, on the other hand, that we have a population of 100 newborn guinea pigs, another 78 who are exactly 1 year old, 80 more who are exactly 2 years old, and 45 more who are exactly 3 years old. Suppose that we observe these four groups for exactly 1 year and observe the number of deaths, shown in Table A-6, among the four groups.

Notice that we spent only 1 year (as opposed to 4 years for the cohort method) in obtaining the data of Table A-6. We can then construct what is known as a *current life table* by assuming that a *hypothetical* cohort of newborn guinea pigs are passing through life subject to the same forces of mortality as those seen in the guinea pigs that were observed for 1 year. We do this by entering the observed proportion dying during each year as the entries for the q_x column. If we choose to start with a hypothetical group of 1000 animals, our current life table then would be constructed from the following entries.

Age in years	q_x	l_x	d_x	L_x	T_x	$\overset{o}{e}_x$
0	0.20	1000				
1	0.50					
2	0.75					
3	1.00					

Under the assumption of equally spaced deaths within an interval, the remaining entries in the table can now be computed. The completed life table has the following appearance:

Age in years	q_x	l_x	d_x	L_x	T_x	$\overset{\circ}{e}_x$
0	0.20	1000	200	900	1800	1.800
1	0.50	800	400	600	900	1.125
2	0.75	400	300	250	300	0.750
3	1.00	100	100	50	50	0.500

The computations for the entries in the l_x and d_x columns were explained in the second paragraph of this section. The computations and meanings for the entries in the last three columns are the same as for the corresponding entries in cohort tables. You will have a chance to test your understanding of these computations in an exercise that follows.

The complete life table for the United States for 1959 to 1961 (Table A-4) is a current life table rather than a cohort life table. Since human beings can live to be over 100 years old, it would take over a century to actually complete a cohort life table for a human population. In interpreting current life tables, one should keep in mind that these represent the expected mortality experience of a hypothetical cohort under the assumption that current death rates will continue to hold for the cohort as it passes through life.

You will now be given an opportunity to complete life tables by the current method as an exercise for this lesson. It is assumed that if you understand the construction of a current life table, you will have no difficulty with the construction of a cohort life table. In the next lesson, we will look at a special form of life table called a follow-up life table.

LESSON 4 EXERCISE

The following data represent the life experience of a certain type of moth. All of a group of moths ranging in age from newly emerged to 5 weeks after emergence are observed for a period of 1 week. The number of moths of each age and the number of moths of each age that died during the week of observation is shown in the table. Complete a current life table for the moths, using 10,000 as the initial l_x.

Age in weeks	Number observed	Number dying
0	1000	200
1	1000	250
2	1000	300
3	1000	500
4	1000	800
5	1000	1000

Lesson 5 □ Follow-up life tables
PERSON-TIME UNITS

In our discussion of life tables, we have been assuming that all persons have been observed for their entire lives for cohort studies or for the entire period for current studies. In addition, we have assumed that the event being studied is one that can occur only once to any person. Let us examine these assumptions one at a time.

First, requiring that all persons be observed during the same period of time often results in having to reject a large amount of potentially useful data. For example, suppose that you were interested in measuring the useful life of a battery used for powering pacemakers. The $\overset{o}{e}_x$ column of a life table for the batteries would certainly be valuable information for physicians and patients. Using either the cohort or current methods described in Lesson 3 to construct a life table for the batteries would require a data-gathering process that would be nearly impossible to implement with any sizable group. Pacemakers are implanted on a schedule that is spread throughout the year, rather than all at one time. In addition, patients are frequently lost to further observation before the batteries fail. Some move to new locations, others die. Dropping from the study those individuals for whom the data are incomplete or those who entered the study after some designated starting period would severely restrict the usable data.

The second assumption we have made is that the event of interest can happen only once to any person. Although this assumption is valid for many events, such as births, deaths, or vision loss, there are many events we might wish to examine that are not "one time" events. Some examples are onset of certain diseases, failure of a tooth crown, rejection of an intrauterine device (IUD), and occurrence of an epileptic seizure. Techniques for measuring any of these events must make provision for multiple occurrences for a single individual.

Table A-7. Data illustrating the use of person-time units to measure the frequency of colds in a population

Individual number	Date of entry into study	Date of exit from study	Length of observation time	Number of colds
1	January 1	December 31	365 days	2
2	July 1	December 31	184 days	0
3	September 15	December 31	108 days	1
4	May 10	October 1	145 days	0
5	January 1	July 1	182 days	0
6	January 1	August 15	227 days	3
7	August 15	December 31	139 days	0
8	February 20	September 30	223 days	4
9	October 10	December 31	83 days	1
10	January 1	May 1	121 days	2
11	November 27	December 31	35 days	0
TOTAL			1812 days	13

Both the problem of differing periods of observation and the problem of multiple occurrence of events can be solved through the use of person-time units. We might use person-days, person-weeks, or person-years, and we might translate one into another. For example, one person is observed for 3 days and a second person for 4 days. Between them, there have been 7 person-days of observation, or 1 person-week. Actually, we have already used the concept of person-years, in the $_nL_x$ and T_x columns of life tables discussed in the last lesson, when we talked about guinea pig–years.

If we record the number of events experienced by each person during the period in which that person is observed, we can later get totals both for the number of events and the accumulated length of observation for all persons. We can then make a ratio between the number of events and the number of person-time units (for example, 2.7 colds per person per year, 1.2 pregnancies per person per decade, or 0.15 IUD rejections per person per month).

The data in Table A-7 illustrate the use of person-time units to measure the frequency of colds in a population. Notice that the raw data recorded for each person are: the time when observation began, the time it ended, and the number of events the person experienced. At the bottom of the table, the totals show 1812 person-days of observation and 13 colds. Dividing 1812 by 365, we get 4.96 person-years. Then, dividing 13 by 4.96, we get 2.62 colds per person per year.

Ratios based on the person-time concept are extremely useful. It must be noted, however, that the validity of such ratios rests on the assumption that the probability of the event does not change in any systematic fashion over time. Failing to recognize this requirement can lead to wrong conclusions. Suppose, for example, that one were looking at the event "failure of an automobile transmission." Obviously, the probabililty of transmission failure increases as the mileage adds up. However, suppose that 200 new cars were observed for various parts of the first year of operation and that there was one failure. If on the average the new cars were observed for $1/2$ year, the total car-years of observation would be 100, and the ratio would be one failure per car per century of operation.

USE AND CONSTRUCTION OF FOLLOW-UP LIFE TABLES

There is a solution for the case where the probability of the event may change over time. The technique also makes use of the person-time concept; it is called a follow-up life table. Unlike ordinary life tables, which result in an $\overset{o}{e}_x$ column reporting average expected remaining life at any point in time, follow-up life tables report the probability of "escaping the event" up to any given point in time. For example, a follow-up life table might report the probability of a tooth crown's lasting 5 or more years, that is, of its not experiencing failure for at least 5 years.

It is not difficult to understand the rationale that serves as a basis for follow-up life tables. It is the bookkeeping procedure for data gathering that is sometimes confusing. Let us begin, therefore, by looking first at a complete table; let us then consider how the data for the table would be gathered. Remember that follow-up life tables are appropriate when

(1) different persons are observed for different periods of time (2) the probability of experiencing the event may change over time. The latter condition implies that there is some starting point, for example, restoration of a tooth, insertion of an IUD, or implantation of a pacemaker. In certain instances, follow-up life tables may be appropriate for events that can happen more than once. The requirement is that the probability of the event return to its initial level once the event occurs.

Now, let us turn to a worked-out example. Suppose that a manufacturer wants to study the durability of a certain high-grade nylon stringing of tennis racquets. As a promotional gimmick, a tennis supply house agrees to provide one stringing free for all members of a certain tennis club during a 1-year period of study. To be sure that data will be reported, the member pays the usual fee at the time of the stringing, but the fee is refunded (1) if the strings break during the study period, (2) if the person leaves the community during the study period, or (3) when the study period ends. The study starts in January, but certain members join the club during the year and thus enter the study late. Others leave the club during the period of the study. Thus the two conditions calling for a follow-up life table are met. Not all persons are observed for the same length of time, and the probability of string failure is likely to increase with use. Our follow-up table will provide us with information as to the probability of the strings' lasting (escaping failure) for various lengths of time. The completed table might look like Table A-8. As with regular life tables, it is the last column, cumulative probability of survival, that interests us. It gives the probability of the event's *not* happening before the *start* of each designated period. For example, the probability of the string's lasting at least 4 months is 0.875. This means that less than 13% of the strings will break during the first 4 months of use. The probability of having a failure before the stringing is, of course, zero, so that P_x for the first entry will always be 1.00.

DETAILS OF DATA GATHERING

As stated earlier, the data-gathering process for construction of a follow-up life table can be confusing unless it is approached in a systematic fashion. A data sheet might be used that has three headings and one line for each person. The headings are: date of entry, date of exit, and date of event. The data sheet for our study, with a few sample entries, might look like Table A-9.

Once a data sheet such as Table A-9 has been completed, with information on all persons in the study, we are ready to begin the analysis. The first decision to be made is how to divide up the time period being studied. Sometimes data-gathering techniques will dictate that the period be divided into calendar years, months, or whatever, because the data are accurate only to a certain degree. In other instances, one arbitrarily will choose a time period resulting in a reasonable number of categories. In some instances, intervals may vary in length. For our study, we may decide to use months made up of 30 calendar days. Our analysis of the data now is similar to the process of making up two frequency distributions; one for those who experienced the event, and one for those who did not. Our work sheet might look like Table A-10. The two columns to the right in Table A-10 would

Table A-8. Example of a follow-up life table for tennis racquet strings

Months after stringing (x)	Initial number observed (O$_x$)	With-drawals (W$_x$)	String failures (d$_x$)	Number observed throughout (N$_x$)	Period proportion failures (q$_x$)	Period proportion survivals (s$_x$)	Cumulative proportion survivals (P$_x$)
0	101	2	1	100	0.010	0.990	1.000
1	98	2	3	97	0.031	0.969	0.990
2	93	2	4	92	0.043	0.957	0.959
3	87	4	4	85	0.047	0.953	0.918
4	79	6	5	76	0.066	0.934	0.875
5	68	2	7	67	0.104	0.896	0.817
6	59	6	6	56	0.107	0.893	0.732
7	47	6	5	44	0.114	0.886	0.654
8	36	10	4	31	0.129	0.871	0.579
9	22	8	3	18	0.167	0.833	0.504
10	11	6	2	8	0.250	0.750	0.420
11	3	2	1	2	0.500	0.500	0.315
12	0						0.158

Details of table construction

1. The x column designates months after stringing. Since 1-month periods are used, it starts at 0 and ends at 12.
2. The O_x column starts in the first row with the total number of persons observed. To find the number for the next row, subtract the number of withdrawals and failures ($101 - 2 - 1 = 98$).
3. The W_x column and the d_x column are the number of withdrawals and the number of failures respectively.
4. The N_x column is the number observed throughout the period to see if the event occurs. Those who withdraw during the period are assumed to do so evenly throughout, and each is counted for half the period. In contrast to the procedures for ordinary life tables, those that experience the event are counted for the whole period. This is because the N_x column will be used to compute the probability of the event, not the person-years of experience (as with the L_x column in life tables). Since our only interest is in whether the event occurred during the period, we, in a sense, quit our observation once the event occurs; but this is only because we already know what happened. If we observed the total period, we would still report a string failure. The N_x column is therefore $O_x - (W_x/2)$; for the first row, $101 - (2/2) = 100$.
5. The q_x column is the probabilty of experiencing the event during the period, and it is found by dividing d_x by N_x. For the first row, this is $1/100 = 0.010$.
6. The s_x column is the probability of not experiencing the event during the period, and it is obtained by subtracting q_x from 1.00. For the first row, this is $1.00 - 0.010 = 0.990$.
7. The entries in the P_x column can be confusing. You remember that they indicate the probability of avoiding the event until the start of any designated period. In our example, this means the probability of the strings' lasting until the start of any period of interest to us. To do so, the strings must last throughout the period immediately preceding the one in which we are interested, as well as through all the periods before that. If we look at the entries for the period immediately preceding our period of interest, we will find the information we need. The s_x column gives the probability of surviving that particular period, and the P_x column gives the probability of surviving all prior periods. Our only problem then is in knowing how to combine this information to find the probability of surviving both all prior periods and the one immediately before the one in which we are interested. Chapter 4 describes the multiplicative rule and its application in this situation. For our purposes here, all we need to know is that we successively multiply the entries in the s_x and P_x columns to get each successive P_x value. For the first row, P_x is 1.00, as stated earlier. For the second row, P_x is found by multiplying s_x and P_x from the preceding row; its value is $(0.990)(1.00)$, or 0.990.

Table A-9. Sample data sheet for a follow-up life table for nylon tennis string

Name	Date of stringing	String failure Date	String failure Number of days after stringing	End of observation Date	End of observation Number of days after stringing
Jones	1/12	5/3	111		
Anderson	1/15			12/31	350
Meyers	4/17			11/10	270
Walters	6/2	7/7	35		
Smith	7/13			12/31	177
Carter	11/14			12/31	47

Table A-10. Sample work sheet for a follow-up life table for nylon tennis strings

Period	End of observation	String failure
0-29		
30-59	Carter	Walters
60-89		
90-119		Jones
120-149		
150-179	Smith	
180-209		
210-239		
240-269		
270-299	Meyers	
300-329		
330-359	Anderson	

Table A-11. Frequency distributions used in the construction of a follow-up life table (Table A-8)

Period	End of observation	String failure
0-29	2	1
30-59	2	3
60-89	2	4
90-119	4	4
120-149	6	5
150-179	2	7
180-209	6	6
210-239	6	5
240-269	10	4
270-299	8	3
300-329	6	2
330-359	2	1
TOTAL	56	45

be the frequency columns in a frequency distribution. In an actual study, these columns would contain check marks (✔) instead of names, with each person in the study being represented by a check mark in one of the two columns. Here, the names are used to make certain the process is clear. Notice that only one check mark is made for each person. The entries in "end of observation" do not include those who experience the event, but only those who are lost to follow-up or have not yet experienced the event when the study ends. The complete data (with counts replacing check marks) might be as in Table A-11.

The data in Table A-11 are now in a form that permits the construction of the life table appearing earlier as Table A-8. The total number of persons observed (101) is the initial value in the O_x column. The "string failure" column becomes the d_x column, and the "end of observation" column becomes the W_x column. It should be noted that in all of the symbols, the small x stands for the number of months after stringing.

In the exercise that follows this lesson, much of the early work has been done, because you will be given data already organized for you in a format similar to that of Table A-11. You are then required to complete the construction of the follow-up life table. It will serve as a good memory aid if you will first review in your mind the steps that would be taken to arrive at the data in the form in which it is presented.

SUMMARY

This concludes our study of public health statistics. In a sense, our discussion remains incomplete, because we have not considered the public health statistics presented in this appendix from the point of view of inferential statistics. Such a discussion would be beyond the scope of this book. You should be aware, however, that rates, the entries in the $\overset{o}{e}_x$ columns of ordinary life tables, and P_x values of the follow-up life tables are usually sample statistics, subject to chance fluctuations based on the groups being studied.

LESSON 5 EXERCISE

The following data represent survival experience with cadaver kidney grafts.

Period (6 months)	Withdrawals	Graft failure
0	20	11
1	21	28
2	19	37
3	38	34
4	59	47
5	22	64
6	63	62
7	67	51
8	92	37
9	84	28
10	63	21
11	22	10
TOTAL	570	430

Construct a follow-up life table and answer the following question: What is the probability of a cadaver kidney graft's still functioning 6 years after the operation?

APPENDIX B

Answers to exercises

1.1   EXERCISE A: **[1]** −13. **[2]** −12. **[3]** 8. **[4]** −5. **[5]** +12. **[6]** 36. **[7]** 36. **[8]** −88. **[9]** 0.01.
[10] −9. **[11]** 30. **[12]** 43.1 **[13]** 81. **[14]** −5. **[15]** −3. **[16]** True. **[17]** False. **[18]**
False. **[19]** ZY. **[20]** $4Y$. **[21]** 6. **[22]** 210. **[23]** 0.3. **[24]** 60.

EXERCISE B: **[1]** 1522. **[2]** 39.788. **[3]** 24,691,155. **[4]** 7385. **[5]** −74. **[6]** 645. **[7]** 7504.
[8] 77.7. **[9]** 0.45632. **[10]** 21. **[11]** 0.00205. **[12]** 0.02. **[13]** 20449. **[14]** 0.015. **[15]**
$\Sigma X = 234. \ \Sigma X^2 = 7258.$ **[16]** $\Sigma X = 402. \ \Sigma X^2 = 9580.$

2.1

Class limits	f	Class limits	f
100-149	1	400-449	7
150-199	3	450-499	6
200-249	3	500-549	4
250-299	6	550-599	0
300-349	8	600-649	1
350-399	11		

2.2

[1] Class limits	f	%	cf	c%
3.60-3.69	2	2.44	2	2.44
3.70-3.79	0	0	2	2.44
3.80-3.89	0	0	2	2.44
3.90-3.99	2	2.44	4	4.88
4.00-4.09	7	8.54	11	13.41
4.10-4.19	7	8.54	18	21.95
4.20-4.29	11	13.41	29	35.37
4.30-4.39	9	10.98	38	46.34
4.40-4.49	13	15.85	51	62.20
4.50-4.59	7	8.54	58	70.73
4.60-4.69	10	12.20	68	82.93
4.70-4.79	6	7.32	74	90.24
4.80-4.89	6	7.32	80	97.56
4.90-4.99	2	2.44	82	100.00

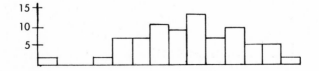

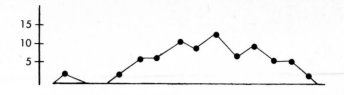

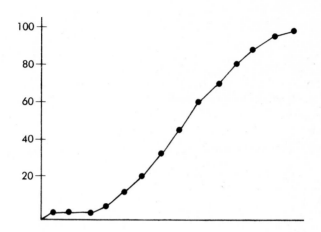

[2]

Continued.

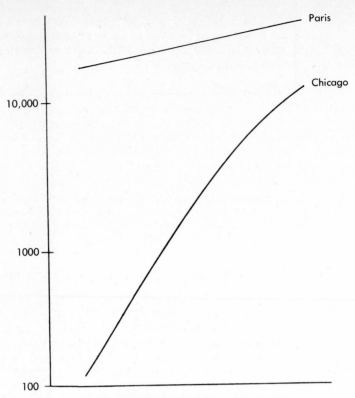

The absolute increase has been approximately equal, but the proportional change has been much greater in Chicago.

3.1 [1] 114. [2] 1307. [3] 5596. [4] 116.583. [5] 116.

[6] Class interval	f	cf
40-49	1	1
50-59	0	1
60-69	0	1
70-79	0	1
80-89	1	2
90-99	4	6
100-109	9	15
110-119	12	27
120-129	9	36
130-139	8	44
140-149	3	47
150-159	1	48

[7] 116.79. [8] 117.

3.2 [1] 105. [2] 9.6. [3] 313.12. [4] 17.7. [5] 15.18. [6] 342.67; 18.51. [7] 129.5 − 106.17 = 23.33.

3.3 [1] $n = 50$; $\Sigma X = 360.5$; $\Sigma X^2 = 2649.77$; $\Sigma Y = 991.6$; $\Sigma Y^2 = 19,695.8$; $\Sigma XY = 7182.67$. [2] $\Sigma x^2 = 50.565$; $\Sigma y^2 = 30.389$; $\Sigma xy = 33.234$. [3] $\mu_X = 7.21$; $\mu_Y = 19.832$; $\sigma_X = 1.006$; $\sigma_Y = 0.780$. [4] $+1.5$; -1.19; 93.32; 11.7; 0.8162. [5] 0.848; there is a fairly strong, positive correlation.

3.4 [1a] -1; -1; $+1$; $+1$. [1b] 1; 1; 19; 19. [2] Subtract 100,101; divide by 100; converted scores are -1, 0, and $+1$; mean and σ of converted scores, 0 and $\sqrt{2/3}$; mean of original scores is $(0)(100) + 100,101$, or 100,101; σ of original scores is $100\sqrt{2/3}$, or 81.65. [3] For X scores, code by subtracting 1100, and then divide by 50; for Y scores, code by subtracting 115, and then divide by 5. Both coded score distributions are identical, so $r = 1.00$ for both the coded and the original scores.

4.1 [1a] $^3/_8$. [1b] $^1/_8$. [1c] By subtraction, $^1/_2$. [1d] By addition, $^1/_2 + ^3/_8 = ^7/_8$. [2a] $(^1/_2)^5$. [2b]$(^1/_2)^5$. [2c] By subtraction, $1 - (^1/_2)^5 = ^{31}/_{32}$. [3] P (A or B or both) $= P(A) + P(B) - P(A$ and $B) = (0.40) + (0.70) - (0.40 \times 0.70) = 0.82$. [4a] $^{145}/_{400}$. [4b] $^{50}/_{400}$. [4c] $^{200}/_{400}$. [4d] $^{25}/_{400}$. [4e] $^{15}/_{60}$. [4f] $^{90}/_{400}$. [4g] $^{55}/_{400}$. [4h] $^{55}/_{235}$.

4.2 [1] $\dfrac{1}{{}_5C_2}$, or $\dfrac{1}{10_i}$. [2] $\dfrac{1}{{}_5P_2}$, or $\dfrac{1}{20}$. [3a] Trait A is a correct guess. [3b] 0.5. [3c] 0.5. [3d] 8. [3e] Yes. [3f] 0, 1, 2, 3, 4, 5, 6, 7, 8. [3g] $28(^1/_2)^8$, or 0.109. [3h] $P(0) = 1(^1/_2)^8$; $P(1) = 8(^1/_2)^8$; $P(2) = 28(^1/_2)^8$; $P(3) = 56(^1/_2)^8$; $P(4) = 70(^1/_2)^8$; $P(5) = 56(^1/_2)^8$; $P(6) = 28(^1/_2)^8$; $P(7) = 8(^1/_2)^8$; $P(8) = 1(^1/_2)^8$; [3i] $37(^1/_2)^8$, or 0.144. [3j] You are not prepared to become a believer yet, but the test deserves further investigation. [4] $P(A) = 0.625$; $P(NA) = 0.375$; $P(B\,|\,A) = 0.60$; $P(B\,|\,NA) = 0.333$; $P(A\,|\,B) = 0.75$. [5] Box 1 = 3750; box 2 = 1250; box 3 = 2500; box 4 = 2500; $P(A\,|\,B) = 3750/(3750 + 1250) = 0.75$.

5.1 [1] $\pi = 0.5$; $\mu = 40.17$; $\sigma = 24.6$.

5.2 [1] Empirically: You could flip 2 coins 25 times and consider a head a positive outcome, or you could take 25 pairs of numbers from a table of random numbers, calling an even number a positive outcome; the possible results are 0, 1, or 2 positive outcomes for each sample. Theoretically: $P(0) = (^1/_2)^2$; $P(2) = (^1/_2)^2$; and by subtraction, $P(1) = 1 - (^1/_2)^2 - (^1/_2)^2 = ^1/_2$. [2] Both $(64)(0.8)$ and $(64)(0.2) \geqslant 5$; therefore, shape is approximately normal, central tendency $= \pi$ (or 0.8), and variability $= SE_p = \sqrt{(0.8)(0.2)/64} = 0.05$. [3] For each sample, choose ten single-digit numbers from a table of random numbers. Consider an even number a positive outcome. Count the even numbers in each sample. Repeat many times. [4] A person who is guessing has a probability of 0.5 of getting each item right. The sample size is 100. The sampling distribution of p, the proportion right, for all persons who are guessing will be normally distributed, with central tendency $= 0.5$ and $SE_p = \sqrt{(0.5)(0.5)/100} = 0.05$; $z = \dfrac{0.70 - 0.50}{0.05} = +4$. The probability of guessing 70 or more right is 3 in 100,000.

5.3 [1] The theoretical distribution is approximately normal, with a mean = 500 and $SE_x = 100/\sqrt{36} = 16.67$. [2] Choose random samples of size 36 from all persons taking the exam in a given year. For each sample, compute $\overline{X}$. Make a frequency polygon of $\overline{X}$'s to determine shape; find the mean of $\overline{X}$'s to determine central tendency; find the standard deviations of $\overline{X}$'s to determine variability. [3] The distribution would still be normal, even though sample size is less than 25, because the population is normal. Central tendency would still be 500. $SE_{\bar{x}} = 100/\sqrt{16} = 25$. **[4]** Sampling distribution is normal, with central tendency of 60 and $SE_{\bar{x}} = 15/\sqrt{16} = 3.75$. To include 99% of the normal distribution, you need to go 2.576 standard deviations from the mean: $60 \pm (2.576)$ $(3.75) = 50.34$ to 69.66.

6.1   **[1a]** $^{60}/_{80} = 0.75$. **[1b]** Yes, because $(80)(0.75)$ and $(80)(0.25) \geq 5$. **[1c]** 2.576; **[1d]** 0.0484. **[1e]** 0.625 to 0.875. **[1f]** An infinite number of confidence intervals calculated in this way would include the true proportion *approximately* 99% of the time. **[1g]** The confidence interval will be narrower. **[2]** $\sqrt{(0.5)(0.5)/80} = 0.056$; 0.606 to 0.89; An infinite number of confidence intervals calculated in this way will include the true proportion *at least* 99% of the time.

6.2   **[1]** $69 \pm (2.576)(2)$, or 63.85 to 74.15. **[2]** Since population is normal and s is used to estimate σ, use the t distribution with 8 *df:* $380 \pm (1.397)(20)$, or 352.06 to 407.94. **[3]** The population is skewed, but the sampling distribution is approximately normal, since $n > 25$; σ known, so use normal curve: $22 \pm (1.645)(2)$, or 18.71 to 25.29. **[4]** The population is skewed, σ is not known, and n is only 25; s is not a good estimate of σ. You will not be able to solve this problem with the present information.

7.1   **[1]** π, the proportion of prisoners involved in violent incidents. **[2]** The prisoners in TM classes. **[3]** All prisoners at the institution not in TM classes. **[4]** $\pi_S = 0.07$. **[5]** One-tail test, because you are only interested in providing the program if violence is reduced. **[6]** H_A: $\pi_T < 0.07$. **[7]** H_0: $\pi_T \geq 0.07$. **[8]** 0.10. **[9]** $n = 200$. **[10]** Normal, central tendency $= 0.07$, $SE_p = 0.018$. **[11]** $p = {}^{9}/_{200} = 0.045$. **[12]** Using approach *A*, you determine the critical region is made up of all sample proportions ≤ 0.047. The observed statistic is in the critical region and has a probability of less than 0.10 when the null hypothesis is true. Using approach *B*, you find that the z score for 0.045 is -1.389. The probability is approximately 0.08 if the null hypothesis is true. **[13]** In both cases, you reject the null hypothesis. **[14]** You might be making a Type I error. This type of error would lead you to claim that the program reduces violent incidents among participants when it actually does not.

7.2   **[1]** H_0: $\mu \leq 14.8$; $\sigma = 2.4$; $SE_{\bar{x}} = \sigma/\sqrt{n} = 0.6$; $z = (\bar{X} - \mu)/SE_{\bar{x}} = (16.4 - 14.8)/0.6 = 2.67$. A value this extreme in one direction happens less than 4 times in 1000. Therefore, you will reject the null hypothesis. Alternatively, you could find that the critical region is 2.326 SE's away from μ: $14.8 + 2.326(0.6)$, or 16.19. The sample mean is in the critical region. Again, you will reject null. **[2]** You are probably interested in differences in either direction. H_0: $\mu = 82$. For the old method, you know that the scores are normal. Evidence suggests that the scores are also approximately normal when the new method of teaching is used. σ is unknown. Use the t distribution with 11 *df;* $SE_{\bar{x}} \doteq s/\sqrt{n}$, or 2.627. Critical region boundaries are $82 \pm (2.201)(2.627)$, or 76.22 and 87.78. The sample mean, 79.4, is not in critical region. There is not sufficient evidence to reject the null hypothesis. **[3]** H_0: $\mu \leq 0.20$ rem per year. A Type I error would lead you to say that exposure is higher near plant when it actually is not. It would cause unnecessary alarm. A Type II error would be a failure to find that exposure is higher when in fact it is. The result would be a failure to give warning. Here, you would probably be more concerned about a Type II error. You would not guard stringently against Type I error. You would probably set an alpha level of 0.20 or even 0.30.

7.3 **[1]** The standard population is composed of noninfected people. The test population is those people infected with rabies. **[2]** $\mu_S = 50$, $\mu_T = 70$, $\sigma_S = 10$, $\sigma_T = 10$. **[3]** For samples of size 1, both sampling distributions are normal, with $SE_{\bar{x}} = 10$. For the standard population, the central tendency is 50; for the test population, it is 70. **[4]** $50 + 1.645(10)$, or 66.45.
[5]

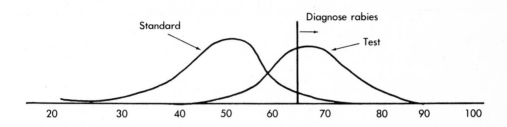

[6] 5%. **[7]** $z = (66.45 - 70)/10 = 0.36$; approximately 36% of the sampling distribution of the test population < 66.45. **[8]** $1.00 - 0.36$, or approximately 64%. **[9]** Yes, because there are far too many Type II errors. Since death results when a Type II error is made, you will want to make the test more powerful, even though more people will needlessly undergo the painful rabies shots. **[10]** The $SE_{\bar{x}}$ would now be $10/\sqrt{4} = 5$. Both sampling distributions would be less variable. Rabies would be diagnosed if the average for the four children is above 58.22. This is 2.36 SE's below the mean for infected persons. Now a Type II error would only have a probability of about 9 in 1000.

8.1 **[1]** The sampling distribution of differences will be approximately normal, because π, the proportion of even numbers, is 0.5 for each table. Therefore, $n\pi$ and $n(1 - \pi) \geqslant 5$ in both cases. The central tendency of the sampling distribution of differences will be 0.00, since, on the average, the difference between p_A and p_B can be expected to be 0.00. The variability of the sampling distribution of differences will be:

$$SE_{p_A - p_B} = \sqrt{\frac{(0.5)(0.5)}{25} + \frac{(0.5)(0.5)}{25}} \qquad \text{or} \qquad 0.1414$$

[2] The shape will be normal, since n_A and $n_B = 25$. For each population, $\mu = 4.5$ and $\sigma^2 = 8.25$. Therefore, the sampling distribution of differences between means will have a central tendency of 0.00 and a variability as follows:

$$SE_{\bar{x}_A - \bar{x}_B} = \sqrt{SE_{\bar{x}_A}^2 + SE_{\bar{x}_B}^2} = \sqrt{\frac{\sigma^2}{n_A} + \frac{\sigma^2}{n_B}} = \sqrt{\frac{8.25}{25} + \frac{8.25}{25}} = 0.812$$

[3] If we assume that $\pi_A = \pi_B$, the best estimate of π is $(32 + 48)/(36 + 64)$, or 0.80. The best estimate of $SE_{p_A - p_B}$ is:

$$SE_{p_A - p_B} \approx \sqrt{\frac{(0.8)(0.2)}{36} + \frac{(0.8)(0.2)}{64}} = \sqrt{0.0069} = 0.083$$

If we do not assume that $\pi_A = \pi_B$, the best estimate of $SE_{p_A - p_B}$ is:

$$SE_{p_A - p_B} \approx \sqrt{\frac{(32/36)(4/36)}{36} + \frac{(48/64)(16/64)}{64}} = 0.0753$$

[4] If we assume that $\sigma_A = \sigma_B$, the best estimate of σ is:

$$\sigma = \sqrt{\frac{\Sigma x_A^2 + \Sigma x_B^2}{n_A + n_B - 2}} = 6.5$$

If $\sigma_A = \sigma_B$, the best estimate of $SE_{\bar{X}_A - \bar{X}_B}$ is:

$$SE_{\bar{X}_A - \bar{X}_B} \approx \sqrt{\frac{6.5^2}{80} + \frac{6.5^2}{20}} = 1.625$$

If we do not assume that $\sigma_A = \sigma_B$, the best estimate of $SE_{\bar{X}_A - \bar{X}_B}$ is:

$$SE_{\bar{X}_A - \bar{X}_B} \approx \sqrt{\frac{s_A^2}{n_A} + \frac{s_B^2}{n_B}} = \sqrt{\frac{46.39}{80} + \frac{25.04}{20}} = 1.35$$

8.2   **[1]** Do not assume $\pi_A = \pi_B$. Therefore:

$$SE_{p_A - p_B} \approx \sqrt{\frac{p_A(1 - p_A)}{n_A} + \frac{p_B(1 - p_B)}{n_B}} \quad \text{or} \quad \sqrt{\frac{(0.09)(0.91)}{400} + \frac{(0.05)(0.95)}{400}} \quad \text{or} \quad 0.018$$

The sampling distribution of differences is normal. For a 99% confidence interval: $(0.09 - 0.05) \pm (2.576)(0.018)$, or 0.04 ± 0.046, or -0.006 to 0.086.

[2] H_0: $\pi_A \leq \pi_B$. The best estimate of common π is $(22 + 18)/(100 + 100)$, or 0.20; $p_A = 22/100$, or 0.22; $p_B = 18/100$, or 0.18. The distribution of differences is normal. The central tendency of differences is zero if the null hypothesis is correct.

$$SE_{p_A - p_B} = \sqrt{\frac{(0.20)(0.80)}{100} + \frac{(0.20)(0.80)}{100}} = 0.056$$

$$z = \frac{(0.22 - 0.18) - 0}{0.056} = \frac{0.04}{0.056} = 0.714$$

Values this extreme in one direction happen about 23% of the time. Therefore, we fail to reject the null hypothesis.

[3] H_0: $\mu_A \leq \mu_B$. If we do not assume that $\sigma_A = \sigma_B$, the best estimate of the variability of the sampling distribution of differences between means is:

$$SE_{\bar{X}_A - \bar{X}_B} \approx \sqrt{\frac{s_A^2}{n_A} + \frac{s_B^2}{n_B}} = \sqrt{\frac{(2.1)^2}{100} + \frac{(1.8)^2}{100}} = 0.2765$$

With samples of this size, the sampling distribution can be expected to be normal. We should use the t distribution, since σ's are unknown. However, with samples this large, the z or t distributions give nearly the same result. Since z is more complete, use it.

$$z = \frac{(5.1 - 4.8) - 0}{0.2765} = 1.08$$

Values this extreme in one direction happen about 14% of the time. Therefore, we fail to reject the null hypothesis.

[4] H_0: $\mu_A \leq \mu_B$. If we assume that $\sigma_A = \sigma_B$, the best estimate of σ^2 is:

$$\sigma^2 = \frac{\Sigma x_A^2 + \Sigma x_B^2}{n_A + n_B - 2} = \frac{1007.5 + 1803.6}{8 + 10 - 2} = 175.69$$

$$SE_{\bar{X}_A - \bar{X}_B} = \sqrt{\frac{175.69}{8} + \frac{175.69}{10}} = 6.287$$

$$\bar{X}_A = 90.25 \text{ and } \bar{X}_B = 80.8$$

$$t = \frac{(90.25 - 80.8) - 0}{6.287} = 1.503 \text{ with } 16 \, df$$

The t value needed for significance is 1.746. The observed t value falls in the noncritical region. Therefore, we fail to reject the null hypothesis.

[5] For this problem, $\Sigma D = 119, \Sigma D^2 = 2151, \bar{D} = \dfrac{\Sigma D}{n} = \dfrac{119}{10} = 11.9$:

$$S_D = \sqrt{\frac{\Sigma D^2 - \dfrac{(\Sigma D)^2}{n}}{n - 1}} = 9.036$$

$$SE_{\bar{D}} \approx \frac{s_D}{\sqrt{n}} = \frac{9.036}{\sqrt{10}} = 2.857$$

$$t = \frac{11.9 - 0}{2.857} = 4.1645 \text{ with } 9 \, df$$

The result is significant beyond 0.005.

9.1 The logic of this problem calls for a directional test. However, direction of differences cannot be easily determined for a 1×4 table. Usually, a nondirectional test is made first. If this turns out to be significant, categories can be grouped into two classes, and directional tests can then be made.

$$X_p^2 = \sum \frac{(f_o - f_e)^2}{f_e} = \frac{(65 - 50)^2}{50} + \frac{(112 - 100)^2}{100} + \frac{(359 - 350)^2}{350} + \frac{(464 - 500)^2}{500}$$

$$X_p^2 = 8.76 \text{ with } 3 \, df$$

Values of X_p^2 this extreme will happen less than 5% of the time by chance. We will probably reject the null hypothesis. Note that if the three categories—death, paralysis, and concussion—are grouped and compared with "other," the result will again be a significant value for X_p^2. With just two classes, a directional test is readily made. We first note that the differences are in the expected direction. We then compare our observed value of X_p^2, 5.184, with the table of χ^2 with 1 df. Values this extreme happen less than 2.5% of the time. For a directional test, we halve the probability. Our result is significant at the 0.0125 level. In Chapter 9, Lesson 2, you will learn of a modification in calculations that is made for the case of 1 df.

9.2 [1] The receiver has a one-in-five chance of guessing the correct object. The null hypothesis is therefore that $\pi \leqslant 0.20$. We calculate X_p^2 corrected for continuity:

$$X_p^2 = \frac{(|28 - 20| - 0.5)^2}{20} + \frac{(|72 - 80| - 0.5)^2}{80} = 3.516 \text{ with } 1 \, df$$

$$z = \sqrt{3.516} \quad \text{or} \quad 1.87$$

Values this extreme in one direction occur only about 3% of the time. There appears to be something other than chance operating.

Using the method of a one-sample test of proportions and correcting for continuity, we calculate as follows:

$$z = \frac{27.5/100 - 20/100}{\sqrt{\dfrac{(0.2)(0.8)}{100}}} = 1.87$$

[2] With $\overline{X} = 116.5$ and $s = 13.2$, division points become: 103.3, 109.6, 116.5, 123.1, and 129.7. The number of sample values less than 103.3 is two, the number between 103.3 and 109.6 is three, and so forth. These are the observed frequencies. From the table of the normal curve, we find that the proportion expected to be below -1 or 103.3 is 0.1587. Multiplying 0.1587 by the sample size of 20 gives the expected frequency of 3.174 for the first class. For the second class, the expected frequency is $(0.1498)(20)$, or 2.996. For the third class, the expected frequency is $(0.1915)(20)$, or 3.83. Since the normal curve is symmetrical, these same expected frequencies are appropriate for the classes above the mean.

$$X_\nu^2 = \frac{(2 - 3.174)^2}{3.174} + \frac{(3 - 2.996)^2}{2.996} + \frac{(5 - 3.83)^2}{3.83} + \frac{(7 - 3.83)^2}{3.83} + \frac{(0 - 2.996)^2}{2.996} +$$

$$\frac{(3 - 3.174)^2}{3.174} = 6.42 \text{ with } 3 \, df$$

Values this extreme happen about 10% of the time by chance. If we have made this test to decide whether the assumption of normality is reasonable before using a parametric test, the result will be disappointing. To be on the safe side, we will probably find a method of analyzing the data that does not depend on the assumption of normality for the population from which the sample is chosen.

9.3   **[1]** Once again, we have a situation in which a directional test is called for, but determining direction for the differences in a 3×4 table is difficult. We first make an overall test to see whether there is evidence of a relationship. Marginal totals are used to find expected frequencies for each cell. Expected frequencies by row from the top are: 8, 6, 10, 16, 8, 6, 10, 16, 4, 3, 5, and 8. $X_p^2 = \dfrac{(14 - 8)^2}{8} + \dfrac{(12 - 6)^2}{6}$ and so forth. $X_p^2 = 30.64$ with 6 df. Values of χ^2 this extreme are very rare. We will decide that there is a significant relationship. We then examine the table and see that the big differences between observed and expected frequencies are in the top row. If we wish to, we can collapse classes for the table to get a 2×2 table in which the direction of the relationship is apparent.

	One pack per day	One or more packs per day
Mild	26	14
Moderate or severe	9	51

[2] For a two-sample test of proportions, the null hypothesis is that π favoring is the same in both age groups. Best estimate of π is found by combining samples.

$$\pi = (10 + 5)/(15 + 15) = 0.5$$

The sampling distribution of differences is approximately normal, with a central tendency of 0.00 and an *SE* as follows:

$$SE_{\nu_A - \nu_B} = \sqrt{\frac{(0.5)(0.5)}{15} + \frac{(0.5)(0.5)}{15}} \quad \text{or} \quad 0.182$$

$$p_A = 10/15 \quad \text{and} \quad p_B = 5/15$$

$$z = \frac{(9.5/15 - 5.5/15) - 0}{0.182} \quad \text{or} \quad 1.46$$

Using $X_p{}^2$, all cells have an expected value of 7.5.

$$X_p{}^2 = \frac{(9.5 - 7.5)^2}{7.5} + \frac{(5.5 - 7.5)^2}{7.5} + \frac{(5.5 - 7.5)^2}{7.5} + \frac{(9.5 - 7.5)^2}{7.5} = 2.133 \text{ with } 1 \, df$$

$$z = \sqrt{2.133} \quad \text{or} \quad 1.46$$

In both solutions, z equals 1.46. You probably will make a directional test, expecting more older people to be in favor. The one-tail P value is approximately 0.0721.

[3] The probabilities are as follows:

$$P \text{ of result shown:} \quad \frac{(_8C_7)(_3C_1)}{_{11}C_8} = \frac{24}{165}$$

$$P \text{ of result} \begin{array}{|c|c|} \hline 8 & 0 \\ \hline 0 & 3 \\ \hline \end{array} : \frac{(_8C_8)(_3C_0)}{_{11}C_8} = \frac{1}{165}$$

$$\text{Sum of probabilities} = \frac{25}{165} \quad \text{or} \quad 0.1515$$

This is a directional probability. It is not an extreme result. We will probably not reject the null hypothesis.

10.1 [1] The null hypothesis is that the variance is the same as for prior test periods, or 0.8. This is another case in which you would like to prove the null hypothesis to be true. Traditional hypothesis testing with a stringent alpha level will not be appropriate. Instead, find the P value of the test statistic. The nearer it is to 1.00, the better. The procedure is a one-sample test of variance using χ^2. Observed sample variance is $8/9$, or 0.889.

$$\chi^2 = (n - 1)\frac{s^2}{\sigma^2} = (9)\frac{0.889}{0.8} = 10 \text{ with } 9 \, df$$

Reference to the table of χ^2 shows that the result is very common, with values this *great* occurring more than 25% of the time. Values this *extreme* in either direction occur more than 50% of the time. Although the results do not prove the null hypothesis to be true, they are at least somewhat reassuring.

[2] Again, if there were a way to do so, he would like to prove the null hypothesis to be true. The appropriate technique is the two-sample test of variance. We make an F ratio by making the larger sample variance the numerator.

$$F = \frac{200}{100} \quad \text{or} \quad 2.0 \text{ with } 40 \text{ and } 40 \, df$$

Reference to tables of F indicates that when $S_A{}^2$ is divided by $S_B{}^2$, an F value this *great* can be expected somewhere between 1% and 5% of the time. To make a nondirectional test, we double these probabilities and conclude that if the null hypothesis is true, sample variances will differ this much approximately 2% to 10% of the time. The t test that does not assume equal values for population variances should be used.

11.1 Means: $\overline{X}_A = 7.8$, $\overline{X}_B = 9.83$, $\overline{X}_C = 13.5$, and $\overline{X}_D = 14.33$. The ANOVA table is as follows.

Source	SS	df	ms	F
Between	183.37	3	61.12	6.12*
Within	239.63	24	9.98	
TOTAL	423.00	27		

*$P < 0.01$.

11.2   Means: $\overline{X}_A = 96.8$, $\overline{X}_B = 75.375$, $\overline{X}_C = 101.71$, $\overline{X}_D = 112.8$ and $\overline{X}_E = 80.7$. The ANOVA table is as follows.

Source	SS	df	ms	F
Between	6,355.57	4	1588.89	3.86*
Within	12,345.00	30	411.5	
TOTAL	18,700.57	34		

*$P < 0.05$.

Planned comparison of $\overline{X}_D$ with others in a nondirectional test.

Observed difference: $112.8 - \frac{1}{4}(96.8) - \frac{1}{4}(75.375) - \frac{1}{4}(101.71) - \frac{1}{4}(80.7) = 24.15$

Tabled value: two-tail value of t for an alpha level of 0.05 with 30 $df = 2.042$

$$SE_{difference} = \sqrt{411.5\left[\frac{(1)^2}{5} + \frac{(-0.25)^2}{5} + \frac{(-0.25)^2}{8} + \frac{(-0.25)^2}{7} + \frac{(-0.25)^2}{10}\right]} = 9.856$$

Interval $= 24.15 \pm (2.042)(9.856)$ or   4.02 to 44.27

We therefore conclude that there is a significant difference.
Unplanned comparison of schools A, C, and D with schools B and E.

Observed difference: $\frac{1}{3}(96.8) - \frac{1}{2}(75.375) + \frac{1}{3}(101.71) + \frac{1}{3}(112.8) - \frac{1}{2}(80.7) = 25.73$.

Tabled value: $\sqrt{(F_{4,30})(k - 1)} = \sqrt{2.69(4)} = 3.28$

$$SE_{difference} = \sqrt{411.5\left[\frac{(0.33)^2}{5} + \frac{(-0.5)^2}{8} + \frac{(0.33)^2}{7} + \frac{(0.33)^2}{5} + \frac{(-0.5)^2}{10}\right]} = 6.89$$

Interval $= 25.73 \pm (3.28)(6.89)$ or   3.13 to 48.33

We therefore conclude that there is a significant difference.

11.3   [1] The means of the groups are 72.6, 69.7, 75, and 83.5 respectively. To get a positive difference, we will use the coefficients $-\frac{1}{2}$, $-\frac{1}{2}$, $\frac{1}{2}$, and $\frac{1}{2}$ for the comparison we desire. The observed difference is: $(-\frac{1}{2})(72.6) + (-\frac{1}{2})(69.7) + (\frac{1}{2})(75) + (\frac{1}{2})(83.5)$, or 8.1. The appropriate tabled value is a modified F value obtained from the F distribution for 3 and 72 df. Since the table does not give the value, we go down to 3 and 60 df. The value in the table is 2.76. To modify this for the number of tests possible and to make it comparable to a value of t, we multiply by $k - 1$, or 3, and then take the square root. We get a value of 2.88 for the first parentheses of the confidence interval. For the SE of the difference we calculate as follows:

$$SE = \sqrt{95.42\left[\frac{(-0.5)^2}{20} + \frac{(-0.5)^2}{20} + \frac{(0.5)^2}{20} + \frac{(0.5)^2}{20}\right]} = 2.184$$

Our confidence interval is: $8.1 \pm (2.88)(2.184)$, or 1.81 to 14.39. Since the interval does not include 0.00, we decide that there is a significant difference.
[2] The ANOVA table is as follows:

Source	SS	df	ms	F
Rows	1476.22	1	1476.22	4.81
Columns	1428.02	1	1428.02	4.65
Interaction	1288.22	1	1288.22	4.2
Between cells	4192.46	3		
Within cells	11,038.30	36	306.61	
TOTAL	15,230.76	39		

The table of F (Table E-9 in Appendix E) does not list values for 1 and 36 *df*. To be conservative, we use the value for 1 and 30 *df*. The value is 4.17 for an alpha level of 0.05. Since all of the F values calculated exceed this value, we conclude that all are statistically significant. We conclude that the mean for all cats treated by method A is higher than the mean for all cats treated by method B. We also conclude that the mean for all cats treated within 48 hours of exposure is higher than the mean for all cats treated more than 48 hours after exposure. Finally, we conclude that the advantage of method A over method B changes significantly depending on whether treatment is started before or after 48 hours following exposure.

11.4 [1] The ANOVA table is as follows:

Source	SS	df	ms	F
Rows	23.058	2	11.52	1.60
Columns	40.534	1	40.53	5.60
Interaction	14.467	2	7.23	5.51
Between cells	78.06	5		
Within cells	23.63	18	1.31	
TOTAL	101.69	23		

For this random-effects design, we first test the significance of the interaction effect. The value 5.51 is found to be significant at the 0.05 level. We therefore use the $ms_{interaction}$ to test the two main effects. The resulting F ratios yield values of F of 1.6 and 5.6. Neither is found to be significant at the 0.05 level with 2 and 2 and 1 and 2 *df* respectively.

[2] The ANOVA table is as follows:

Source	SS	df	ms	F
Rows	12.69	2	6.34	2.43
Columns	9.00	1	9.0	3.45
Residual	5.20	2	2.6	
TOTAL	26.89	5		

Neither of the main effects is significant at the 0.05 level.

[3] The ANOVA table is as follows:

Source	SS	df	ms	F
Layers *(A)*	23.06	2	11.53	7.95*
Slices *(B)*	40.53	1	40.53	27.95*
Sections *(C)*	0.28	1	0.28	0.193
$A \times B$	14.47	2	7.23	4.99†
$A \times C$	1.31	2	0.65	0.45
$B \times C$	4.58	1	4.58	3.16
$A \times B \times C$	0.05	2	0.025	0.02
Between cells	84.28	11		
Within cells	17.42	12	1.45	
TOTAL	101.70	23		

*$P < 0.01$.
†$P < 0.05$.

We conclude that there is a significant effect for factors A and B. There is not a significant difference for sex. There is a significant $A \times B$ interaction effect. An examination of the means reveals that the difference for levels of factor A is much greater in combination with level 1 of factor B than with level 2 of factor B.

[4] The ANOVA table is as follows:

Source	SS	df	ms	F
Rows	6.49	2	3.24	7.7*
Columns	9.28	1	9.28	22.1*
Interaction	4.3	2	2.15	5.12†
Between cells	20.07	5		
Within cells		15	0.42	

*$P < 0.01$.
†$P < 0.05$.

The interpretation is similar to that of problem 3.

12.1   **[1a]** $SS_T = 12$; $SS_{bgroups} = 6$; $SS_{wgroups} = 6$. **[1b]** $SS_{bgroups}/SS_T = 0.50$; 50% of the total variability of scores can be attributed to differences between groups. **[1c]** $r = 0.7071$. **[1d]** $r^2 = 0.50$; 50% of the total variability of Y scores can be explained on the basis of differences in X. **[1e]** They both represent the same concept: the portion of the total variability of Y scores that can be attributed to differences between groups.

[2]
$$t = \frac{0.7071\sqrt{7}}{\sqrt{1 - 0.5}} = 2.64 \text{ with } 7 \; df$$

For a one-tail test, the t value must be 2.998 to be significant at the 0.01 level. We therefore fail to reject the null hypothesis that $R = 0.00$.

[3]
$$z = \frac{0.707 - 0}{\dfrac{1}{\sqrt{37 - 1}}} = 4.242$$

Values greater than 2.326 are significant at the 0.01 level. We would therefore reject the null hypothesis.

[4] Using Fisher's transformation we get:
$$z = \frac{1.020 - 0.510}{\sqrt{\dfrac{1}{18} + \dfrac{1}{18}}} = 1.53$$

We conclude that there is not a significant difference.

[5] We convert 0.68 to a transformed r of 0.829 and then find transformed boundaries of $0.829 \pm (1.96)(1/\sqrt{25})$, or 0.437 to 1.221. From the table, we convert these boundaries to ordinary correlation coefficients of 0.41 to 0.84.

12.2   **[1]** $r_{xy} = 0.572$; $t = (0.572\sqrt{8})/\sqrt{1 - 0.572^2}$, or 1.97 with 8 df. We fail to reject the null hypothesis that $R = 0.00$, using an alpha level of 0.05 for the two-tail test.
[2] $r_{pb} = 0.597$; $t = (0.597\sqrt{8})/\sqrt{1 - 0.597^2}$, or 2.10 with 8 df. We fail to reject the null hypothesis that $R_{pb} = 0.00$.
[3] $\phi = 0.60$; $\chi^2 = \dfrac{1^2}{2.5} + \dfrac{1^2}{2.5} + \dfrac{1^2}{2.5} + \dfrac{1^2}{2.5}$, or 1.6 with 1 df; $\sqrt{\chi^2}$ z; therefore, $z = 1.26$. We fail to reject the null hypothesis that ϕ in the population is 0.00.
[4] $r_S = 0.594$; $t = (0.594\sqrt{8})/\sqrt{1 - 0.594^2}$, or 2.09 with 8 df. We fail to reject the null hypothesis that $R_S = 0.00$.

[5]

$$w = \frac{12\left[3290.5 - \dfrac{165^2}{10}\right]}{10(3)^2(99)} = \frac{6816}{8910} = 0.765$$

$$\text{Average } r_A = \frac{(3)(0.765) - 1}{3 - 1} = \frac{1.295}{2} = 0.648$$

[6]

$$\text{Eta} = \sqrt{\frac{11}{125.6}} = \sqrt{0.088} = 0.297$$

$$F = \frac{[0.088 - (-0.21)^2]/2}{(1 - 0.088)/36} = 0.866 \text{ with 2 and 36 } df$$

We decide that there is not sufficient evidence against linearity.

12.3 **[1]** $r = 0.473$. **[2]** $\hat{Y}_i = 539.61 + (0.46)(X_i)$. **[3]** To find SE_{est} by the long way, use the formula from answer 2 to predict each Y score. Then find the difference between predicted and actual Y scores. Square these differences and add. The result should be 370,120.5. $SE_{est} = \sqrt{370,120.6/23} = 126.85$. Using computational formula, $SE_{est} = (\sqrt{24/23})$ $(140.9)(\sqrt{1 - 0.473^2})$, or 126.87. Note that s had to be calculated using $n - 1$. **[4]** To find the coefficient of determination, we square r; $r^2 = 0.473^2 = 0.224$. The interpretation is that approximately 22% of the observed variance of Y can be explained on the basis of differences in variable X. **[5]** In the example in the text, the predictor variable was a pure variable that had a 50% overlap with the criterion variable, and r^2 was approximately 50%. In this exercise, the predictor has two components, one of which has no relationship to the criterion. Half of the predictor variance makes up half of the criterion variance. As a result, r^2 is near 25%.

12.4 **[1]** Means: 4.064, 77.4, 68.95. **[2]** Standard deviations: 0.439, 23.55, 11.52. **[3]** Correlations: X and $Y = 0.484$, X and $Z = 0.553$, Y and $Z = 0.75$. **[4]** The formula is $Z_i = 18.498 + 6.454(X_i) + 0.308(Y_i)$. **[5]** Multiple correlation is 0.781; multiple correlation squared is 0.610. **[6]** $SE_{est} = 7.61$. **[7]** Partial correlations are: for X, 0.329 and for Y, 0.661.

13.1 **[1]**

Source	SS	df	ms	F
Between	244.45	2	122.22	1.62
Within	1282.1	17	75.42	
TOTAL	1526.55	19		

The F value of 1.62 with 2 and 17 df would not be significant at an alpha level of 0.10. **[2]** Method 1 residuals = 479.2; method 2 residuals = 574.79; method 3 residuals = 819.2; $F_1 = 1.397$ with 2 and 14 df. It is reasonable to conclude that slopes are equal. $F_2 = 3.4$ with 2 and 16 df. We conclude that adjusted means differ. In this case, where $\overline{X}$'s are all equal, adjustment has no effect; so adjusted means are the same as original means: 12.2, 16.625, and 21.285. Confidence intervals are: $(\mu_A - \mu_B)$, -10.39 to 1.54; $(\mu_A - \mu_C)$, -15.2 to -2.95; $(\mu_B - \mu_C)$, -10.08 to 0.75. We conclude that there is a significant difference between μ_A and μ_C. The use of ANCOVA has provided a more powerful test than ANOVA, even though all groups had equal means for the covariate.

[3] Method 1 residuals = 479.17; method 2 residuals = 574.79; method 3 residuals = 691.69; F_1 = 1.39 with 2 and 14 *df*; F_2 = 1.63 with 2 and 16 *df*. Neither *F* is significant. The conclusion is that the observed differences in means for *Y* can be explained on the basis of differences in means for *X*.

14.1 [1] For the ten scores, three are above 75 and seven are below. With the correction for continuity, $X_p{}^2$ = 0.9, z = $\sqrt{0.9}$ = 0.95. Since this is not an extreme value, we do not reject the null hypothesis that the median is 75. [2] The sum of the positive ranks is 7. The expected sum of ranks is 27.5. The standard error is 9.81. The distribution is approximately normal. z = −2.09. The two-tail probability is 0.0366. We will reject the null hypothesis that the median is 75. [3] The median of the combined samples falls between 65.3 and 69. For sample *C*, six scores are above 67.15 and four are below. For sample *D*, just the reverse is true. With the correction for continuity, $X_p{}^2$ = 0.2. z = $\sqrt{0.2}$ = 0.45. Since this is not an extreme value, we do not reject the null hypothesis that the populations have a common median. [4] U_D = 40, the expected value of U is 50, SE_U = 13.23, the distribution of *U* is approximately normal, and z = −0.76. Although the value of z is more extreme than it was for the two-sample median test, it is still not very unusual. We will probably not reject the null hypothesis that the two samples come from populations having identical distributions. [5] If the measurement for *D* is subtracted from that of *C*, there are three negative differences and seven positive differences. With the correction for continuity, $X_p{}^2$ = 0.9, z = $\sqrt{0.9}$, or 0.95. The value is more extreme than that found for the unmatched data of the Mann-Whitney *U* Test, but it is still not very unusual. We will probably not reject the null hypothesis that the samples are from populations with identical distributions. [6] The absolute value of negative signed ranks is 6, the expected value of *T* is 27.5, and SE_T = 9.81. The distribution is approximately normal, z = 2.19, and the two-tail probability is 0.0285. We will reject the null hypothesis that the samples come from populations with identical distributions.

14.2 [1] The median of the combined samples falls between 76.3 and 77.3. For sample *A*, seven scores are above 76.8 and three are below. For samples *B* and *C*, the corresponding figures are five and five, and eight and twelve respectively. The value of $X_p{}^2$ is 2.4 with 2 *df*. Since the test statistic has a probability greater than 0.25, we will not reject the null hypothesis that the samples come from populations with a common median. [2] T_C = 500, T_D = 185.5, T_E = 134.5, T = 820; $\bar{R}_C$ = 25, $\bar{R}_D$ = 18.55, $\bar{R}_E$ = 13.45, $\bar{R}_T$ = 20.5; H = 6.88 with 2 *df*. From the table of χ^2 we find the value to be significant at the 0.05 level. There are three possible comparisons. Each confidence interval is set at 0.933. In each case, the tabled value is 1.84. For the comparisons involving *C*, *SE* is 4.53; for the comparison involving *D* and *E*, *SE* is 5.23. The confidence intervals are: ($\bar{R}_C$ − $\bar{R}_D$), −1.88 to 14.78; ($\bar{R}_C$ − $\bar{R}_E$), 3.22 to 19.88; and ($\bar{R}_D$ − $\bar{R}_E$), −4.52 to 14.72. We conclude that there is a significant difference between the distributions of populations *C* and *E*. [3] The average ranks are: 4.8, 4.0, 2.9, 2.1, and 1.2. $\bar{R}_T$ = 3.0; S = 33.2 with 4 *df*. From the table of χ^2, we find that the result is significant at the 0.001 level. If we are making only comparisons involving *A*, there are four that are possible. Individual confidence intervals will be set at $1 - \dfrac{0.2}{4}$, or 0.95. The tabled value is 1.96. *SE* is 0.71. The confidence intervals are: ($\bar{R}_A$ − $\bar{R}_B$), −0.59 to 2.19; ($\bar{R}_A$ − $\bar{R}_C$), 0.51 to 3.29; ($\bar{R}_A$ − $\bar{R}_D$), 1.31 to 4.09; and ($\bar{R}_A$ − $\bar{R}_C$), 2.21 to 4.99. We conclude that the distribution of population *A* is different from the distributions of populations *C*, *D*, and *E*.

A.2 [1] City A: $^{478}/_{9100} = 0.0525$, or 52.5 per 1000. City B: $^{402}/_{10000} = 0.0402$, or 40.2 per 1000. [2] Event: pregnancy among high school girls. Related variable: age of girl. City A has relatively more individuals in the high age groups, which are the high-risk categories. We would expect city A to have an adjusted crude rate lower than its crude rate and city B to have an adjusted crude rate higher than its crude rate.

[3]

		City A		City B	
Age	Standard population	Age-specific rate	Expected pregnancies	Age-specific rate	Expected pregnancies
15	6500	0.022	143	0.018	117
16	5300	0.04	212	0.04	212
17	4200	0.055	231	0.06	252
18	3100	0.10	310	0.09	279
		TOTAL	896	TOTAL	860

Age-adjusted crude rate

City A 896/19,100 × 1000 = 46.9 per 1000
City B 860/19,100 × 1000 = 45.0 per 1000

[4] It appears that the rates of pregnancies among high-school girls in the two cities are about the same, after taking into account the different age distributions.

A.3 **Note:** Standardized mortality ratio *(SMR)* and force of mortality are demographic terms that were developed for mortality studies. In the context of studies of other events, such as pregnancy and morbidity, these terms are still used, rather than constructing new labels for each application.
[1] $^{300}/_{5000} = 0.06$, or 60 per 1000. [2] City C has proportionately more students in the high-risk group (high ages). Hence, we expect the handicap factor to be less than 1.

[3]

Age-specific pregnancy rates (standard population)	Expected pregnancies in city C (study population)
0.022	22
0.04	40
0.055	55
0.10	200
TOTAL	317

$SMR = {}^{300}/_{317} = 0.9464$
Index rate $= {}^{317}/_{5000} \times 1000 = 63.4$ per 1000
Handicap factor $= {}^{52.5}/_{63.4} = 0.8281$
Adjusted rate $=$ (crude rate for the study population)(handicap factor)
           $= 0.06(0.8281) = 0.0497$, or 49.7 per 1000
or adjusted rate $=$ (crude rate for the standard population)*(SMR)*
           $= 0.0525(0.9464) = 0.0497$, or 49.7 per 1000

An *SMR* of less than 1 implies fewer pregnancies for girls in city C than for girls of similar ages in city A. The index rate indicates that the crude rate in city C would be 63.4 per 1000 if the "forces for pregnancy" were the same in city C as in city A. The handicap factor of 0.8281 is indicative of the fact that a large proportion of the girls in city C are in the high-risk categories (higher ages).
[4] It appears that after adjusting for age, the pregnancy rates in city A and city C are about the same (*SMR* is about 1).

A.4

x	Number observed	Number dying	q_x	l_x	d_x	L_x	T_x	$\overset{o}{e}_x$
0	1000	200	0.2	10,000	2000	9000	25,720	2.57
1	1000	250	0.25	8000	2000	7000	16,720	2.09
2	1000	300	0.3	6000	1800	5100	9720	1.62
3	1000	500	0.5	4200	2100	3150	4620	1.10
4	1000	800	0.8	2100	1680	1260	1470	0.70
5	1000	1000	1.0	420	420	210	210	0.50

A.5

Period	O_x	W_x	d_x	N_x	q_x	s_x	P_x
0	1000	20	11	990.0	.0111	.9889	1.0
1	969	21	28	958.5	.0292	.9708	.9889
2	920	19	37	910.5	.0406	.9594	.96
3	864	38	34	845.0	.0402	.9598	.9210
4	792	59	47	762.5	.0616	.9384	.8839
5	686	22	64	675.0	.0948	.9052	.8294
6	600	63	62	568.5	.1091	.8909	.7508
7	475	67	51	441.5	.1155	.8845	.6689
8	357	92	37	311.0	.1190	.8810	.5917
9	228	84	28	186.0	.1505	.8495	.5213
10	116	63	21	84.5	.2485	.7515	.4428
11	32	22	10	21.0	.4762	.5238	.3327
12							.1743

The probability of a kidney graft still functioning 6 years after the operation is $P_{12} = 0.1743$.

Comprehensive example of hypothesis testing

KINGS AND COINS
Problem statement

Suppose for a moment that you are the Imperial Wizard of a mythical land with a numbering system that extends only to 100. The land contains 100 provinces, each governed by an overseer who brings annual taxes to the king at a 1-day festival. This is a rich kingdom, and each overseer is expected to bring three bags of coins, each equal to his or her own weight. One bag is to contain all gold coins, the second all silver, and the third is to be equally divided between gold and silver.

It happens that the king is an inveterate gambler who even gambles on tax collections. As each overseer brings the bags to him, he draws a single coin from the mixed bag. If it is a silver coin, he keeps the mixed bag and the bag of silver coins, and the overseer gets to keep the gold coins. If he draws a gold coin, he keeps the mixed bag and the bag of gold coins, and the overseer must be satisfied with the silver coins.

The king understands human nature well enough to know that without some safeguards, an overseer might be inclined to improve his or her chances of keeping the gold coins by putting a higher percentage of silver coins in the mixed bag. He calls upon you as Imperial Wizard to tell him if any overseer tries to cheat him. Anyone you say is cheating will be beheaded. Describe in detail the process you will use to make your decision. Remember, you are unable to count beyond 100 and you must decide about the honesty of all 100 overseers on the same day. Use your knowledge of sampling and hypothesis testing to set up a decision rule.

Relevant questions

1. What is the test popuation and what parameter are you interested in?
2. What is the standard population and what is the parameter value?
3. What is the null hypothesis and is it directional or nondirectional? What is the alternative hypothesis?
4. What is a Type I error in this problem and what is its consequence?

5. What is a Type II error and what is its consequence?
6. What alpha level will you choose and what does your selection of alpha level imply?
7. What is the minimum-size sample you will choose in order to use the normal curve? Why?
8. How would you describe the sampling distribution of sample p values taken from the standard population?
9. How would you describe the critical region you will establish, and how would you translate this description into a decision rule, such as, "If more than nine out of ten coins chosen from the mixed bag are silver, I will decide that the overseer is cheating"?

Additional information

It happens that there are three overseers who are cheating. The percentages of silver coins in the mixed bags are 0.51, 0.60, and 0.80 respectively.

Additional relevant questions

10. What is the probability that each of the cheats will be detected by your decision rule? (Hint: What will be the sampling distribution of p values from samples taken from each of their bags, and what percentage of the sample values will fall in the region you have called critical on your sampling distribution from the standard population?)
11. What is the probability of a Type I and Type II error for an alpha level equal to 0.001 for an innocent overseer and for each of the cheating overseers with sample sizes of both 50 and 100?
12. Finally, judging by this problem, what affects the probability of Type I errors and what affects the probability of Type II errors?

Answers

1. You are interested in π, the percentage of silver coins in the mixed bag provided by each overseer. Each overseer's mixed bag represents a test population that is to be compared with the standard population. We will call the percentage of silver coins in each overseer's bag π_T, where the T stands for test population.
2. The standard population is a hypothetical bag of mixed coins with exactly half silver and half gold coins. The parameter for the standard population is therefore known to be 0.50. In symbols, $\pi_S = 0.50$, where S stands for standard population.
3. Since the king will be cheated only if the percentage of silver coins is greater than 0.5, the test will be directional. The null hypothesis will be that $\pi_T \leq \pi_S$, or $\pi_T \leq 0.5$. The alternative is that $\pi_T > \pi_S$, or $\pi_T > 0.5$.
4. You make a Type I error when you decide that the null hypothesis is false when in fact it is not. In other words, a Type I error occurs when you erroneously reject the null hypothesis and conclude that the alternative is correct. In this problem, a Type I error means that you decide that an overseer has stacked the bet in his or her favor by having a majority of silver coins in the bag when in fact the overseer is innocent and

the percentage of silver coins in the bag is 0.50. The consequence of such an error is that an innocent person will be executed.

5. You make a Type II error when you fail to reject a false null hypothesis. In other words, a Type II error occurs when you fail to identify a difference that does exist between the parameter of the test population and the parameter of the standard population. In this instance, a crooked overseer gets away with improving his or her chances of keeping the gold coins.

6. The alpha level chosen refers to the probability of making a Type I error when honest overseers are considered. In this problem, it is the probability that an innocent person will be decided to be guilty. Notice that this is a conditional probability. If the alpha level is chosen to be 0.05, you will not necessarily make a Type I error 5% of the time. If all the overseers are cheating, you will make no Type I errors, since the null hypothesis is always false.

Traditionally, the alpha level is chosen to be 0.05 or 0.01. This is a tradition that needs to be examined in the context of each experiment. As we will learn, guarding stringently against Type I errors increases the risk of Type II errors. There-fore, the comparative cost of each type of error must be considered when you choose the alpha level. Suppose that the king were going to check up on your decisions to see whether you had allowed anyone to cheat him. Suppose further that you would be executed as a penalty for such an error. You would be very concerned about Type II errors. On the other hand, suppose that the penalty for Type II errors were small and you were a humane wizard who would not want to be responsible for the death of an innocent person. In this case, you would guard against Type I errors.

Let us assume that the penalty for a Type II error is small and that you do not want to cause the death of an innocent person. Assume that each year you make a decision about each of 100 overseers, most of whom are honest. If you choose an alpha level of 0.05, about five innocent overseers will be executed yearly. For an alpha level of 0.01, about one innocent overseer will be executed per year. For an alpha level of 0.001, about one innocent overseer will die every 10 years. Finally, for an alpha level of 0.0001, about one innocent overseer will be killed every 100 years. Let us assume that we want to guard very stringently against being responsible for the death of an innocent person, so we choose an alpha level of 0.0001.

7. If you wish to make use of the normal curve for the sampling distribution of p, both $n\pi$ and $n(1 - \pi)$ must be ≥ 5. Since π for the standard population is 0.5, the minimum sample you could choose would be ten coins. Later in the analysis of this problem we will need to make use of the sampling distribution of p from a population in which $\pi = 0.80$. Since in this instance $(1 - \pi) = 0.2$, the minimum sample we can choose and still make use of the normal curve is 25. To show the effect of sample size on the results of hypothesis testing, we will later compare samples of 50 and 100. For now, let us decide to use a sample of 100, which is as high as our numbering system goes.

8. With $n = 100$ and π equal to 0.5, the sampling distribution of p's will be normal

with an average equal to π, or 0.5, and a standard error equal to $\sqrt{\pi(1-\pi)/n}$, or $\sqrt{(0.5)(0.5)/100}$, or 0.05.

9. We are using a one-tail test with an alpha level equal to 0.0001. Looking at the table of the normal curve, we find that the z value that is exceeded only one time in 10,000 is 3.719. In other words, the probability of a z value greater than 3.719 is 0.0001. Since a z score is the number of standard deviations above the mean, we need to find the sample proportion that is 3.719 standard deviations above the mean of our sampling distribution. The value is $\pi + 3.719\sqrt{\pi(1-\pi)/n}$, or 0.5 + (3.719)(0.05), or 0.686. The interpretation is that if we take samples of size 100 from the mixed bags of innocent overseers, only one time in 10,000 would we expect to get a $p \geq 0.686$. Since we have been pretty conservative in choosing an alpha level of 0.0001, let us make the following our decision rule: "When an overseer presents the bag of mixed coins, I will choose a sample of 100 coins at random. If there are 68 or more silver coins, I will inform the king that the overseer is cheating."

10. Since we are using 68 as the dividing point, we need to know how often a sample will have 68 or more silver coins when taken from each of the mixed bags of the cheating overseers. For this, we need to know the sampling distribution of p for samples of size 100 from populations in which π equal 0.51, 0.60, and 0.80 respectively.

    Let us start with the mixed bag containing 0.80 silver coins. With samples of size 100, the sample p's will be normally distributed, with an average equal to π, or 0.80, and a standard error equal to $\sqrt{(0.80)(0.20)/100}$, or 0.04. To find the probability of 68 or more silver coins, we find the z score for 0.68. The z scores equal (observed score − expected score)/standard deviation, or (0.68 − 0.80)/0.04, or −3.0. Looking at the table of the normal curve, we find that 99.86% of all z scores are above −3.0. Translated, this means that when we take samples of size 100 from the mixed bag in which $\pi = 0.80$, 99.86% of the time we will get 68 or more silver coins, and the guilty overseer will be found out. Alternatively, the probability of failing to detect the cheating (a Type II error) is 0.0014.

    Using the same procedure, the z score for $p = 0.68$ when samples of size 100 are chosen from a population where $\pi = 0.60$ is $+1.63$. Using the normal curve table, we find that samples of 68 or more silver coins will occur only 5.15% of the time. In other words, about 5.15% of the time the overseer with 60% silver coins will be discovered. The rest of the time (94.88%) the overseer will be undetected (Type II error).

    For $\pi = $ to 0.51, the z score is $+3.4$. This means that samples with 68 or more silver coins will occur only 0.03% of the time. The probability of detecting cheating by an overseer with 51% silver coins is only 0.0003. The probability that the overseer will go undetected (Type II error) is 0.9997.

11. Since we have already done the problem for $n = 100$, we need only repeat it for $n = 50$. The sampling distribution of p values for samples of size 50 from the standard

population in which $\pi = 0.50$ will be normal, with an average equal to 0.5 and a standard error equal to $\sqrt{(0.5)(0.5)/50}$, or 0.0707.

The p value that is 3.719 standard deviations above the mean would be 0.763. For a sample of 50 coins, 38 silver coins would be the closest equivalent to a p of 0.763. Our decision rule would be: Take a sample of 50 coins; if 38 or more are silver, inform the king that the overseer is cheating.

When samples of size 50 are taken from a bag in which $\pi = 0.80$, the distribution of p is again normal, with an average of 0.80 and a standard error of $\sqrt{0.0032}$. The cutoff point already determined is 38 or more silver coins, which is 76%. The z score for $0.76 = -0.7071$. This means that samples with 38 or more silver coins will occur about 76.11% of the time, and the guilty overseer will be caught. About 23.89% of the time, the difference will go undetected (Type II error).

The z score for 76% silver coins from the bag in which $\pi = 0.60$ is 2.3. This means that only about 1.04% of the samples will have 38 or more silver coins. About 98.96% of the guilty overseers with 60% silver coins will go undetected (Type II error).

The z score for 76% silver coins from the bag in which $\pi = 0.51$ is about 3.54. Samples with 38 or more silver coins will be very rare (0.02%). Almost all (99.98%) of the overseers with 51% silver coins will be undetected (Type II error).

12. In your solution to the problem, you probably chose some alpha level other than 0.0001. Your solution, along with the two solutions just provided, allows you to see how each of several factors affects the probability of Type I and Type II errors. In general, the problem should confirm in your mind the following statements:
   a. The probability of making a Type I error is set by the experimenter when he or she decides on the alpha level to be used. This probability will hold any time there is actually no difference between the parameter values of the standard population and of the test population. It will not be affected by sample size, variability in the population, or any other factor.
   b. The probability of a Type II error depends on:
      (1) The chosen alpha level.
      (2) The size of sample.
      (3) The actual difference between the parameter of the standard and the parameter of the best population.
      (4) The variability in the population.

You should be able to look at the example and find how each of these factors affects the probability of a Type II error. The summary table that follows will help. It shows the probability of Type I and Type II errors with two different sample sizes, three different alpha levels, and four different values of π in the overseers' bags. Notice that when π equals 0.50, the probability of a Type II error is zero and that when π is any value greater than 0.50, the probability of a Type I error is zero. It should be obvious from the table that the most important factor in determining the probability of a Type II error is the size of the actual difference between π_T and π_S.

The last item listed as a factor in determining the probability of a Type II error is variability in the population. It is a bit tricky to understand. You should have seen that the standard error of p gets smaller as the population gets closer and closer to all silver coins. The biggest standard error of p for any given sample size occurs when the population contains the most variability, that is, when half the coins are silver and half are gold. This concept is easier to see when we consider $\overline{X}$'s rather than p's, because then the $SE_{\overline{x}}$ is affected directly by σ, the standard deviation in the population.

Probability of Type I and Type II errors

	Alpha level	$\pi = 0.50$ Type I	Type II	$\pi = 0.51$ Type I	Type II	$\pi = 0.60$ Type I	Type II	$\pi = 0.80$ Type I	Type II
$n = 50$	0.01	0.01	0	0	0.9830	0	0.8078	0	0.0067
	0.001	0.001	0	0	0.9985	0	0.9582	0	0.0793
	0.0001	0.0001	0	0	0.9998	0	0.9896	0.	0.2399
$n = 100$	0.01	0.01	0	0	0.9772	0	0.5793	0	0.0000
	0.001	0.001	0	0	0.9974	0	0.8461	0	0.0001
	0.0001	0.0001	0	0	0.9997	0	0.9485	0	0.0014

Final note: Have you wondered how you, as Imperial Wizard, are expected to know the meaning of a number like 0.0004 when your numbering system only goes up to 100? Unfortunately, that is one of the mysteries forever lost in antiquity.

Important formulas appearing in the text

3.1-1   Population mean:

$$\mu = \frac{\Sigma X}{N}$$

3.1-2   Population mean from a frequency distribution:

$$\mu = \frac{\Sigma fM}{N}$$

3.2-1   Deviation score:

$$x_i = X_i - \mu$$

where i = any individual

3.2-2   Average deviation:

$$\text{Average deviation} = \frac{\Sigma |x|}{N}$$

3.2-3   Population variance:

$$\sigma^2 = \frac{\Sigma x^2}{N}$$

3.2-4   Population standard deviation:

$$\sigma = \sqrt{\frac{\Sigma x^2}{N}}$$

3.2-5   Sum of squared deviations:

$$\Sigma x^2 = \Sigma X^2 - \frac{(\Sigma X)^2}{N}$$

3.2-7   Computational formula, population standard deviation:

$$\sigma = \sqrt{\frac{\Sigma X^2 - \dfrac{(\Sigma X)^2}{N}}{N}}$$

3.2-9 Coefficient of variation:

$$\text{Coefficient of variation} = \frac{\text{standard deviation}}{\text{mean}} \times 100$$

3.3-1 Standard score:

$$z_i = \frac{X_i - \mu}{\sigma}$$

3.3-2 Correlation coefficient, definitional formula:

$$r_{XY} = \frac{\Sigma z_X z_Y}{n}$$

3.3-5 Correlation coefficient, computational formula:

$$r_{XY} = \frac{\Sigma XY - \dfrac{(\Sigma X)(\Sigma Y)}{n}}{\sqrt{\left[\Sigma X^2 - \dfrac{(\Sigma X)^2}{n}\right]\left[\Sigma Y^2 - \dfrac{(\Sigma Y)^2}{n}\right]}}$$

4.1-1 Probability of event *A*:

$$P(A) = \frac{N_A}{N}$$

4.1-2 Probability of events *A* and *B*:

$$P(A \text{ and } B) = P(A) \times P(B|A)$$

4.1-3 Probability of event *A* for all instances:

$$P(\text{all } A) = [P(A)]^n$$

4.1-4 Probability of event *A* or *B* or both:

$$P(A \text{ or } B \text{ or both}) = P(A) + P(B) - P(A \text{ and } B)$$

4.2-1 Number of possible combinations:

$$_nC_k = \frac{n!}{(n-k)!\,k!}$$

4.2-2 Number of possible permutations:

$$_nP_k = \frac{n!}{(n-k)!}$$

4.2-3 Probability of exactly *k* out of *n* positive outcomes:

$$P(\text{exactly } k \text{ out of } n) = [_nC_k][P(A)]^k[P(NA)]^{n-k}$$

4.2-4 Poisson probabilities:

$$P(k \text{ occurrences}) = \frac{(\mu)^k}{k!e^\mu}$$

4.2-5 Bayes' theorem:

$$P(A|B) = \frac{P(A) \cdot P(B|A)}{[P(A) \cdot P(B|A)] + [P(NA) \cdot P(B|NA)]}$$

5.1-1 Sample standard deviation:

$$s = \sqrt{\frac{\Sigma x^2}{n-1}}$$

5.2-2 Standard error of a proportion:

$$SE_p = \sqrt{\frac{\pi(1-\pi)}{n}}$$

5.3-1 Standard error of a mean:

$$SE_{\bar{X}} = \frac{\sigma}{\sqrt{n}}$$

5.3-2 Standard error of a mean, sampling without replacement:

$$SE_{\bar{X}} = \left(\sqrt{\frac{N-n}{N-1}}\right)\left(\frac{\sigma}{\sqrt{n}}\right)$$

6.1-2 Confidence interval:

Parameter = observed statistic ±
(appropriate tabled value)(standard error of the statistic)

6.2-1 Standard score for a sample mean:

$$z_{\bar{X}} = \frac{\bar{X} - \mu}{\sigma/\sqrt{n}}$$

6.2-2 *t* score for a sample mean:

$$t_{\bar{X}} = \frac{\bar{X} - \mu}{s/\sqrt{n}}$$

8.1-1 Standard error of a difference between two statistics:

$$SE_{\text{stat } A - \text{stat } B} = \sqrt{(SE_{\text{stat } A})^2 + (SE_{\text{stat } B})^2}$$

8.1-2 Standard error of a difference between sample proportions:

$$SE_{p_A - p_B} = \sqrt{\frac{\pi_A(1-\pi_A)}{n_A} + \frac{\pi_B(1-\pi_B)}{n_B}}$$

8.1-3 Standard error of a difference between sample means:

$$SE_{\bar{X}_A - \bar{X}_B} = \sqrt{\frac{\sigma_A{}^2}{n_A} + \frac{\sigma_B{}^2}{n_B}}$$

8.1-4 Pooled estimated standard deviation:

$$s = \sqrt{\frac{\Sigma x_A^2 + \Sigma x_B^2}{n_A + n_B - 2}}$$

8.1-5 Variance of sum of variables A and B:

$$\sigma_{A+B}^2 = \sigma_A^2 + \sigma_B^2 + 2r_{AB}\sigma_A\sigma_B$$

8.1-6 Variance of difference of variables A and B:

$$\sigma_{A-B}^2 = \sigma_A^2 + \sigma_B^2 - 2r_{AB}\sigma_A\sigma_B$$

8.2-1 Degrees of freedom for t test with unequal variances:

$$df = \frac{[(SE_{\bar{x}_A})^2 + (SE_{\bar{x}_B})^2]^2}{\dfrac{(SE_{\bar{x}_A})^4}{n_A} + \dfrac{(SE_{\bar{x}_B})^4}{n_B}}$$

8.2-2 Standard normal score for a difference between sample means, σ known:

$$z = \frac{(\bar{X}_A - \bar{X}_B) - 0}{\sqrt{\dfrac{\sigma_A^2}{n_A} + \dfrac{\sigma_B^2}{n_B}}}$$

8.2-3 t score for a difference between sample means, σ unknown but assumed equal:

$$t = \frac{(\bar{X}_A - \bar{X}_B) - 0}{\sqrt{\dfrac{s^2}{n_A} + \dfrac{s^2}{n_B}}}$$

8.2-4 t score for a difference between sample means, σ unknown but assumed unequal:

$$t = \frac{(\bar{X}_A - \bar{X}_B) - 0}{\sqrt{\dfrac{s_A^2}{n_A} + \dfrac{s_B^2}{n_B}}}$$

8.2-5 t score for a difference between means, correlated samples:

$$t = \frac{\bar{D} - 0}{s_D/\sqrt{n}}$$

9.1-1 Pearson's chi-square statistic:

$$X_p^2 = \sum \frac{(f_o - f_e)^2}{f_e}$$

9.2-1 Pearson's chi-square statistic with correction for continuity:

$$X_p^2 = \sum \frac{(|f_o - f_e| - 0.5)^2}{f_e}$$

9.3-1 Pearson's chi-square statistic with correction for continuity, 2×2 table:

$$X_p^2 = \frac{(|ad - bc| - n/2)^2 \, n}{(a + b)(a + c)(b + d)(c + d)}$$

9.3-2 Probability for Fisher's exact test:

$$P = \frac{[_{m_1}C_{f_1}][_{m_2}C_{f_2}]}{_{T}C_{m_3}}$$

10.1-1 Definition of chi-square:

$$\chi^2 = \sum \left(\frac{X_i - \mu}{\sigma}\right)^2$$

10.1-2 One-sample test of variance:

$$\chi^2 = (n - 1)\frac{s^2}{\sigma^2}$$

10.1-3 Two-sample test of variances:

$$F = s_A^2/s_B^2$$

11.1-1 SS_{wcells} for ANOVA:

$$SS_{wcells} = SS_T - SS_{bcells}$$

11.1-2 SS_T for ANOVA:

$$SS_T = \Sigma X^2 - \frac{T^2}{n}$$

11.1-3 SS_{bcells} for ANOVA:

$$SS_{bcells} = \left(\sum \frac{T_j^2}{n_j}\right) - C$$

11.2-1 SE for individual comparisons in ANOVA:

$$SE = \sqrt{ms_{wcells} \left(\sum \frac{c_j^2}{n_j}\right)}$$

12.1-1 Coefficient of determination:

$$\text{Coefficient of determination} = r^2$$

12.1-2 Standard error of r:

$$SE_r = \frac{1 - R^2}{\sqrt{n - 1}}$$

12.1-3 Test of $R = 0$, small samples:

$$t = \frac{r\sqrt{n - 2}}{\sqrt{1 - r^2}}$$

12.1-4 Test of $R = $ nonzero value:

$$z = \frac{r_F - R_F}{1/\sqrt{n - 3}}$$

12.2-1   Phi coefficient:

$$\phi = \frac{ad - bc}{\sqrt{(a + b)(a + c)(c + d)(b + d)}}$$

12.2-3   Coefficient of concordance:

$$w = \frac{12\left[\Sigma T^2 - \frac{(\Sigma T)^2}{n}\right]}{nm^2(n^2 - 1)}$$

12.2-5   Eta coefficient:

$$\text{Eta} = \sqrt{\frac{SS_{bgroups}}{SS_{total}}}$$

12.2-6   Test for linearity:

$$F = \frac{(\text{eta}^2 - r^2)/(k - 2)}{(1 - \text{eta}^2)/(n - k)}$$

12.3-2   One-variable prediction of *Y:*

$$\hat{Y}_i = a + bX_i$$

12.3-3   Slope of regression line, one predictor:

$$b = \frac{\Sigma xy}{\Sigma x^2}$$

12.3-5   Intercept of regression line, one predictor:

$$a = \text{mean}_Y - b(\text{mean}_X)$$

12.3-8   Sum of residuals from regression:

$$\Sigma(Y - \hat{Y})^2 = \Sigma y^2 - \frac{(\Sigma xy)^2}{\Sigma x^2}$$

12.3-9   Sample SE_{est}, conceptual:

$$SE_{est} = \sqrt{\frac{\Sigma(Y_i - \hat{Y})^2}{n - 1 - k}}$$

12.3-10   Sample SE_{est}, computational:

$$SE_{est} = \sqrt{\frac{n - 1}{n - 1 - k}}\ s_Y\sqrt{1 - r^2}$$

12.4-1   Two-variable prediction of *Y:*

$$\hat{Z}_i = a + b_X X_i + b_Y Y_i$$

12.4-2   Weight for predictor 1, two-predictor case:

$$b_X = \frac{s_Z}{s_X}\left[\frac{r_{ZX} - r_{ZY}r_{XY}}{1 - r_{XY}^2}\right]$$

12.4-3   Weight for predictor 2, two-predictor case:

$$b_Y = \frac{s_Z}{s_Y}\left[\frac{r_{ZY} - r_{ZX}r_{XY}}{1 - r_{XY}^2}\right]$$

12.4-4 Y = intercept; two-predictor case:

$$a = \bar{Z} - b_X\bar{X}_i - b_Y\bar{Y}_i$$

12.4-5   Multiple correlation coefficient:

$$r_{Z\cdot XY} = \sqrt{\frac{r_{ZX}^2 + r_{ZY}^2 - 2r_{ZX}r_{ZY}r_{XY}}{1 - r_{XY}^2}}$$

12.4-6   Partial correlation coefficient:

$$r_{ZX\cdot Y} = \frac{r_{ZX} - r_{ZY}r_{XY}}{\sqrt{1 - r_{ZY}^2}\sqrt{1 - r_{XY}^2}}$$

12.4-7   Partial correlation coefficient:

$$r_{ZY\cdot X} = \frac{r_{ZY} - r_{ZX}r_{XY}}{\sqrt{1 - r_{ZX}^2}\sqrt{1 - r_{XY}^2}}$$

13.1-1   ANCOVA test for equal slopes:

$$F_1 = \frac{\dfrac{Res_2 - Res_1}{k - 1}}{\dfrac{Res_1}{n_T - 2k}}$$

13.1-2   ANCOVA test for equal intercepts:

$$F_2 = \frac{\dfrac{Res_3 - Res_2}{k - 1}}{\dfrac{Res_2}{n_T - k - 1}}$$

13.1-3   Adjusted means in ANCOVA:

$$\text{adj }\bar{Y}_j = \bar{Y}_j + b_w(\bar{X}_T - \bar{X}_j)$$

13.1-4   Number of possible pairwise comparisons:

$$m = \frac{k(k - 1)}{2}$$

13.1-5   Individual alpha levels:

$$\text{Individual alpha level} = 1.00 - \frac{\text{overall alpha level}}{m}$$

14.1-1   Expected T in signed-rank median test:

$$\text{Mean}_T = n(n + 1)/4$$

14.1-2 SE_T in signed-rank median test:

$$SE_T = \sqrt{\frac{n(n + 1)(2n + 1)}{24}}$$

14.1-3   Expected U in Mann-Whitney U Test:

$$\text{Mean}_U = (n_1)(n_2)/2$$

14.1-4 SE_U in Mann-Whitney U Test:

$$SE_U = \sqrt{n_1 n_2(n_1 + n_2 + 1)/12}$$

14.2-1   Test statistic for Kruskal-Wallis ANOVA by ranks:

$$H = \frac{12}{n_T(n_T + 1)} \sum_{j=1}^{k} n_j(\bar{R}_j - \bar{R}_T)^2$$

14.2-4   Test statistic for the Friedman test for multiple matched groups:

$$S = \frac{12n}{k(k + 1)} \sum_{j=1}^{k} (\bar{R}_j - \bar{R}_T)^2$$

Useful tables

Table E-1. Random numbers

Line	1	2	3	4	5	6	7	8	9	10	11	12	13	14
													Column	
1	10480	15011	01536	02011	81647	91646	69179	14194	62500	36207	20969	99570	91291	90700
2	22368	46573	25595	85393	30995	89198	27982	53402	93965	34095	52666	19174	39615	99505
3	24130	48360	22527	97265	76393	64809	15179	24830	49340	32081	30680	19655	63348	58629
4	42167	93093	06243	61680	07856	16376	39440	53537	71341	57004	00849	74917	97758	16379
5	37370	39975	81837	16656	06121	91782	60468	81305	49684	60672	14110	06927	01263	54613
6	77921	06907	11008	42751	27756	53498	18602	70659	90655	15053	21916	81825	44394	42880
7	99562	72905	56420	69994	98872	31016	71194	18738	44013	48840	63213	21069	10634	12952
8	96301	91977	05463	07972	18876	20922	94595	56869	69014	60045	18425	84903	42508	32307
9	89579	14342	63661	10281	17453	18103	57740	84378	25331	12566	58678	44947	05585	56941
10	85475	36857	53342	53988	53060	59533	38867	62300	08158	17983	16439	11458	18593	64952
11	25918	69578	88231	33276	70997	79936	56865	05859	90106	31595	01547	85590	91610	78188
12	63553	40961	48235	03427	49626	60445	18663	72695	52180	20847	12234	90511	33703	90322
13	09429	93969	52636	92737	88974	33488	36320	17617	30015	08272	84115	27156	30613	74952
14	10365	61129	87529	85689	48237	52267	67689	93394	01511	26358	85104	20285	29975	89868
15	07119	97336	71048	08178	77233	13916	47564	81056	97735	85977	29372	74461	28551	90707
16	51085	12765	51821	51259	77452	16308	60756	92144	49442	53900	70960	63990	75601	40719
17	02368	21382	52404	60268	89368	19885	55322	44819	01188	65255	64835	44919	05944	55157
18	01011	54092	33362	94904	31273	04146	18594	29852	71585	85030	51132	01915	92747	64951
19	52162	53916	46369	53586	23216	14513	83149	98736	23495	64350	94738	17752	35156	35749
20	07056	97628	33787	09998	42698	06691	76988	13602	51851	46104	88916	19509	25625	58104
21	48663	91245	85828	14346	09172	30168	90229	04734	59193	22178	30421	61666	99904	32812
22	54104	58492	22421	74103	47070	25306	76468	26384	58151	06646	21524	15227	96909	44592
23	32639	32363	05597	24200	13363	38005	94342	28728	35806	06912	17012	64161	18296	22851
24	29334	27001	87637	87308	58731	00256	45834	15398	46557	41135	10367	07684	36188	18510
25	02488	33062	28834	07351	19731	92420	60952	61280	50001	67658	32586	86679	50720	94953
26	81525	72295	04839	96423	24878	82651	66566	14778	76797	14780	13300	87074	79866	95725
27	29676	20591	68086	26432	46901	20849	89768	81536	86645	12659	92259	57102	80428	25280
28	00742	57392	39064	66432	84673	40027	32832	61362	98947	96067	64760	64584	96096	98253
29	05306	04213	25669	26422	44407	44048	37937	63904	45766	66134	75470	66520	34693	90449
30	91021	26418	64117	94305	26766	25940	39972	22209	71500	64568	91402	42416	07844	69618

31	00582	04711	87917	77341	42206	35126	74087	99547	81817	42607	43808	76655	62028	76630
32	00725	69884	62797	56170	86324	88072	76222	36086	84637	93161	76038	65855	77919	88006
33	69011	65795	93876	55293	18988	27354	26575	08625	40801	59920	29841	80150	12777	48501
34	25976	57948	29888	88604	67917	48708	18912	82271	65424	69774	33611	54262	85963	03547
35	09763	83473	73577	12908	30883	18317	28290	35797	05998	41688	34952	37888	38917	88050
36	91567	42595	27958	30134	04024	86385	29880	99730	55536	84855	29080	09250	79656	73211
37	17955	56349	90999	49127	20044	59931	06115	20542	18059	02008	73708	83517	36103	42791
38	46503	18584	18845	49618	02304	51038	20655	58727	28168	15475	56942	53389	20562	87338
39	92157	89634	94824	78171	84610	82834	09922	25417	44137	48413	25555	21246	35509	20468
40	14577	82765	35606	81263	39667	47358	56873	56307	61607	49518	89656	20103	77490	18062
41	98427	07523	33362	64270	01638	92477	66969	98420	04880	45585	46565	04102	46880	45709
42	34914	63976	88720	82765	34476	17032	87589	40836	32427	70002	70663	88863	77775	69348
43	70060	28277	39475	46473	23219	53416	94970	25832	69975	94884	19661	72828	00102	66794
44	53978	54914	06990	67245	68350	82948	11308	42878	80287	88267	47363	46634	06541	97809
45	76072	29515	40980	07391	58745	25774	22987	80059	39911	96189	41151	14222	60697	59583
46	90725	52210	83974	29992	65831	38857	50490	83765	55657	14361	31720	57375	56228	41546
47	64364	67412	33339	31926	14882	24413	59744	92351	97473	89286	35931	04110	23726	51900
48	08962	00358	31662	25388	61642	34072	81249	35648	56891	69352	48373	45578	78547	81788
49	95012	68379	93526	70765	10592	04542	76463	54328	02349	17247	28865	14777	62730	92277
50	15664	10493	20492	38391	91132	21999	59516	81652	27195	48223	46751	22923	32261	85653
51	16408	81899	04153	53381	79401	21438	83035	92350	36693	31238	59649	91754	72772	02338
52	18629	81953	05520	91962	04739	13092	97662	24822	94730	06496	35090	04822	86774	98289
53	73115	35101	47498	87637	99016	71060	88824	71013	18735	20286	23153	72924	35165	43040
54	57491	16703	23167	49323	45021	33132	12544	41035	80780	45393	44812	12515	98931	91202
55	30405	83946	23792	14422	15059	45799	22718	19792	09983	74353	68668	30429	70735	25499
56	16631	35006	85900	98275	32388	52390	16815	69298	82732	38480	73817	38523	41961	44437
57	96773	20206	42559	78985	05300	22164	24389	54224	35083	19687	11052	91491	60383	19746
58	38935	64202	14349	82674	66523	44133	00697	35552	35970	19124	63318	29686	03387	59846
59	31624	76384	17403	53363	44167	64486	64758	75368	76554	31601	12614	33072	60332	92325
60	78919	19474	23632	27889	47914	02584	37680	20801	72152	39339	34806	08930	85001	87820

From Walker, H. M., and Lev, J.: Statistical inference, New York, 1953, Holt, Rinehart and Winston, Inc.

Table E-2. Number of possible combinations of n things taken k at a time, $_nC_k$

n	0	1	2	3	4	5	6	7	8	9	10	11	12	13	14	15
									k							
1	1	1	0	0	0	0	0	0	0	0	0	0	0	0	0	0
2	1	2	1	0	0	0	0	0	0	0	0	0	0	0	0	0
3	1	3	3	1	0	0	0	0	0	0	0	0	0	0	0	0
4	1	4	6	4	1	0	0	0	0	0	0	0	0	0	0	0
5	1	5	10	10	5	1	0	0	0	0	0	0	0	0	0	0
6	1	6	15	20	15	6	1	0	0	0	0	0	0	0	0	0
7	1	7	21	35	35	21	7	1	0	0	0	0	0	0	0	0
8	1	8	28	56	70	56	28	8	1	0	0	0	0	0	0	0
9	1	9	36	84	126	126	84	36	9	1	0	0	0	0	0	0
10	1	10	45	120	210	252	210	120	45	10	1	0	0	0	0	0
11	1	11	55	165	330	462	462	330	165	55	11	1	0	0	0	0
12	1	12	66	220	495	792	924	792	495	220	66	12	1	0	0	0
13	1	13	78	286	715	1287	1716	1716	1287	715	286	78	13	1	0	0
14	1	14	91	364	1001	2002	3003	3432	3003	2002	1001	364	91	14	1	0
15	1	15	105	455	1365	3033	5005	6435	6435	5005	3003	1365	455	105	15	1

Table E-3. Number of possible permutations of n things taken k at a time, $_nP_k$

n	0	1	2	3	4	5	6	7	8	9	10
						k					
1	1	1	0	0	0	0	0	0	0	0	0
2	1	2	1	0	0	0	0	0	0	0	0
3	1	3	6	1	0	0	0	0	0	0	0
4	1	4	12	24	1	0	0	0	0	0	0
5	1	5	20	60	120	1	0	0	0	0	0
6	1	6	30	120	360	720	1	0	0	0	0
7	1	7	42	210	840	2520	5040	1	0	0	0
8	1	8	56	336	1680	6720	20,160	40,320	1	0	0
9	1	9	72	504	3024	15,120	60,480	181,440	362,880	1	0
10	1	10	90	720	5040	30,240	151,200	604,800	1,814,400	3,628,800	1

Table E-4. Individual binomial probabilities (the probability of exactly k out of n at various values of π)

n	0	1	2	3	4	5	6	7	8	9	10	11	12	13	14	15
When π = .1																
1	.9000	.1000	0	0	0	0	0	0	0	0	0	0	0	0	0	0
2	.8100	.1800	.0100	0	0	0	0	0	0	0	0	0	0	0	0	0
3	.7290	.2430	.0270	.0010	0	0	0	0	0	0	0	0	0	0	0	0
4	.6561	.2916	.0486	.0036	.0001	0	0	0	0	0	0	0	0	0	0	0
5	.5904	.3280	.0729	.0081	.0004	0	0	0	0	0	0	0	0	0	0	0
6	.5314	.3542	.0984	.0145	.0012	0	0	0	0	0	0	0	0	0	0	0
7	.4782	.3720	.1240	.0229	.0025	.0001	0	0	0	0	0	0	0	0	0	0
8	.4304	.3826	.1488	.0330	.0045	.0004	0	0	0	0	0	0	0	0	0	0
9	.3874	.3874	.1721	.0446	.0074	.0008	0	0	0	0	0	0	0	0	0	0
10	.3486	.3874	.1937	.0573	.0111	.0014	.0001	0	0	0	0	0	0	0	0	0
11	.3138	.3835	.2130	.0710	.0157	.0024	.0002	0	0	0	0	0	0	0	0	0
12	.2824	.3765	.2301	.0852	.0213	.0037	.0004	0	0	0	0	0	0	0	0	0
13	.2541	.3671	.2447	.0997	.0277	.0055	.0008	0	0	0	0	0	0	0	0	0
14	.2287	.3558	.2570	.1142	.0349	.0077	.0012	.0001	0	0	0	0	0	0	0	0
15	.2058	.3431	.2668	.1285	.0428	.0104	.0019	.0002	0	0	0	0	0	0	0	0
When π = .2																
1	.8000	.2000	0	0	0	0	0	0	0	0	0	0	0	0	0	0
2	.6400	.3200	.0400	0	0	0	0	0	0	0	0	0	0	0	0	0
3	.5120	.3840	.0960	.0080	0	0	0	0	0	0	0	0	0	0	0	0
4	.4096	.4096	.1536	.0256	.0016	0	0	0	0	0	0	0	0	0	0	0
5	.3276	.4096	.2048	.0512	.0064	.0003	0	0	0	0	0	0	0	0	0	0
6	.2621	.3932	.2457	.0819	.0153	.0015	0	0	0	0	0	0	0	0	0	0
7	.2097	.3670	.2752	.1146	.0286	.0043	.0003	0	0	0	0	0	0	0	0	0
8	.1677	.3355	.2936	.1468	.0458	.0091	.0011	0	0	0	0	0	0	0	0	0
9	.1342	.3019	.3019	.1761	.0660	.0165	.0027	.0002	0	0	0	0	0	0	0	0
10	.1073	.2684	.3019	.2013	.0880	.0264	.0055	.0007	0	0	0	0	0	0	0	0

Continued.

440 *Appendix E*

Table E-4. Individual binomial probabilities (the probability of exactly *k* out of *n* at various values of π)—cont'd

n	0	1	2	3	4	5	6	7	8	9	10	11	12	13	14	15
When π = .2—cont'd																
11	.0858	.2362	.2952	.2214	.1107	.0387	.0096	.0017	.0002	0	0	0	0	0	0	0
12	.0687	.2061	.2834	.2362	.1328	.0531	.0155	.0033	.0005	0	0	0	0	0	0	0
13	.0549	.1786	.2680	.2456	.1535	.0690	.0230	.0057	.0010	.0001	0	0	0	0	0	0
14	.0439	.1539	.2501	.2501	.1719	.0859	.0322	.0092	.0020	.0003	0	0	0	0	0	0
15	.0351	.1319	.2308	.2501	.1876	.1031	.0429	.0138	.0034	.0006	.0001	0	0	0	0	0
When π = .3																
1	.7000	.3000														
2	.4900	.4200	.0900													
3	.3430	.4410	.1890	.0270												
4	.2401	.4116	.2646	.0756	.0081											
5	.1680	.3601	.3087	.1323	.0283	.0024										
6	.1176	.3025	.3241	.1852	.0595	.0102	.0007	0	0	0	0	0	0	0	0	0
7	.0823	.2470	.3176	.2268	.0972	.0250	.0035	.0002	0	0	0	0	0	0	0	0
8	.0576	.1976	.2964	.2541	.1361	.0466	.0100	.0012	0	0	0	0	0	0	0	0
9	.0403	.1556	.2668	.2668	.1715	.0735	.0210	.0038	.0004	0	0	0	0	0	0	0
10	.0282	.1210	.2334	.2668	.2001	.1029	.0367	.0090	.0014	.0001	0	0	0	0	0	0
11	.0197	.0932	.1997	.2568	.2201	.1320	.0566	.0173	.0037	.0005	0	0	0	0	0	0
12	.0138	.0711	.1677	.2397	.2311	.1584	.0792	.0291	.0077	.0014	.0001	0	0	0	0	0
13	.0096	.0539	.1388	.2181	.2337	.1802	.1030	.0441	.0141	.0033	.0005	.0001	0	0	0	0
14	.0067	.0406	.1133	.1943	.2290	.1963	.1262	.0618	.0231	.0066	.0014	.0002	0	0	0	0
15	.0047	.0305	.0915	.1700	.2186	.2061	.1472	.0811	.0347	.0155	.0029	.0005	0	0	0	0
When π = .4																
1	.6000	.4000														
2	.3600	.4800	.1600													
3	.2160	.4320	.2880	.0640												
4	.1296	.3456	.3456	.1536	.0256	0	0	0	0	0	0	0	0	0	0	0
5	.0777	.2592	.3456	.2304	.0768	.0102	0	0	0	0	0	0	0	0	0	0

n	0	1	2	3	4	5	6	7	8	9	10	11	12	13	14	15
6	.0466	.1866	.3110	.2764	.1382	.0368	.0040	0	0	0	0	0	0	0	0	0
7	.0279	.1306	.2612	.2903	.1935	.0774	.0172	.0016	0	0	0	0	0	0	0	0
8	.0167	.0895	.2090	.2786	.2322	.1238	.0412	.0078	.0006	0	0	0	0	0	0	0
9	.0100	.0604	.1612	.2508	.2508	.1672	.0743	.0212	.0035	.0003	0	0	0	0	0	0
10	.0060	.0403	.1209	.2149	.2508	.2006	.1114	.0424	.0106	.0016	.0001	0	0	0	0	0
11	.0036	.0266	.0886	.1773	.2364	.2207	.1471	.0700	.0233	.0051	.0006	0	0	0	0	0
12	.0021	.0174	.0638	.1418	.2128	.2270	.1765	.1009	.0420	.0124	.0024	.0003	0	0	0	0
13	.0013	.0113	.0452	.1106	.1844	.2213	.1967	.1311	.0655	.0242	.0064	.0011	.0001	0	0	0
14	.0007	.0073	.0316	.0845	.1549	.2065	.2065	.1574	.0918	.0408	.0136	.0032	.0005	.0001	0	0
15	.0004	.0047	.0219	.0633	.1267	.1859	.2065	.1770	.1180	.0612	.0244	.0074	.0016	.0002	0	0

When $\pi = .5$

n	0	1	2	3	4	5	6	7	8	9	10	11	12	13	14	15
1	.5000	.5000	0	0	0	0	0	0	0	0	0	0	0	0	0	0
2	.2500	.5000	.2500	0	0	0	0	0	0	0	0	0	0	0	0	0
3	.1250	.3750	.3750	.1250	0	0	0	0	0	0	0	0	0	0	0	0
4	.0625	.2500	.3750	.2500	.0625	0	0	0	0	0	0	0	0	0	0	0
5	.0312	.1562	.3125	.3125	.1562	.0312	0	0	0	0	0	0	0	0	0	0
6	.0156	.0937	.2343	.3125	.2343	.0937	.0156	0	0	0	0	0	0	0	0	0
7	.0078	.0546	.1640	.2734	.2734	.1640	.0546	.0078	0	0	0	0	0	0	0	0
8	.0039	.0312	.1093	.2187	.2734	.2187	.1093	.0312	.0039	0	0	0	0	0	0	0
9	.0019	.0175	.0703	.1640	.2460	.2460	.1640	.0703	.0175	.0019	0	0	0	0	0	0
10	.0009	.0097	.0439	.1171	.2050	.2460	.2050	.1171	.0439	.0097	.0009	0	0	0	0	0
11	.0004	.0053	.0268	.0805	.1611	.2255	.2255	.1611	.0805	.0268	.0053	.0004	0	0	0	0
12	.0002	.0029	.0161	.0537	.1208	.1933	.2255	.1933	.1208	.0537	.0161	.0029	.0002	0	0	0
13	.0001	.0015	.0095	.0349	.0872	.1571	.2094	.2094	.1571	.0872	.0349	.0095	.0015	.0001	0	0
14	0	.0008	.0055	.0222	.0610	.1221	.1832	.2094	.1832	.1221	.0610	.0222	.0055	.0008	0	0
15	0	.0004	.0032	.0138	.0416	.0916	.1527	.1963	.1963	.1527	.0916	.0416	.0138	.0032	.0004	0

Table E-5. Sum of binomial probabilities (the probability of at least k out of n at various values of π)

k

n	0	1	2	3	4	5	6	7	8	9	10	11	12	13	14	15
When π = .1																
1	1.0000	.1000	0	0	0	0	0	0	0	0	0	0	0	0	0	0
2	1.0000	.1900	.0100	0	0	0	0	0	0	0	0	0	0	0	0	0
3	1.0000	.2710	.0280	.0010	0	0	0	0	0	0	0	0	0	0	0	0
4	1.0000	.3439	.0523	.0037	.0001	0	0	0	0	0	0	0	0	0	0	0
5	1.0000	.4095	.0814	.0085	.0004	0	0	0	0	0	0	0	0	0	0	0
6	1.0000	.4685	.1142	.0158	.0012	0	0	0	0	0	0	0	0	0	0	0
7	1.0000	.5217	.1496	.0256	.0027	.0001	0	0	0	0	0	0	0	0	0	0
8	1.0000	.5695	.1868	.0380	.0050	.0004	0	0	0	0	0	0	0	0	0	0
9	1.0000	.6125	.2251	.0529	.0083	.0008	0	0	0	0	0	0	0	0	0	0
10	1.0000	.6513	.2639	.0701	.0127	.0016	.0001	0	0	0	0	0	0	0	0	0
11	1.0000	.6861	.3026	.0895	.0185	.0027	.0002	0	0	0	0	0	0	0	0	0
12	1.0000	.7175	.3409	.1108	.0256	.0043	.0005	0	0	0	0	0	0	0	0	0
13	1.0000	.7458	.3786	.1338	.0341	.0064	.0009	0	0	0	0	0	0	0	0	0
14	1.0000	.7712	.4153	.1583	.0441	.0092	.0014	.0001	0	0	0	0	0	0	0	0
15	1.0000	.7941	.4509	.1840	.0555	.0127	.0022	.0003	0	0	0	0	0	0	0	0
When π = .2																
1	1.0000	.2000	0	0	0	0	0	0	0	0	0	0	0	0	0	0
2	1.0000	.3600	.0400	0	0	0	0	0	0	0	0	0	0	0	0	0
3	1.0000	.4880	.1040	.0080	0	0	0	0	0	0	0	0	0	0	0	0
4	1.0000	.5904	.1808	.0272	.0016	0	0	0	0	0	0	0	0	0	0	0
5	1.0000	.6723	.2627	.0579	.0067	.0003	0	0	0	0	0	0	0	0	0	0
6	1.0000	.7378	.3446	.0988	.0169	.0016	0	0	0	0	0	0	0	0	0	0
7	1.0000	.7902	.4232	.1480	.0333	.0046	.0003	0	0	0	0	0	0	0	0	0
8	1.0000	.8322	.4966	.2030	.0562	.0104	.0012	0	0	0	0	0	0	0	0	0
9	1.0000	.8657	.5637	.2618	.0856	.0195	.0030	.0003	0	0	0	0	0	0	0	0
10	1.0000	.8926	.6241	.3222	.1208	.0327	.0063	.0008	0	0	0	0	0	0	0	0
11	1.0000	.9141	.6778	.3825	.1611	.0504	.0116	.0019	.0002	0	0	0	0	0	0	0
12	1.0000	.9312	.7251	.4416	.2054	.0725	.0194	.0039	.0005	0	0	0	0	0	0	0
13	1.0000	.9450	.7663	.4983	.2526	.0991	.0300	.0070	.0012	.0001	0	0	0	0	0	0
14	1.0000	.9560	.8020	.5519	.3018	.1298	.0438	.0116	.0023	.0003	0	0	0	0	0	0
15	1.0000	.9648	.8328	.6019	.3518	.1642	.0610	.0180	.0042	.0007	.0001	0	0	0	0	0

When $\pi = .3$

n	0	1	2	3	4	5	6	7	8	9	10	11	12	13
1	1.0000	.3000	0	0	0	0	0	0	0	0	0	0	0	0
2	1.0000	.5100	.0900	0	0	0	0	0	0	0	0	0	0	0
3	1.0000	.6570	.2160	.0270	0	0	0	0	0	0	0	0	0	0
4	1.0000	.7599	.3483	.0837	.0081	0	0	0	0	0	0	0	0	0
5	1.0000	.8319	.4717	.1630	.0307	.0024	0	0	0	0	0	0	0	0
6	1.0000	.8823	.5798	.2556	.0704	.0109	.0007	0	0	0	0	0	0	0
7	1.0000	.9176	.6705	.3529	.1260	.0287	.0037	.0002	0	0	0	0	0	0
8	1.0000	.9423	.7447	.4482	.1941	.0579	.0112	.0012	.0001	0	0	0	0	0
9	1.0000	.9596	.8039	.5371	.2703	.0988	.0252	.0042	.0004	0	0	0	0	0
10	1.0000	.9717	.8506	.6172	.3503	.1502	.0473	.0105	.0015	.0001	0	0	0	0
11	1.0000	.9802	.8870	.6872	.4304	.2103	.0782	.0216	.0042	.0005	0	0	0	0
12	1.0000	.9861	.9149	.7471	.5074	.2763	.1178	.0386	.0094	.0016	.0002	0	0	0
13	1.0000	.9903	.9363	.7975	.5793	.3456	.1653	.0623	.0182	.0040	.0006	0	0	0
14	1.0000	.9932	.9525	.8391	.6448	.4157	.2194	.0932	.0314	.0082	.0016	.0002	0	0
15	1.0000	.9952	.9647	.8731	.7031	.4845	.2783	.1311	.0500	.0152	.0036	.0006	.0001	0

When $\pi = .4$

n	0	1	2	3	4	5	6	7	8	9	10	11	12	13
1	1.0000	.4000	0	0	0	0	0	0	0	0	0	0	0	0
2	1.0000	.6400	.1600	0	0	0	0	0	0	0	0	0	0	0
3	1.0000	.7840	.3520	.0640	0	0	0	0	0	0	0	0	0	0
4	1.0000	.8704	.5248	.1792	.0256	0	0	0	0	0	0	0	0	0
5	1.0000	.9222	.6630	.3174	.0870	.0102	0	0	0	0	0	0	0	0
6	1.0000	.9533	.7667	.4556	.1792	.0409	.0040	0	0	0	0	0	0	0
7	1.0000	.9720	.8413	.5800	.2897	.0962	.0188	.0016	0	0	0	0	0	0
8	1.0000	.9832	.8936	.6846	.4059	.1736	.0498	.0085	.0006	0	0	0	0	0
9	1.0000	.9899	.9294	.7682	.5173	.2665	.0993	.0250	.0038	.0003	0	0	0	0
10	1.0000	.9939	.9536	.8327	.6177	.3668	.1662	.0547	.0122	.0016	.0001	0	0	0
11	1.0000	.9963	.9697	.8810	.7037	.4672	.2465	.0993	.0292	.0059	.0007	0	0	0
12	1.0000	.9978	.9804	.9165	.7746	.5618	.3347	.1582	.0573	.0152	.0028	.0003	0	0
13	1.0000	.9986	.9873	.9420	.8314	.6469	.4256	.2288	.0976	.0320	.0077	.0013	.0001	0
14	1.0000	.9992	.9919	.9602	.8756	.7207	.5141	.3075	.1501	.0583	.0175	.0039	.0006	.0001
15	1.0000	.9995	.9948	.9728	.9094	.7827	.5967	.3901	.2131	.0950	.0338	.0093	.0019	.0002

Continued.

Table E-5. Sum of binomial probabilities (the probability of at least *k* out of *n* at various values of π)—cont'd

When $\pi = .5$

n	0	1	2	3	4	5	6	7	8	9	10	11	12	13	14	15
1	1.0000	.5000	0	0	0	0	0	0	0	0	0	0	0	0	0	0
2	1.0000	.7500	.2500	0	0	0	0	0	0	0	0	0	0	0	0	0
3	1.0000	.8750	.5000	.1250	0	0	0	0	0	0	0	0	0	0	0	0
4	1.0000	.9375	.6875	.3125	.0625	0	0	0	0	0	0	0	0	0	0	0
5	1.0000	.9687	.8125	.5000	.1875	.0312	0	0	0	0	0	0	0	0	0	0
6	1.0000	.9843	.8906	.6562	.3437	.1093	.0156	0	0	0	0	0	0	0	0	0
7	1.0000	.9921	.9375	.7734	.5000	.2265	.0625	.0078	0	0	0	0	0	0	0	0
8	1.0000	.9960	.9648	.8554	.6367	.3632	.1445	.0351	.0039	0	0	0	0	0	0	0
9	1.0000	.9980	.9804	.9101	.7460	.5000	.2539	.0898	.0195	.0019	0	0	0	0	0	0
10	1.0000	.9990	.9892	.9453	.8281	.6230	.3769	.1718	.0546	.0107	.0009	0	0	0	0	0
11	1.0000	.9995	.9941	.9672	.8867	.7255	.5000	.2744	.1132	.0327	.0058	.0004	0	0	0	0
12	1.0000	.9997	.9968	.9807	.9270	.8061	.6127	.3872	.1938	.0729	.0192	.0031	.0002	0	0	0
13	1.0000	.9998	.9982	.9887	.9538	.8665	.7094	.5000	.2905	.1334	.0461	.0112	.0017	.0001	0	0
14	1.0000	.9999	.9990	.9935	.9713	.9102	.7880	.6047	.3952	.2119	.0897	.0286	.0064	.0009	0	0
15	1.0000	.9999	.9995	.9963	.9824	.9407	.8491	.6963	.4999	.3036	.1508	.0592	.0175	.0036	.0004	0

Table E-6. The standard normal curve

	One tail		Two tail	
z	π Beyond	π Remainder	π Beyond	π Remainder
0.00	0.5000	0.5000	1.0000	0.0000
0.01	0.4960	0.5040	0.9920	0.0080
0.02	0.4920	0.5080	0.9840	0.0160
0.03	0.4880	0.5120	0.9761	0.0239
0.04	0.4840	0.5160	0.9681	0.0319
0.05	0.4801	0.5199	0.9601	0.0399
0.06	0.4761	0.5239	0.9522	0.0478
0.07	0.4721	0.5279	0.9442	0.0558
0.08	0.4681	0.5319	0.9362	0.0638
0.09	0.4641	0.5359	0.9283	0.0717
0.10	0.4602	0.5398	0.9203	0.0797
0.11	0.4562	0.5438	0.9124	0.0876
0.12	0.4522	0.5478	0.9045	0.0955
0.13	0.4483	0.5517	0.8966	0.1034
0.14	0.4443	0.5557	0.8887	0.1113
0.15	0.4404	0.5596	0.8808	0.1192
0.16	0.4364	0.5636	0.8729	0.1271
0.17	0.4325	0.5675	0.8650	0.1350
0.18	0.4286	0.5714	0.8571	0.1429
0.19	0.4247	0.5753	0.8493	0.507
0.20	0.4207	0.5793	0.8415	0.1585
0.21	0.4168	0.5832	0.8337	0.1663
0.22	0.4129	0.5871	0.8259	0.1741
0.23	0.4090	0.5910	0.8181	0.1819
0.24	0.4052	0.5948	0.8103	0.1897
0.25	0.4013	0.5987	0.8026	0.1974
0.26	0.3974	0.6026	0.7949	0.2051
0.27	0.3936	0.6064	0.7872	0.2128
0.28	0.3897	0.6103	0.7795	0.2205
0.29	0.3859	0.6141	0.7718	0.2282
0.30	0.3821	0.6179	0.7642	0.2358
0.31	0.3783	0.6217	0.7566	0.2434
0.32	0.3745	0.6255	0.7490	0.2510
0.33	0.3707	0.6293	0.7414	0.2586
0.34	0.3669	0.6331	0.7339	0.2661
0.35	0.3632	0.6368	0.7263	0.2737
0.36	0.3594	0.6406	0.7188	0.2812
0.37	0.3557	0.6443	0.7114	0.2886
0.38	0.3520	0.6480	0.7039	0.2961
0.39	0.3483	0.6517	0.6965	0.3035

Continued.

Adapted from Pearson, E. S., and Hartley, H. O.: Biometrika tables for statisticians, vol. 1, ed. 3, London, 1966, Cambridge University Press.

Table E-6. The standard normal curve—cont'd

z	One tail		Two tail	
	π Beyond	π Remainder	π Beyond	π Remainder
0.40	0.3446	0.6554	0.6892	0.3108
0.41	0.3409	0.6591	0.6818	0.3182
0.42	0.3372	0.6628	0.6745	0.3255
0.43	0.3336	0.6664	0.6672	0.3328
0.44	0.3300	0.6700	0.6599	0.3401
0.45	0.3264	0.6736	0.6527	0.3473
0.46	0.3228	0.6772	0.6455	0.3545
0.47	0.3192	0.6808	0.6384	0.3616
0.48	0.3156	0.6844	0.6312	0.3688
0.49	0.3121	0.6879	0.6241	0.3759
0.50	0.3085	0.6915	0.6171	0.3829
0.51	0.3050	0.6950	0.6101	0.3899
0.52	0.3015	0.8985	0.6031	0.3969
0.53	0.2981	0.7019	0.5961	0.4039
0.54	0.2946	0.7054	0.5892	0.4108
0.55	0.2912	0.7088	0.5823	0.4177
0.56	0.2877	0.7123	0.5755	0.4245
0.57	0.2843	0.7157	0.5687	0.4313
0.58	0.2810	0.7190	0.5619	0.4381
0.59	0.2776	0.7224	0.5552	0.4448
0.60	0.2743	0.7257	0.5485	0.4515
0.61	0.2709	0.7291	0.5419	0.4581
0.62	0.2676	0.7324	0.5353	0.4647
0.63	0.2643	0.7357	0.5287	0.4713
0.64	0.2611	0.7389	0.5222	0.4778
0.65	0.2578	0.7422	0.5157	0.4843
0.66	0.2546	0.7454	0.5093	0.4907
0.67	0.2514	0.7486	0.5029	0.4971
0.6745	0.25	0.75	0.50	0.50
0.68	0.2483	0.7517	0.4965	0.5035
0.69	0.2451	0.7549	0.4902	0.5098
0.70	0.2420	0.7580	0.4839	0.5161
0.71	0.2389	0.7611	0.4777	0.5223
0.72	0.2358	0.7642	0.4715	0.5285
0.73	0.2327	0.7673	0.4654	0.5346
0.74	0.2296	0.7704	0.4593	0.5407
0.75	0.2266	0.7734	0.4533	0.5467
0.76	0.2236	0.7764	0.4473	0.5527
0.77	0.2206	0.7794	0.4413	0.5587
0.78	0.2177	0.7823	0.4354	0.5646

Table E-6. The standard normal curve—cont'd

	One tail		Two tail	
z	π Beyond	π Remainder	π Beyond	π Remainder
0.79	0.2148	0.7852	0.4295	0.5705
0.80	0.2119	0.7881	0.4237	0.5763
0.81	0.2090	0.7910	0.4179	0.5821
0.82	0.2061	0.7939	0.4122	0.5878
0.83	0.2033	0.7967	0.4065	0.5935
0.84	0.2005	0.7995	0.4009	0.5991
0.8416	0.20	0.80	0.40	0.60
0.85	0.1977	0.8023	0.3953	0.6047
0.86	0.1949	0.8051	0.3898	0.6102
0.87	0.1922	0.8078	0.3843	0.6157
0.88	0.1894	0.8106	0.3789	0.6211
0.89	0.1867	0.8133	0.3735	0.6265
0.90	0.1841	0.8159	0.3681	0.6319
0.91	0.1814	0.8186	0.3628	0.6372
0.92	0.1788	0.8212	0.3576	0.6424
0.93	0.1762	0.8238	0.3524	0.6476
0.94	0.1736	0.8264	0.3472	0.6528
0.95	0.1711	0.8289	0.3421	0.6579
0.96	0.1685	0.8315	0.3371	0.6629
0.97	0.1660	0.8340	0.3320	0.6680
0.98	0.1635	0.8365	0.3271	0.6729
0.99	0.1611	0.8389	0.3222	0.6778
1.00	0.1587	0.8413	0.3173	0.6827
1.01	0.1562	0.8438	0.3125	0.6875
1.02	0.1539	0.8461	0.3077	0.6923
1.03	0.1515	0.8485	0.3030	0.6970
1.04	0.1492	0.8508	0.2983	0.7017
1.05	0.1469	0.8531	0.2937	0.7063
1.06	0.1446	0.8554	0.2891	0.7109
1.07	0.1423	0.8577	0.2846	0.7154
1.08	0.1401	0.8599	0.2801	0.7199
1.09	0.1379	0.6621	0.2757	0.7243
1.10	0.1357	0.8643	0.2713	0.7287
1.11	0.1335	0.8665	0.2670	0.7330
1.12	0.1314	0.8686	0.2627	0.7373
1.13	0.1292	0.8708	0.2585	0.7415
1.14	0.1271	0.8729	0.2543	0.7457
1.15	0.1251	0.8749	0.2501	0.7499
1.16	0.1230	0.8770	0.2460	0.7540
1.17	0.1210	0.8790	0.2420	0.7580

Continued.

Table E-6. The standard normal curve—cont'd

z	One tail		Two tail	
	π **Beyond**	π **Remainder**	π **Beyond**	π **Remainder**
1.18	0.1190	0.8810	0.2380	0.7620
1.19	0.1170	0.8830	0.2340	0.7660
1.20	0.1151	0.8049	0.2301	0.7699
1.21	0.1131	0.8869	0.2263	0.7737
1.22	0.1112	0.8888	0.2225	0.7775
1.23	0.1093	0.8907	0.2187	0.7813
1.24	0.1075	0.8925	0.2150	0.7890
1.25	0.1056	0.8444	0.2113	0.7887
1.26	0.1038	0.8962	0.2077	0.7923
1.27	0.1020	0.8980	0.2041	0.7959
1.28	0.1003	0.8997	0.2005	0.7995
1.282	0.10	0.90	0.20	0.80
1.29	0.0985	0.9015	0.1971	0.8029
1.30	0.0968	0.9032	0.1936	0.8064
1.31	0.0951	0.9049	0.1902	0.8098
1.32	0.0934	0.9066	0.1868	0.8132
1.33	0.0918	0.9082	0.1835	0.8165
1.34	0.0901	0.9099	0.1802	0.8198
1.35	0.0885	0.9115	0.1770	0.8230
1.36	0.0869	0.9131	0.1738	0.8202
1.37	0.0853	0.9147	0.1707	0.8293
1.38	0.0838	0.9162	0.1676	0.8324
1.39	0.0823	0.9177	0.1645	0.8355
1.40	0.0808	0.9192	0.1615	0.8385
1.41	0.0793	0.9207	0.1585	0.8415
1.42	0.0778	0.9222	0.1556	0.8444
1.43	0.0764	0.9286	0.1527	0.8473
1.44	0.0749	0.9251	0.1499	0.8501
1.45	0.0735	0.9265	0.1471	0.8529
1.46	0.0721	0.9279	0.1443	0.8567
1.47	0.0708	0.9292	0.1416	0.8584
1.48	0.0694	0.9306	0.1389	0.8611
1.49	0.0681	0.9319	0.1362	0.8638
1.50	0.0668	0.9332	0.1336	0.8664
1.51	0.0655	0.9345	0.1310	0.8690
1.52	0.0643	0.9357	0.1285	0.8715
1.53	0.0630	0.9370	0.1260	0.8740
1.54	0.0618	0.9382	0.1236	0.8764
1.55	0.0606	0.9394	0.1211	0.8789
1.56	0.0594	0.9406	0.1188	0.8812

Table E-6. The standard normal curve—cont'd

	One tail		Two tail	
z	π **Beyond**	π **Remainder**	π **Beyond**	π **Remainder**
1.57	0.0582	0.9418	0.1164	0.8836
1.58	0.0571	0.9429	0.1141	0.8859
1.59	0.0559	0.9441	0.1118	0.8882
1.60	0.0548	0.9452	0.1096	0.8904
1.61	0.0537	0.9463	0.1074	0.8926
1.62	0.0526	0.9474	0.1052	0.8948
1.63	0.0516	0.9484	0.1031	0.8969
1.64	0.0505	0.9495	0.1010	0.8990
1.645	0.05	0.95	0.10	0.90
1.65	0.0495	0.9505	0.0989	0.9011
1.66	0.0485	0.9515	0.0969	0.9031
1.67	0.0475	0.9525	0.0949	0.9051
1.68	0.0465	0.9535	0.0930	0.9070
1.69	0.0455	0.9545	0.0910	0.9090
1.70	0.0446	0.9554	0.0891	0.9109
1.71	0.0436	0.9564	0.0873	0.9127
1.72	0.0427	0.9573	0.0854	0.9146
1.73	0.0418	0.9582	0.0836	0.9164
1.74	0.0409	0.9591	0.0819	0.9181
1.75	0.0401	0.9599	0.0801	0.9199
1.76	0.0392	0.9608	0.0784	0.9216
1.77	0.0384	0.9616	0.0767	0.9233
1.78	0.0375	0.9625	0.0751	0.9249
1.79	0.0367	0.9633	0.0734	0.9266
1.80	0.0359	0.9641	0.0719	0.9281
1.81	0.0352	0.9649	0.0703	0.9297
1.82	0.0344	0.9656	0.0688	0.9312
1.83	0.0336	0.9664	0.0672	0.9328
1.84	0.0329	0.9671	0.0658	0.9342
1.85	0.0322	0.9678	0.0643	0.9357
1.86	0.0314	0.9686	0.0629	0.9371
1.87	0.0307	0.9693	0.0615	0.9385
1.88	0.0301	0.9699	0.0601	0.9399
1.89	0.0294	0.9706	0.0588	0.9412
1.90	0.0287	0.9713	0.0574	0.9426
1.91	0.0281	0.9719	0.0561	0.9439
1.92	0.0274	0.9726	0.0549	0.9451
1.93	0.0268	0.9732	0.0536	0.9464
1.94	0.0262	0.9738	0.0524	0.9476
1.95	0.0256	0.9744	0.0512	0.9488

Continued.

Table E-6. The standard normal curve—cont'd

z	One tail		Two tail	
	π **Beyond**	π **Remainder**	π **Beyond**	π **Remainder**
1.960	0.025	0.975	0.05	0.95
1.97	0.0244	0.9756	0.0488	0.9512
1.98	0.0239	0.9761	0.0477	0.9523
1.9	0.0233	0.9767	0.0466	0.9534
2.00	0.0228	0.9772	0.0455	0.9545
2.01	0.0222	0.9778	0.0444	0.9556
2.02	0.0217	0.9783	0.0434	0.9566
2.03	0.0212	0.9788	0.0424	0.9576
2.04	0.0207	0.9793	0.0414	0.9586
2.05	0.0202	0.9798	0.0404	0.9596
2.054	0.02	0.98	0.04	0.96
2.06	0.0197	0.9803	0.0394	0.9606
2.07	0.0192	0.9808	0.0385	0.9615
2.08	0.0188	0.9812	0.0375	0.9625
2.09	0.0183	0.9817	0.0366	0.9634
2.10	0.0179	0.9821	0.0357	0.9643
2.11	0.0174	0.9826	0.0349	0.9651
2.12	0.0170	0.9830	0.0340	0.9660
2.13	0.0166	0.9834	0.0332	0.9668
2.14	0.0162	0.9838	0.0324	0.9676
2.15	0.0158	0.9842	0.0316	0.9684
2.16	0.0154	0.9846	0.0308	0.9692
2.17	0.0150	0.9850	0.0300	0.9700
2.18	0.0146	0.9854	0.0293	0.9707
2.19	0.0143	0.9857	0.0285	0.9715
2.20	0.0139	0.9661	0.0278	0.9722
2.21	0.0136	0.9864	0.0271	0.9729
2.22	0.0132	0.9868	0.0264	0.9736
2.23	0.0129	0.9871	0.0257	0.9743
2.24	0.0125	0.9875	0.0251	0.9749
2.25	0.0122	0.9878	0.0244	0.9756
2.26	0.0119	0.9881	0.0238	0.9762
2.27	0.0116	0.9884	0.0232	0.9768
2.28	0.0113	0.9887	0.0226	0.9774
2.29	0.0110	0.9890	0.0220	0.9780
2.30	0.0107	0.9893	0.0214	0.9786
2.31	0.0104	0.9896	0.0209	0.9791
2.32	0.0102	0.9898	0.0203	0.9797
2.326	0.01	0.99	0.02	0.98
2.33	0.0099	0.9901	0.0198	0.9802

Table E-6. The standard normal curve—cont'd

	One tail		Two tail	
z	π Beyond	π Remainder	π Beyond	π Remainder
2.34	0.0096	0.9904	0.0193	0.9807
2.35	0.0094	0.9906	0.0188	0.9812
2.36	0.0091	0.9909	0.0183	0.9817
2.37	0.0089	0.991	0.0178	0.9822
2.38	0.0087	0.9913	0.0173	0.9827
2.39	0.0084	0.9916	0.0168	0.9832
2.40	0.0082	0.9918	0.0164	0.9836
2.41	0.0080	0.9920	0.0160	0.9840
2.42	0.0078	0.9922	0.0155	0.9845
2.43	0.0075	0.9925	0.0151	0.9849
2.44	0.0073	0.9927	0.0147	0.9853
2.45	0.0071	0.9929	0.0143	0.9857
2.46	0.0069	0.9931	0.0139	0.9861
2.47	0.0068	0.9932	0.0135	0.9865
2.48	0.0066	0.9934	0.0131	0.9869
2.49	0.0064	0.9936	0.0128	0.9872
2.50	0.0062	0.9938	0.0124	0.9876
2.51	0.0060	0.9940	0.0121	0.9879
2.52	0.0059	0.9941	0.0117	0.9883
2.53	0.0057	0.9943	0.0114	0.9886
2.54	0.0055	0.9945	0.0111	0.9889
2.55	0.0054	0.9946	0.0108	0.9892
2.56	0.0052	0.9948	0.0105	0.9895
2.57	0.0051	0.9949	0.0102	0.9898
2.576	0.005	0.995	0.01	0.99
2.58	0.0049	0.9951	0.0099	0.9901
2.59	0.0048	0.9952	0.0096	0.9904
2.60	0.0047	0.9953	0.0093	0.9907
2.61	0.0045	0.9955	0.0091	0.9909
2.62	0.0044	0.9956	0.0088	0.9912
2.63	0.0043	0.9957	0.0085	0.9915
2.64	0.0041	0.9959	0.0083	0.9917
2.65	0.0040	0.9960	0.0080	0.9920
2.70	0.0035	0.9965	0.0069	0.9931
2.75	0.0030	0.9970	0.0060	0.9940
2.80	0.0026	0.9974	0.0051	0.9949
2.85	0.0022	0.9978	0.0044	0.9956
2.90	0.0019	0.9981	0.0037	0.9963
2.95	0.0016	0.9984	0.0032	0.9968
3.00	0.0013	0.9987	0.0027	0.9973

Continued.

Table E-6. The standard normal curve—cont'd

	One tail		Two tail	
z	π Beyond	π Remainder	π Beyond	π Remainder
3.05	0.0011	0.9989	0.0023	0.9977
3.090	0.001	0.999	0.002	0.998
3.10	0.0010	0.990	0.0019	0.9981
3.15	0.0008	0.9992	0.0016	0.9984
3.20	0.0007	0.9993	0.0014	0.9988
3.25	0.0006	0.9994	0.0012	0.9986
3.291	0.0005	0.9995	0.001	0.999
3.30	0.0005	0.9995	0.0010	0.9990
3.35	0.0004	0.9996	0.0008	0.9992
3.40	0.0003	0.9997	0.0007	0.9993
3.45	0.0003	0.9997	0.0006	0.9994
3.50	0.0002	0.9998	0.0005	0.9995
3.55	0.0002	0.9998	0.0004	0.9996
3.60	0.0002	0.9998	0.0003	0.9997
3.65	0.0001	0.9999	0.0003	0.9997
3.719	0.0001	0.9999	0.0002	0.9998
3.80	0.0001	0.9999	0.0001	0.9999
3.891	0.00005	0.99995	0.0001	0.9999
4.000	0.00003	0.99997	0.00006	0.99994
4.265	0.00001	0.99999	0.00002	0.99998

Table E-7. Student's *t* distribution

df	.20	.10	.05	.025	.01	.005
	One tail (π beyond t)					
	t values					
1	1.376	3.078	6.314	12.706	31.821	63.657
2	1.061	1.886	2.920	4.303	6.965	9.925
3	.978	1.638	2.353	3.182	4.541	5.841
4	.941	1.533	2.132	2.776	3.747	4.604
5	.920	1.476	2.015	2.571	3.365	4.032
6	.906	1.440	1.943	2.447	3.143	3.707
7	.896	1.415	1.895	2.365	2.998	3.499
8	.889	1.397	1.860	2.306	2.896	3.355
9	.883	1.383	1.833	2.262	2.821	3.250
10	.879	1.372	1.812	2.228	2.764	3.169
11	.876	1.363	1.796	2.201	2.718	3.106
12	.873	1.356	1.782	2.179	2.681	3.055
13	.870	1.350	1.771	2.160	2.650	3.012
14	.868	1.345	1.761	2.145	2.624	2.977
15	.866	1.341	1.753	2.131	2.602	2.947
16	.865	1.337	1.746	2.120	2.583	2.921
17	.863	1.333	1.740	2.110	2.567	2.898
18	.862	1.330	1.734	2.101	2.552	2.878
19	.861	1.328	1.729	2.093	2.539	2.861
20	.860	1.325	1.725	2.086	2.528	2.845
21	.859	1.323	1.721	2.080	2.518	2.831
22	.858	1.321	1.717	2.074	2.508	2.819
23	.858	1.319	1.714	2.069	2.500	2.807
24	.857	1.318	1.711	2.064	2.492	2.797
25	.856	1.316	1.708	2.060	2.485	2.787
26	.856	1.315	1.706	2.056	2.479	2.779
27	.855	1.314	1.703	2.052	2.473	2.771
28	.855	1.313	1.701	2.048	2.467	2.763
29	.854	1.311	1.699	2.045	2.462	2.756
30	.854	1.310	1.697	2.042	2.457	2.750
40	.851	1.303	1.684	2.021	2.423	2.704
60	.848	1.296	1.671	2.000	2.390	2.660
120	.845	1.289	1.658	1.980	2.358	2.617
∞	.842	1.282	1.645	1.960	2.326	2.576
	.40	**.20**	**.10**	**.05**	**.02**	**.01**
	Two tail (π beyond t)					

Adapted from Pearson, E. H., and Hartley, H.O.: Biomerika tables for statisticians, vol. 1, ed. 3, London, 1966, Cambridge University Press.

Table E-8. Chi-square distribution

df	.995	.99	.97	.95	.90	.75	.50	.25	.10	.05	.025	.01	.005	.001
	χ^2 values													
1	.00	.00	.00	.00	.02	.10	.45	1.32	2.70	3.84	5.02	6.63	7.88	10.83
2	.00	.02	.05	.10	.21	.58	1.39	2.77	4.60	5.99	7.38	9.21	10.60	13.82
3	.07	.11	.22	.35	.58	1.21	2.36	4.11	6.25	7.81	9.35	11.34	12.84	16.27
4	.21	.30	.48	.71	1.06	1.92	3.36	5.38	7.78	9.49	11.14	13.28	14.86	18.47
5	.41	.55	.83	1.14	1.61	2.67	4.35	6.62	9.24	11.07	12.83	15.09	16.75	20.52
6	.68	.87	1.24	1.64	2.20	3.45	5.35	7.84	10.64	12.59	14.45	16.81	18.55	22.46
7	.99	1.24	1.69	2.17	2.83	4.25	6.34	9.04	12.02	14.07	16.01	18.47	20.28	24.32
8	1.34	1.65	2.18	2.73	3.49	5.07	7.34	10.22	13.36	15.51	17.53	20.09	21.95	26.12
9	1.73	2.09	2.70	3.32	4.17	5.90	8.34	11.39	14.68	16.92	19.02	21.67	23.59	27.88
10	2.16	2.56	3.25	3.94	4.87	6.74	9.34	12.55	15.99	18.31	20.48	23.21	25.19	29.59
11	2.60	3.05	3.82	4.57	5.58	7.58	10.34	13.70	17.28	19.68	21.92	24.72	26.76	31.26
12	3.07	3.57	4.40	5.23	6.30	8.44	11.34	14.84	18.55	21.03	23.34	26.22	28.30	32.91
13	3.57	4.11	5.01	5.89	7.04	9.30	12.34	15.98	19.81	22.36	24.74	27.69	29.82	34.53
14	4.07	4.66	5.63	6.57	7.79	10.16	13.34	17.12	21.06	23.68	26.12	29.14	31.32	36.12
15	4.60	5.23	6.26	7.26	8.55	11.04	14.34	18.24	22.31	25.00	27.49	30.58	32.80	37.70
16	5.14	5.81	6.91	7.96	9.31	11.91	15.34	19.37	23.54	26.30	28.84	32.00	34.27	39.25
17	5.70	6.41	7.56	8.67	10.08	12.79	16.34	20.49	24.77	27.59	30.19	33.41	35.72	40.79
18	6.26	7.01	8.23	9.39	10.86	13.68	17.34	21.60	25.99	28.87	31.53	34.80	37.16	42.31
19	6.84	7.63	8.91	10.12	11.65	14.56	18.34	22.72	27.20	30.14	32.85	36.19	38.58	43.82
20	7.43	8.26	9.59	10.85	12.44	15.45	19.34	23.83	28.41	31.41	34.17	37.57	39.00	45.31

Proportion greater than χ^2

21	8.03	8.90	10.28	11.59	13.24	16.34	20.34	24.93	29.61	32.67	35.48	38.93	41.40	46.80
22	8.64	9.54	10.98	12.34	14.04	17.24	21.34	26.04	30.81	33.92	36.78	40.29	42.80	48.27
23	9.26	10.20	11.69	13.09	14.85	18.14	22.34	27.14	32.01	35.17	38.07	41.64	44.18	49.73
24	9.89	10.86	12.40	13.85	15.66	19.04	23.34	28.24	33.20	36.41	39.36	42.98	45.56	51.18
25	10.52	11.52	13.12	14.61	16.47	19.94	24.34	29.34	34.38	37.65	40.65	44.31	46.93	52.62
26	11.16	12.20	13.84	15.38	17.29	20.84	25.34	30.43	35.56	38.88	41.92	45.64	48.29	54.05
27	11.81	12.88	14.57	16.15	18.11	21.75	26.34	31.53	36.74	40.11	43.19	49.96	49.64	55.48
28	12.46	13.56	15.31	16.93	18.94	22.66	27.34	32.62	37.92	41.34	44.46	48.28	50.99	56.89
29	13.12	14.26	16.05	17.71	19.77	23.57	28.34	33.71	39.09	42.56	45.72	49.59	52.34	58.30
30	13.79	14.95	16.79	18.49	20.60	24.48	29.34	34.80	40.26	43.77	46.98	50.90	53.67	59.70
40	20.71	22.16	24.43	26.51	29.05	33.66	39.34	45.62	51.80	55.76	59.34	63.69	66.77	73.40
50	27.99	29.71	32.36	34.76	37.69	42.94	49.33	56.33	63.17	67.50	71.42	76.15	79.49	86.66
60	35.53	37.48	40.48	43.19	46.46	52.29	59.33	66.98	74.40	79.08	83.30	88.38	91.95	99.61
70	43.28	45.44	48.76	51.74	55.33	61.70	69.33	77.58	85.53	90.53	95.02	100.42	104.22	112.32
80	51.17	53.54	57.15	60.39	64.28	71.14	79.33	88.13	96.58	101.88	106.63	112.33	116.32	124.84
90	59.20	61.75	65.65	69.13	73.29	80.62	89.33	98.65	107.56	113.14	118.14	124.12	128.30	137.21
100	67.33	70.06	74.22	77.93	82.36	90.13	99.33	109.14	118.50	124.34	129.56	135.81	140.17	149.45

From Pearson, E. S., and Hartley, H. O.: Biometrika tables for statisticians, vol. 1, ed. 3, London, 1966, Cambridge University Press.

Table E-9. Tables of the F distribution for probabilities: 0.25, 0.10, 0.05, and 0.01

F VALUES FOR P = 0.25

df	df								
	1	**2**	**3**	**4**	**5**	**6**	**7**	**8**	**9**
1	5.83	7.50	8.20	8.58	8.82	8.98	9.10	9.19	9.26
2	2.57	3.00	3.15	3.23	3.28	3.31	3.34	3.35	3.37
3	2.02	2.28	2.36	2.39	2.41	2.42	2.43	2.44	2.44
4	1.81	2.00	2.05	2.06	2.07	2.08	2.08	2.08	2.08
5	1.69	1.85	1.88	1.89	1.89	1.89	1.89	1.89	1.89
6	1.62	1.76	1.78	1.79	1.79	1.78	1.78	1.78	1.77
7	1.57	1.70	1.72	1.72	1.71	1.71	1.70	1.70	1.69
8	1.54	1.66	1.67	1.66	1.66	1.65	1.64	1.64	1.63
9	1.51	1.62	1.63	1.63	1.62	1.61	1.60	1.60	1.59
10	1.49	1.60	1.60	1.59	1.59	1.58	1.57	1.56	1.56
11	1.47	1.58	1.58	1.57	1.56	1.55	1.54	1.53	1.53
12	1.46	1.56	1.56	1.55	1.54	1.53	1.52	1.51	1.51
13	1.45	1.55	1.55	1.53	1.52	1.51	1.50	1.49	1.49
14	1.44	1.53	1.53	1.52	1.51	1.50	1.49	1.48	1.47
15	1.43	1.52	1.52	1.51	1.49	1.48	1.47	1.46	1.46
16	1.42	1.51	1.51	1.50	1.48	1.47	1.46	1.45	1.44
17	1.42	1.51	1.50	1.49	1.47	1.46	1.45	1.44	1.43
18	1.41	1.50	1.49	1.48	1.46	1.45	1.44	1.43	1.42
19	1.41	1.49	1.49	1.47	1.46	1.44	1.43	1.42	1.41
20	1.40	1.49	1.48	1.47	1.45	1.44	1.43	1.42	1.41
21	1.40	1.48	1.48	1.46	1.44	1.43	1.42	1.41	1.40
22	1.40	1.48	1.47	1.45	1.44	1.42	1.41	1.40	1.39
23	1.39	1.47	1.47	1.45	1.43	1.42	1.41	1.40	1.39
24	1.39	1.47	1.46	1.44	1.43	1.41	1.40	1.39	1.38
25	1.39	1.47	1.46	1.44	1.42	1.41	1.40	1.39	1.38
26	1.38	1.46	1.45	1.44	1.42	1.41	1.39	1.38	1.37
27	1.38	1.46	1.45	1.43	1.42	1.40	1.39	1.38	1.37
28	1.38	1.46	1.45	1.43	1.41	1.40	1.39	1.38	1.37
29	1.38	1.45	1.45	1.43	1.41	1.40	1.38	1.37	1.36
30	1.38	1.45	1.44	1.42	1.41	1.39	1.38	1.37	1.36
40	1.36	1.44	1.42	1.40	1.39	1.37	1.36	1.35	1.34
60	1.35	1.42	1.41	1.38	1.37	1.35	1.33	1.32	1.31
120	1.34	1.40	1.39	1.37	1.35	1.33	1.31	1.30	1.29
∞	1.32	1.39	1.37	1.35	1.33	1.31	1.29	1.28	1.27

Adapted from Pearson, E. S., and Hartley, H. O.: Biometrika tables for statisticians, vol. 1, ed. 3, London,

df									
10	**12**	**15**	**20**	**24**	**30**	**40**	**60**	**120**	**∞**
9.32	9.41	9.49	9.58	9.63	9.67	9.71	9.76	9.80	9.85
3.38	3.39	3.41	3.43	3.43	3.44	3.45	3.46	3.47	3.48
2.44	2.45	2.46	2.46	2.46	2.47	2.47	2.47	2.47	2.47
2.08	2.08	2.08	2.08	2.08	2.08	2.08	2.08	2.08	2.08
1.89	1.89	1.89	1.88	1.88	1.88	1.88	1.87	1.87	1.87
1.77	1.77	1.76	1.76	1.75	1.75	1.75	1.74	1.74	1.74
1.69	1.68	1.68	1.67	1.67	1.66	1.66	1.65	1.65	1.65
1.63	1.62	1.62	1.61	1.60	1.60	1.59	1.59	1.58	1.58
1.59	1.58	1.57	1.56	1.56	1.55	1.54	1.54	1.53	1.53
1.55	1.54	1.53	1.52	1.52	1.51	1.51	1.50	1.49	1.48
1.52	1.51	1.50	1.49	1.49	1.48	1.47	1.47	1.46	1.45
1.50	1.49	1.48	1.47	1.46	1.45	1.45	1.44	1.43	1.42
1.48	1.47	1.46	1.45	1.44	1.43	1.42	1.42	1.41	1.40
1.46	1.45	1.44	1.43	1.42	1.41	1.41	1.40	1.39	1.38
1.45	1.44	1.43	1.41	1.41	1.40	1.39	1.38	1.37	1.36
1.44	1.43	1.41	1.40	1.39	1.38	1.37	1.36	1.35	1.34
1.43	1.41	1.40	1.39	1.38	1.37	1.36	1.35	1.34	1.33
1.42	1.40	1.39	1.38	1.37	1.36	1.35	1.34	1.33	1.32
1.41	1.40	1.38	1.37	1.36	1.35	1.34	1.33	1.32	1.30
1.40	1.39	1.37	1.36	1.35	1.34	1.33	1.32	1.31	1.29
1.39	1.38	1.37	1.35	1.34	1.33	1.32	1.31	1.30	1.28
1.39	1.37	1.36	1.34	1.33	1.32	1.31	1.30	1.29	1.28
1.38	1.37	1.35	1.34	1.33	1.32	1.31	1.30	1.28	1.27
1.38	1.36	1.35	1.33	1.32	1.31	1.30	1.29	1.28	1.26
1.37	1.36	1.34	1.33	1.32	1.31	1.29	1.28	1.27	1.25
1.37	1.35	1.34	1.32	1.31	1.30	1.29	1.28	1.26	1.25
1.36	1.35	1.33	1.32	1.31	1.30	1.28	1.27	1.26	1.24
1.36	1.34	1.33	1.31	1.30	1.29	1.28	1.27	1.25	1.24
1.35	1.34	1.32	1.31	1.30	1.29	1.27	1.26	1.25	1.23
1.35	1.34	1.32	1.30	1.29	1.28	1.27	1.26	1.24	1.23
1.33	1.31	1.30	1.28	1.26	1.25	1.24	1.22	1.21	1.19
1.30	1.29	1.27	1.25	1.24	1.22	1.21	1.19	1.17	1.15
1.28	1.26	1.24	1.22	1.21	1.19	1.18	1.16	1.13	1.10
1.25	1.24	1.22	1.19	1.18	1.16	1.14	1.12	1.08	1.00

1966, Cambridge University Press. *Continued.*

Table E-9. Tables of the *F* distribution for probabilities: 0.25, 0.10, 0.05, and
F VALUES FOR P = 0.10

df	df 1	2	3	4	5	6	7	8	9
1	39.86	49.50	53.59	55.83	57.24	58.20	58.91	59.44	59.86
2	8.53	9.00	9.16	9.24	9.29	9.33	9.35	9.37	9.38
3	5.54	5.46	5.39	5.34	5.31	5.28	5.27	5.25	5.24
4	4.54	4.32	4.19	4.11	4.05	4.01	3.98	3.95	3.94
5	4.06	3.78	3.62	3.52	3.45	3.40	3.37	3.34	3.32
6	3.78	3.46	3.29	3.18	3.11	3.05	3.01	2.98	2.96
7	3.59	3.26	3.07	2.96	2.88	2.83	2.78	2.75	2.72
8	3.46	3.11	2.92	2.81	2.73	2.67	2.62	2.59	2.56
9	3.36	3.01	2.81	2.69	2.61	2.55	2.51	2.47	2.44
10	3.29	2.92	2.73	2.61	2.52	2.46	2.41	2.38	2.35
11	3.23	2.86	2.66	2.54	2.45	2.39	2.34	2.30	2.27
12	3.18	2.81	2.61	2.48	2.39	2.33	2.28	2.24	2.21
13	3.14	2.76	2.56	2.43	2.35	2.28	2.23	2.20	2.16
14	3.10	2.73	2.52	2.39	2.31	2.24	2.19	2.15	2.12
15	3.07	2.70	2.49	2.36	2.27	2.21	2.16	2.12	2.09
16	3.05	2.67	2.46	2.33	2.24	2.18	2.13	2.09	2.06
17	3.03	2.64	2.44	2.31	2.22	2.15	2.10	2.06	2.03
18	3.01	2.62	2.42	2.29	2.20	2.13	2.08	2.04	2.00
19	2.99	2.61	2.40	2.27	2.18	2.11	2.06	2.02	1.98
20	2.97	2.59	2.38	2.25	2.16	2.09	2.04	2.00	1.96
21	2.96	2.57	2.36	2.23	2.14	2.08	2.02	1.98	1.95
22	2.95	2.56	2.35	2.22	2.13	2.06	2.01	1.97	1.93
23	2.94	2.55	2.34	2.21	2.11	2.05	1.99	1.95	1.92
24	2.93	2.54	2.33	2.19	2.10	2.04	1.98	1.94	1.91
25	2.92	2.53	2.32	2.18	2.09	2.02	1.97	1.93	1.89
26	2.91	2.52	2.31	2.17	2.08	2.01	1.96	1.92	1.88
27	2.90	2.51	2.30	2.17	2.07	2.00	1.95	1.91	1.87
28	2.89	2.50	2.29	2.16	2.06	2.00	1.94	1.90	1.87
29	2.89	2.50	2.28	2.15	2.06	1.99	1.93	1.89	1.86
30	2.88	2.49	2.28	2.14	2.05	1.98	1.93	1.88	1.85
40	2.84	2.44	2.23	2.09	2.00	1.93	1.87	1.83	1.79
60	2.79	2.39	2.18	2.04	1.95	1.87	1.82	1.77	1.74
120	2.75	2.35	2.13	1.99	1.90	1.82	1.77	1.72	1.68
∞	2.71	2.30	2.08	1.94	1.85	1.77	1.72	1.67	1.63

0.01—cont'd

				df					
10	**12**	**15**	**20**	**24**	**30**	**40**	**60**	**120**	**∞**
60.19	60.71	61.22	61.74	62.00	62.26	62.53	62.79	63.06	63.33
9.39	9.41	9.42	9.44	9.45	9.46	9.47	9.47	9.48	9.49
5.23	5.22	5.20	5.18	5.18	5.17	5.16	5.15	5.14	5.13
3.92	3.90	3.87	3.84	3.83	3.82	3.80	3.79	3.78	3.76
3.30	3.27	3.24	3.21	3.19	3.17	3.16	3.14	3.12	3.10
2.94	2.90	2.87	2.84	2.82	2.80	2.78	2.76	2.74	2.72
2.70	2.67	2.63	2.59	2.58	2.56	2.54	2.51	2.49	2.47
2.54	2.50	2.46	2.42	2.40	2.38	2.36	2.34	2.32	2.29
2.42	2.38	2.34	2.30	2.28	2.25	2.23	2.21	2.18	2.16
2.32	2.28	2.24	2.20	2.18	2.16	2.13	2.11	2.08	2.06
2.25	2.21	2.17	2.12	2.10	2.08	2.05	2.03	2.00	1.97
2.19	2.15	2.10	2.06	2.04	2.01	1.99	1.96	1.93	1.90
2.14	2.10	2.05	2.01	1.98	1.96	1.93	1.90	1.88	1.85
2.10	2.05	2.01	1.96	1.94	1.91	1.89	1.86	1.83	1.80
2.06	2.02	1.97	1.92	1.90	1.87	1.85	1.82	1.79	1.76
2.03	1.99	1.94	1.89	1.87	1.84	1.81	1.78	1.75	1.72
2.00	1.96	1.91	1.86	1.84	1.81	1.78	1.75	1.72	1.69
1.98	1.93	1.89	1.84	1.81	1.78	1.75	1.72	1.69	1.66
1.96	1.91	1.86	1.81	1.79	1.76	1.73	1.70	1.67	1.63
1.94	1.89	1.84	1.79	1.77	1.74	1.71	1.68	1.64	1.61
1.92	1.87	1.83	1.78	1.75	1.72	1.69	1.66	1.62	1.59
1.90	1.86	1.81	1.76	1.73	1.70	1.67	1.64	1.60	1.57
1.89	1.84	1.80	1.74	1.72	1.69	1.66	1.62	1.59	1.55
1.88	1.83	1.78	1.73	1.70	1.67	1.64	1.61	1.57	1.53
1.87	1.82	1.77	1.72	1.69	1.66	1.63	1.59	1.56	1.52
1.86	1.81	1.76	1.71	1.68	1.65	1.61	1.58	1.54	1.50
1.85	1.80	1.75	1.70	1.67	1.64	1.60	1.57	1.53	1.49
1.84	1.79	1.74	1.69	1.66	1.63	1.59	1.56	1.52	1.48
1.83	1.78	1.73	1.68	1.65	1.62	1.58	1.55	1.51	1.47
1.82	1.77	1.72	1.67	1.64	1.61	1.57	1.54	1.50	1.46
1.76	1.71	1.66	1.61	1.57	1.54	1.51	1.47	1.42	1.38
1.71	1.66	1.60	1.54	1.51	1.48	1.44	1.40	1.35	1.29
1.65	1.60	1.55	1.48	1.45	1.41	1.37	1.32	1.26	1.19
1.60	1.55	1.49	1.42	1.38	1.34	1.30	1.24	1.17	1.00

Continued.

Table E-9. Tables of the *F* distribution for probabilities: 0.25, 0.10, 0.05, and

F VALUES FOR P = 0.05

df	df 1	2	3	4	5	6	7	8	9
1	161.4	199.5	215.7	224.6	230.2	234.0	236.8	238.9	240.5
2	18.51	19.00	19.16	19.25	19.30	19.33	19.35	19.37	19.38
3	10.13	9.55	9.28	9.12	9.01	8.94	8.89	8.85	8.81
4	7.71	6.94	6.59	6.39	6.26	6.16	6.09	6.04	6.00
5	6.61	5.79	5.41	5.19	5.05	4.95	4.88	4.82	4.77
6	5.99	5.14	4.76	4.53	4.39	4.28	4.21	4.15	4.10
7	5.59	4.74	4.35	4.12	3.97	3.87	3.79	3.73	3.68
8	5.32	4.46	4.07	3.84	3.69	3.58	3.50	3.44	3.39
9	5.12	4.26	3.86	3.63	3.48	3.37	3.29	3.23	3.18
10	4.96	4.10	3.71	3.48	3.33	3.22	3.14	3.07	3.02
11	4.84	3.98	3.59	3.36	3.20	3.09	3.01	2.95	2.90
12	4.75	3.89	3.49	3.26	3.11	3.00	2.91	2.85	2.80
13	4.67	3.81	3.41	3.18	3.03	2.92	2.83	2.77	2.71
14	4.60	3.74	3.34	3.11	2.96	2.85	2.76	2.70	2.65
15	4.54	3.68	3.29	3.06	2.90	2.79	2.71	2.64	2.59
16	4.49	3.63	3.24	3.01	2.85	2.74	2.66	2.59	2.54
17	4.45	3.59	3.20	2.96	2.81	2.70	2.61	2.55	2.49
18	4.41	3.55	3.16	2.93	2.77	2.66	2.58	2.51	2.46
19	4.38	3.52	3.13	2.90	2.74	2.63	2.54	2.48	2.42
20	4.35	3.49	3.10	2.87	2.71	2.60	2.51	2.45	2.39
21	4.32	3.47	3.07	2.84	2.68	2.57	2.49	2.42	2.37
22	4.30	3.44	3.05	2.82	2.66	2.55	2.46	2.40	2.34
23	4.28	3.42	3.03	2.80	2.64	2.53	2.44	2.37	2.32
24	4.26	3.40	3.01	2.78	2.62	2.51	2.42	2.36	2.30
25	4.24	3.39	2.99	2.76	2.60	2.49	2.40	2.34	2.28
26	4.23	3.37	2.98	2.74	2.59	2.47	2.39	2.32	2.27
27	4.21	3.35	2.96	2.73	2.57	2.46	2.37	2.31	2.25
28	4.20	3.34	2.95	2.71	2.56	2.45	2.36	2.29	2.24
29	4.18	3.33	2.93	2.70	2.55	2.43	2.35	2.28	2.22
30	4.17	3.32	2.92	2.69	2.53	2.42	2.33	2.27	2.21
40	4.08	3.23	2.84	2.61	2.45	2.34	2.25	2.18	2.12
60	4.00	3.15	2.76	2.53	2.37	2.25	2.17	2.10	2.04
120	3.92	3.07	2.68	2.45	2.29	2.17	2.09	2.02	1.96
∞	3.84	3.00	2.60	2.37	2.21	2.10	2.01	1.94	1.88

0.01—cont'd

				df					
10	**12**	**15**	**20**	**24**	**30**	**40**	**60**	**120**	**∞**
241.9	243.9	245.9	248.0	249.1	250.1	251.1	252.2	253.3	254.3
19.40	19.41	19.43	19.45	19.45	19.46	19.47	19.48	19.49	19.50
8.79	8.74	8.70	8.66	8.64	8.62	8.59	8.57	8.55	8.53
5.96	5.91	5.86	5.80	5.77	5.75	5.72	5.69	5.66	5.63
4.74	4.68	4.62	4.56	4.53	4.50	4.46	4.43	4.40	4.36
4.06	4.00	3.94	3.87	3.84	3.81	3.77	3.74	3.70	3.67
3.64	3.57	3.51	3.44	3.41	3.38	3.34	3.30	3.27	3.23
3.35	3.28	3.22	3.15	3.12	3.08	3.04	3.01	2.97	2.93
3.14	3.07	3.01	2.94	2.90	2.86	2.83	2.79	2.75	2.71
2.98	2.91	2.85	2.77	2.74	2.70	2.66	2.62	2.58	2.54
2.85	2.79	2.72	2.65	2.61	2.57	2.53	2.49	2.45	2.40
2.75	2.69	2.62	2.54	2.51	2.47	2.43	2.38	2.34	2.30
2.67	2.60	2.53	2.46	2.42	2.38	2.34	2.30	2.25	2.21
2.60	2.53	2.46	2.39	2.35	2.31	2.27	2.22	2.18	2.13
2.54	2.48	2.40	2.33	2.29	2.25	2.20	2.16	2.11	2.07
2.49	2.42	2.35	2.28	2.24	2.19	2.15	2.11	2.06	2.01
2.45	2.38	2.31	2.23	2.19	2.15	2.10	2.06	2.01	1.96
2.41	2.34	2.27	2.19	2.15	2.11	2.06	2.02	1.97	1.92
2.38	2.31	2.23	2.16	2.11	2.07	2.03	1.98	1.93	1.88
2.35	2.28	2.20	2.12	2.08	2.04	1.99	1.95	1.90	1.84
2.32	2.25	2.18	2.10	2.05	2.01	1.96	1.92	1.87	1.81
2.30	2.23	2.15	2.07	2.03	1.98	1.94	1.89	1.84	1.78
2.27	2.20	2.13	2.05	2.01	1.96	1.91	1.86	1.81	1.76
2.25	2.18	2.11	2.03	1.98	1.94	1.89	1.84	1.79	1.73
2.24	2.16	2.09	2.01	1.96	1.92	1.87	1.82	1.77	1.71
2.22	2.15	2.07	1.99	1.95	1.90	1.85	1.80	1.75	1.69
2.20	2.13	2.06	1.97	1.93	1.88	1.84	1.79	1.73	1.67
2.19	2.12	2.04	1.96	1.91	1.87	1.82	1.77	1.71	1.65
2.18	2.10	2.03	1.94	1.90	1.85	1.81	1.75	1.70	1.64
2.16	2.09	2.01	1.93	1.89	1.84	1.79	1.74	1.68	1.62
2.08	2.00	1.92	1.84	1.79	1.74	1.69	1.64	1.58	1.51
1.99	1.92	1.84	1.75	1.70	1.65	1.59	1.53	1.47	1.39
1.91	1.83	1.75	1.66	1.61	1.55	1.50	1.43	1.35	1.25
1.83	1.75	1.67	1.57	1.52	1.46	1.39	1.32	1.22	1.00

Continued.

Table E-9. Tables of the F distribution for probabilities: 0.25, 0.10, 0.05, and
F VALUES FOR P = 0.01

df	df								
	1	2	3	4	5	6	7	8	9
1	4052	4999.5	5403	5625	5764	5859	5928	5981	6022
2	98.50	99.00	99.17	99.25	99.30	99.33	99.36	99.37	99.39
3	34.12	30.82	29.46	28.71	28.24	27.91	27.67	27.49	27.35
4	21.20	18.00	16.69	15.98	15.52	15.21	14.98	14.80	14.66
5	16.26	13.27	12.06	11.39	10.97	10.67	10.46	10.29	10.16
6	13.75	10.92	9.78	9.15	8.75	8.47	8.26	8.10	7.98
7	12.25	9.55	8.45	7.85	7.46	7.19	6.99	6.84	6.72
8	11.26	8.65	7.59	7.01	6.63	6.37	6.18	6.03	5.91
9	10.56	8.02	6.99	6.42	6.06	5.80	5.61	5.47	5.35
10	10.04	7.56	6.55	5.99	5.64	5.39	5.20	5.06	4.94
11	9.65	7.21	6.22	5.67	5.32	5.07	4.89	4.74	4.63
12	9.33	6.93	5.95	5.41	5.06	4.82	4.64	4.50	4.39
13	9.07	6.70	5.74	5.21	4.86	4.62	4.44	4.30	4.19
14	8.86	6.51	5.56	5.04	4.69	4.46	4.28	4.14	4.03
15	8.68	6.36	5.42	4.89	4.56	4.32	4.14	4.00	3.89
16	8.53	6.23	5.29	4.77	4.44	4.20	4.03	3.89	3.78
17	8.40	6.11	5.18	4.67	4.34	4.10	3.93	3.79	3.68
18	8.29	6.01	5.09	4.58	4.25	4.01	3.84	3.71	3.60
19	8.18	5.93	5.01	4.50	4.17	3.94	3.77	3.63	3.52
20	8.10	5.85	4.94	4.43	4.10	3.87	3.70	3.56	3.46
21	8.02	5.78	4.87	4.37	4.04	3.81	3.64	3.51	3.40
22	7.95	5.72	4.82	4.31	3.99	3.76	3.59	3.45	3.35
23	7.88	5.66	4.76	4.26	3.94	3.71	3.54	3.41	3.30
24	7.82	5.61	4.72	4.22	3.90	3.67	3.50	3.36	3.26
25	7.77	5.57	4.68	4.18	3.85	3.63	3.46	3.32	3.22
26	7.72	5.53	4.64	4.14	3.82	3.59	3.42	3.29	3.18
27	7.68	5.49	4.60	4.11	3.78	3.56	3.39	3.26	3.15
28	7.64	5.45	4.57	4.07	3.75	3.53	3.36	3.23	3.12
29	7.60	5.42	4.54	4.04	3.73	3.50	3.33	3.20	3.09
30	7.56	5.39	4.51	4.02	3.70	3.47	3.30	3.17	3.07
40	7.31	5.18	4.31	3.83	3.51	3.29	3.12	2.99	2.89
60	7.08	4.98	4.13	3.65	3.34	3.12	2.95	2.82	2.72
120	6.85	4.79	3.95	3.48	3.17	2.96	2.79	2.66	2.56
∞	6.63	4.61	3.78	3.32	3.02	2.80	2.64	2.51	2.41

0.01—cont'd

				df					
10	**12**	**15**	**20**	**24**	**30**	**40**	**60**	**120**	**∞**
6056	6106	6157	6209	6235	6261	6287	6313	6339	6366
99.40	99.42	99.43	99.45	99.46	99.47	99.47	99.48	99.49	99.50
27.23	27.05	26.87	26.69	26.60	26.50	26.41	26.32	26.22	26.13
14.55	14.37	14.20	14.02	13.93	13.84	13.75	13.65	13.56	13.46
10.05	9.89	9.72	9.55	9.47	9.38	9.29	9.20	9.11	9.02
7.87	7.72	7.56	7.40	7.31	7.23	7.14	7.06	6.97	6.88
6.62	6.47	6.31	6.16	6.07	5.99	5.91	5.82	5.74	5.65
5.81	5.67	5.52	5.36	5.28	5.20	5.12	5.03	4.95	4.86
5.26	5.11	4.96	4.81	4.73	4.65	4.57	4.48	4.40	4.31
4.85	4.71	4.56	4.41	4.33	4.25	4.17	4.08	4.00	3.91
4.54	4.40	4.25	4.10	4.02	3.94	3.86	3.78	3.69	3.60
4.30	4.16	4.01	3.86	3.78	3.70	3.62	3.54	3.45	3.36
4.10	3.96	3.82	3.66	3.59	3.51	3.43	3.34	3.25	3.17
3.94	3.80	3.66	3.51	3.43	3.35	3.27	3.18	3.09	3.00
3.80	3.67	3.52	3.37	3.29	3.21	3.13	3.05	2.96	2.87
3.69	3.55	3.41	3.26	3.18	3.10	3.02	2.93	2.84	2.75
3.59	3.46	3.31	3.16	3.08	3.00	2.92	2.83	2.75	2.65
3.51	3.37	3.23	3.08	3.00	2.92	2.84	2.75	2.66	2.57
3.43	3.30	3.15	3.00	2.92	2.84	2.76	2.67	2.58	2.49
3.37	3.23	3.09	2.94	2.86	2.78	2.69	2.61	2.52	2.42
3.31	3.17	3.03	2.88	2.80	2.72	2.64	2.55	2.46	2.36
3.26	3.12	2.98	2.83	2.75	2.67	2.58	2.50	2.40	2.31
3.21	3.07	2.93	2.78	2.70	2.62	2.54	2.45	2.35	2.26
3.17	3.03	2.89	2.74	2.66	2.58	2.49	2.40	2.31	2.21
3.13	2.99	2.85	2.70	2.62	2.54	2.45	2.36	2.27	2.17
3.09	2.96	2.81	2.66	2.58	2.50	2.42	2.33	2.23	2.13
3.06	2.93	2.78	2.63	2.55	2.47	2.38	2.29	2.20	2.10
3.03	2.90	2.75	2.60	2.52	2.44	2.35	2.26	2.17	2.06
3.00	2.87	2.73	2.57	2.49	2.41	2.33	2.23	2.14	2.03
2.98	2.84	2.70	2.55	2.47	2.39	2.30	2.21	2.11	2.01
2.80	2.66	2.52	2.37	2.29	2.20	2.11	2.02	1.92	1.80
2.63	2.50	2.35	2.20	2.12	2.03	1.94	1.84	1.73	1.60
2.47	2.34	2.19	2.03	1.95	1.86	1.76	1.66	1.53	1.38
2.32	2.18	2.04	1.88	1.79	1.70	1.59	1.47	1.32	1.00

Index